Neo-Aristotelian Perspectives on Contemporary Science

The last two decades have seen two significant trends emerging within the philosophy of science: the rapid development and focus on the philosophy of the specialised sciences, and a resurgence of Aristotelian metaphysics, much of which is concerned with the possibility of emergence, as well as the ontological status and indispensability of dispositions and powers in science. Despite these recent trends, few Aristotelian metaphysicians have engaged directly with the philosophy of the specialised sciences. Additionally, the relationship between fundamental Aristotelian concepts—such as "hylomorphism", "substance", and "faculties"—and contemporary science has yet to receive a critical and systematic treatment. *Neo-Aristotelian Perspectives on Contemporary Science* aims to fill this gap in the literature by bringing together essays on the relationship between Aristotelianism and science that cut across interdisciplinary boundaries. The chapters in this volume are divided into two main sections covering the philosophy of physics and the philosophy of the life sciences. Featuring original contributions from distinguished and early career scholars, this book will be of interest to specialists in analytical metaphysics and the philosophy of science.

William M.R. Simpson is a Research Associate at the University of St Andrews, UK, and a postgraduate student of philosophy at the University of Cambridge, UK. He was formerly a Research Fellow in Theoretical Physics at the Weizmann Institute, Israel.

Robert C. Koons is a Professor of Philosophy at the University of Texas at Austin, USA.

Nicholas J. Teh is an Assistant Professor of Philosophy at the University of Notre Dame, USA. He was formerly a Postdoctoral Research Fellow at the University of Cambridge, UK.

Routledge Studies in the Philosophy of Science

7 Popper's Critical Rationalism
A Philosophical Investigation
Darrell Rowbottom

8 Conservative Reductionism
Michael Esfeld and Christian Sachse

9 Models, Simulations, and Representations
Paul Humphreys and Cyrille Imbert

10 Platonism, Naturalism, and Mathematical Knowledge
James Robert Brown

11 Thought Experiments in Science, Philosophy, and the Arts
Edited by Mélanie Frappier, Letitia Meynell, and James Robert Brown

12 Kuhn's The Structure of Scientific Revolutions Revisited
Edited by Vasso Kindi and Theodore Arabatzis

13 Contemporary Philosophical Naturalism and Its Implications
Edited by Bana Bashour and Hans D. Muller

14 Science after the Practice Turn in Philosophy, History, and the Social
Studies of Science
*Edited by Léna Soler, Sjoerd Zwart, Vincent Israel-Jost, and Michael
Lynch*

15 Causation, Evidence, and Inference
Julian Reiss

16 Conceptual Change and the Philosophy of Science
Alternative Interpretations of the A Priori
David J. Stump

17 Neo-Aristotelian Perspectives on Contemporary Science
Edited by William M.R. Simpson, Robert C. Koons and Nicholas J. Teh

Neo-Aristotelian Perspectives on Contemporary Science

Edited by William M. R. Simpson,
Robert C. Koons and Nicholas J. Teh

Routledge
Taylor & Francis Group

LONDON AND NEW YORK

First published 2018 by Routledge

2 Park Square, Milton Park, Abingdon, Oxfordshire OX14 4RN
52 Vanderbilt Avenue, New York, NY 10017

Routledge is an imprint of the Taylor & Francis Group, an informa business

First issued in paperback 2019

Library of Congress Cataloging-in-Publication Data
A catalog record for this book has been requested

ISBN: 978-0-415-79256-1 (hbk)
ISBN: 978-0-367-88515-1 (pbk)

Typeset in Sabon
by Apex CoVantage, LLC

Contents

Acknowledgements   vii
Foreword   ix
JOHN HALDANE

**Introduction: Reflections on Contemporary Science and
the New Aristotelianism**   1
ROBERT C. KOONS, WILLIAM M. R. SIMPSON, AND NICHOLAS J. TEH

PART 1
The Philosophy of Physics   13

1 Dodging the Fundamentalist Threat   15
   XAVI LANAO AND NICHOLAS J. TEH

2 Actuality, Potentiality, and Relativity's Block Universe   35
   EDWARD FESER

3 The Many Worlds Interpretation of QM: A Hylomorphic
   Critique and Alternative   61
   ROBERT C. KOONS

4 A Traveling Forms Interpretation of Quantum Mechanics   105
   ALEXANDER R. PRUSS

5 Half-Baked Humeanism   123
   WILLIAM M. R. SIMPSON

6 Disentangling Nature's Joints   147
   TUOMAS E. TAHKO

PART 2
The Philosophy of the Life Sciences 167

7 Structural Powers and the Homeodynamic Unity
of Organisms 169
CHRISTOPHER J. AUSTIN AND ANNA MARMODORO

8 A Biologically Informed Hylomorphism 185
CHRISTOPHER J. AUSTIN

9 The Great Unifier: Form and the Unity of the Organism 211
DAVID S. ODERBERG

10 Action, Animacy, and Substance Causation 235
JANICE CHIK BREIDENBACH

11 Psychology Without a Mental-Physical Dichotomy 261
WILLIAM JAWORSKI

12 Hylomorphism and the New Mechanist Philosophy in
Biology, Neuroscience, and Psychology 293
DANIEL D. DE HAAN

About the Contributors 327
Index 331

Acknowledgements

William Simpson would like to acknowledge a number of institutions for financial support at different stages of this project, including his college, *Peterhouse* (Cambridge), the *Institute for the Study of Philosophy, Politics, and Religion* (Cambridge), and the *John Templeton Foundation*, which supported him through the Scientists in Congregations programme in Scotland. He would also like to thank the Harvey Fellows programme for financial support during part of his studies.

Robert Koons would like to acknowledge the support during the 2014–15 academic year of the James Madison Program in American Ideals and Institutions at Princeton University (for a Visiting Fellowship) and the University of Texas at Austin (for a faculty research grant). He would also like to thank Anna Marmodoro and the Power Structuralism and Ancient Ontologies project for their support for his visit to Oxford in October of 2016.

Nicholas Teh would like to acknowledge the support during the 2016–17 academic year of the Institute for Scholarship in the Liberal Arts at the University of Notre Dame (for an Integrative Scholarship grant to organize activities at the intersection of philosophy, mathematics, and physics). He would also like to thank Brian Pitts and CamPOS for organizing his visit to Cambridge in spring 2016.

Foreword

John Haldane *

I

Reading the essays in this collection, several thoughts come to mind. First, one of congratulations to the editors for gathering such an interesting, timely, well-informed and rigorously argued set of essays. Second, recognition of the confidence displayed by the authors. I mean, this not as a matter of personal temperament or psychology, but of a shared sense that Aristotelian ideas have much to offer in the effort to understand the things, processes and events that are the focus of the natural sciences, especially physics and biology, and to recognise the scope for, but also the limits of cross-categorical explanations: the inanimate, the non-rational animate, the minded.

Fifty years ago, such confidence would have seemed eccentric, for the tide of reductionism in the philosophy of science was then strong and high, and the presumption was that by stages, and perhaps increasingly rapidly, psychology would reduce to biology, that to chemistry and that, in turn, to physics. Given the prevailing conception of chemical and physical systems this implied a 'mechanistic' view of living organisms. Additionally, it was assumed that causal explanations are law-like, or presuppose law-like generalisations, and that talk of causation as expressing the powers and thereby the essential nature of an agent is at best a *façon de parler* and at worst the effect of fallaciously inferring the existence of non-contingent relations in nature from conceptually linked modes of description in which antecedent conditions are redescribed in terms of consequent ones. Thus, from 'sugar was placed in water and dissolved', one moves to 'water caused the sugar to dissolve' and from that proceeds to 'a solvent caused the sugar to dissolve' thence to 'water has a power to dissolve sugar' and finally to 'it is part of the intrinsic nature of water to dissolve sugar'. By means of analyses of this sort, applied in reverse, it seemed easy to dismiss any talk of natures and causal powers as a projection of certain modes of description, and not as indicating the possibility of non-contingent connections between things themselves. In much the same way, talk of natural processes serving and occurring for the sake of some end, typically an end beneficial to the subject of that process, was thought to be a legacy of something akin, at its core, to animism, but given faux intellectual credibility through schemes that postulated 'final' causes: ends towards processes tend on account of intrinsic principles of order and activity.

It was generally known, of course, that these various ideas had a place in the history of philosophical and scientific theorising, and that the main figure associated with them was Aristotle, but as with the modern theorists of the seventeenth and eighteenth centuries, the empiricists and reductionists of the twentieth century saw themselves as thinking beyond and against both Aristotle and the tradition that followed him into the Middle Ages among Arabic and Western Latin thinkers.

The standard collection of Aristotle's works on logic and methodology was named by his ancient followers the *Organon* and known to the medieval by the Latinised version *Organum* (*tool* or *instrument*, in this case, for developing knowledge by sound reasoning). From the point of view of thinking about reasoning on relation to 'science', the most important of these texts is the *Posterior Analytics*, and this together with Aristotle's *Metaphysics* provided the core of scholastic thinking about the analysis of change and the systematic study of things and their causes.

This history explains the significance of the title of Francis Bacon's 1620 work *Novum Organum Scientiarum—New Tool for Science*—focussed on methods of experiment and records of differences of setups, leading to inductions as to effective factors. Bacon refers to Aristotle 15 times in the *New Organon*, mostly unfavourably and for the same reason, that, according to Bacon, he confuses natural philosophy with a priori reasoning 'logic': '[he] made his natural philosophy a mere bond servant to his logic, thereby rendering it contentious and well-nigh useless'; and later he writes:

> The most conspicuous example of [rational philosophers] was Aristotle, who corrupted natural philosophy by his logic: *fashioning the world out of categories* . . . doing the business of density and rarity (which is to make bodies of greater or less dimensions, that is, occupy greater or less spaces), *by the frigid distinction of act and power*; asserting that single bodies have each a *single and proper motion*, and that if they participate in any other, then this results from an external cause; and *imposing countless other arbitrary restrictions* on the nature of things . . . in the physics of Aristotle you hear hardly anything but the words of logic, which in his metaphysics also, under a more imposing name . . .[1]

Similar contrasts and criticisms of Aristotle are to be found in writings on the methods of natural philosophy by Galileo, Boyle and Newton, and while the effect of such criticism and repudiation was to marginalise and more or less eliminate avowed Aristotelian ways of thinking, the suspicion of unwarranted insertions of metaphysical notions into the sphere of empirical enquiry continued to be associated with 'Aristotelianism' into the twentieth century as in Quine's strictures against postulation of non-contingent modalities:

> Curiously, a philosophical tradition does exist for just such a distinction between necessary and contingent attributes. It lives on in the terms 'essence' and 'accident' . . . It is a distinction that one attributes to Aristotle (subject to contradiction by scholars, such being the penalty for attributions to Aristotle). But, however venerable the distinction, it is surely indefensible . . .[2]

II

Given the modern consensus against the view that to understand the natural world and what occurs in it, we need to apply notions of essences, causal powers, natural tendencies, non-contingent relations and so on, how can it be anything other than quixotic to be discussing 'neo-Aristotelian perspectives on contemporary science', rather as if one were to be exploring astrological perspectives on astronomy? The answer is two-fold. First, there has been growing dissatisfaction with the approaches favoured by reductionist empiricism, and by its anti-realist constructivist and relativistic rivals. Second, serious attention has been given to the possible merits of an Aristotelian approach free of the prejudices of earlier periods.

Early indications of the former included Larry Wright's tentative reinstatement of teleology in his important 1973 article on 'Functions', according to which, in brief:

> 'The function of X is Z' [e.g., the function of oxygen in the blood is . . . the function of the human heart is . . . etc.] *means*
>
> (a) X is there because it does Z,
> (b) Z is a consequence (or result) of X's being there.[3]

And in the same year, Rom Harré and E. H. Madden published one in a sequence of anti-empiricist essays on the idea of natural agency, which led two years later to the book *Causal Powers*. In the former they write:

> The justification of a wholly non-Humean conceptual scheme, based upon the idea of enduring individuals with powers, rests in part on the success of such a scheme in resolving the problems bequeathed to us by the Humean tradition and in part must be achieved by a careful construction of the metaphysics of the new scheme itself. By this we mean a thorough exposition of the meaning and interrelations of the concepts of the new scheme.[4]

In *Causal Powers*, these, along with *natures, natural kinds* and *natural necessity*, are invoked in the development of a recognisably Aristotelian theory of scientific description and explanation. They write:

> The relationship between co-existing properties or successive events or states is naturally necessary when they are understood by scientists to be related in fact by generative mechanisms, whose structure and components constitute the essential nature of the permanent things and materials in the world.[5]

They themselves do not make much of the Aristotelian character of their approach, but significantly, they do relate the causal power element of it to that of Aquinas, as expounded by two prominent analytical philosophers in a work published a decade previously:

> An argument, very similar to ours, can be found in Anscombe and Geach *Three Philosophers*. In giving an account of the philosophy of science of Aquinas they develop a theory of scientific explanation, of the related metaphysics of nature, in which they introduce the concept of a tendency attributable to natural agents, which is closely comparable to our notion of a power.[6]

In fact, the author of the chapter on Aquinas was Peter Geach alone, and this and other writings of his have been an important factor in the revival of a broadly Aristotelian approach presented in engagement with contemporary analytical philosophy. To this has been added the significant revival of modal metaphysics through the work of Kripke and others in consequence of which appeals to essence and to nature-expressive causality are now familiar, if not commonly accepted, in contemporary philosophy of science and metaphysics. For the most part, however, these remain somewhat general, not engaging with the details of the special sciences, and a valuable feature of the essays gathered here is that they do exactly that.

III

I mentioned two thoughts brought to mind in reading the following essays, but let me add a third that begins with an anecdote. Thirty years ago (1987), I went from the University of St Andrews to the University of Pittsburgh as a fellow of its Center for Philosophy of Science. At that time, the Director of the Center was Nicholas Rescher, who was then, and remains, the most prolific author of academic philosophical monographs among English-language philosophers (and probably *all* professional philosophers). At that point, he had authored 50 books; since then, there have been over a hundred more. In anticipation of me going to Pittsburgh, my former PhD supervisor David Hamlyn, who was editor of *Mind* (and who had translated the

De Anima in the Clarendon Aristotle Series), said, adverting to the prodigious scale of Rescherian productivity, 'when you get there I want you to go into the Philosophy Department and call out "How many of you here are Nicholas Rescher?"' Anyone who is familiar with Aristotle's range, output and insight must feel something similar only intensified by the depth of his genius: 'How many of those in the Lyceum were Aristotle?' to which the answer, *mirabile dictu*, appears to be 'just one!' In my time at Pittsburgh, interest in Aristotle was very limited and exclusively historical, but 30 years on, the essays presented here, and many others besides, are testament to the enduring potential of Aristotelian philosophy to illuminate not only issues of human conduct, but the very nature of things inanimate, animate and minded.

Notes

* John Haldane is the J. Newton Rayzor Sr. Distinguished Professor of Philosophy at Baylor University, Professor of Philosophy at St Andrews University, Fellow of the Royal Society Edinburgh and Chair of the Royal Institute of Philosophy London.
1 Francis Bacon, *New Organon* (Aphorisms: Book I), LXIII (*my emphases*).
2 W.V.O. Quine, 'Modality', in *Word and Object* (Cambridge, MA: MIT Press, 1960), 199–200.
3 Larry Wright, 'Function', *Philosophical Review* 82(2) (April 1973):139–68, see 161.
4 R. Harré and E.H. Madden, 'Natural Powers and Powerful Natures', *Philosophy* 48(185) (1973):209–30.
5 R. Harré and E.H. Madden, *Causal Powers: A Theory of Natural Necessity* (Oxford: Blackwell, 1975), 130.
6 Op. cit., 98. For Anscombe and Geach see *Three Philosophers: Aristotle, Aquinas, Frege* (Oxford: Blackwell, 1961).

Introduction

Reflections on Contemporary Science and the New Aristotelianism

Robert C. Koons, William M. R. Simpson, and Nicholas J. Teh

Preliminary Remarks

A recent revival in (neo-)Aristotelian philosophy is beginning to transform the landscape of contemporary analytic philosophy. Ethics, metaphysics, and the philosophy of science are already feeling its influence.[1] In this volume, we are taking some first steps in bringing this revolution to the special sciences of physics, biology, and psychology.

What does it take for a philosophical project to count as *neo-Aristotelian*? We (the editors) suggest five possible criteria of demarcation:

First, neo-Aristotelian philosophers count the concept of *potentiality* as an essential feature of their metaphysics. A neo-Aristotelian philosophy embraces what is commonly called a *causal powers ontology*, in which both active and passive powers are regarded as fundamental features of particular things in the world that bring about change by some kind of natural necessity. In this picture, the passage of time essentially involves intrinsic changes, and these changes are the actualization of prior potentialities. Causation is reducible neither to patterns of categorical fact (as in the neo-Humean project) nor to fundamental, transcendent laws of nature (as in the Armstrongian project). A causal powers ontology entails the reality of some type of *teleology*, since powers exhibit what George Molnar called 'natural intentionality', being directed toward possibly unrealized (future) actualities.

Secondly, a neo-Aristotelian account must include what Jonathan Schaffer has called a 'layered' or 'structured' view of reality,[2] insofar as some entities and properties are regarded as being *more fundamental* than others, and other entities and properties are considered to be *derived*, existing in virtue of the existence of something more fundamental. The fundamental entities, according to some neo-Aristotelians, are basic *substances*; according to others, certain primary *powers*.

Thirdly, a neo-Aristotelian account of reality is not monistic but involves a plurality of entities, both simple and composite. The whole cosmos is not the only substance, nor is everything made from a single set of simple substances, but the world also includes natural unities like biological organisms, which are composite substances.

Fourthly, substances in nature belong to recurring natural kinds, each with its own intelligible nature or essence. These commonalties are not subjective, conventional, or wholly mind-dependent. With the possible exception of artefacts, neo-Aristotelians embrace a 'sparse' theory of real natures or essences (to use David Lewis's terminology). There are no adventitious or arbitrarily constructed kinds of things (like Nelson Goodman's *grue*). It is the task of empirical science to discover which kinds really exist and which causal powers are grounded in each kind.

Finally, neo-Aristotelians reject *extreme realism* or Platonism. Neo-Aristotelian accounts of nature include no appeal to non-immanent, non-natural universals; they instead hold that mathematical models of the physical world should be regarded as idealizations that invariably involve an empirical loss of scientifically relevant information.

For the philosopher or scientist who has yet to explore this burgeoning branch of contemporary philosophy, the existence of such anthologies as this one may at first seem surprising (or even perverse): why this sudden revival of interest in Aristotle's metaphysics and philosophy of science, nearly 600 years after the Scientific Revolution had supposedly discredited them? In our view, there are four developments over the last 50 years that account for this remarkable reversal:

First, there has been a resuscitation of metaphysics within the world of analytic philosophy at large (much of it inspired by Aristotelian and scholastic models)[3] and a growing dissatisfaction with the rigid Humeanism about modality and causation that characterized much of early analytic philosophy. Moreover, with the fall of the deductive-nomological model in the philosophy of science and the rise of causal explanation, new varieties of 'dispositionalism' have emerged that reject reductive Humean analyses of dispositions in favour of causal powers (or capacities) occupying a more fundamental role in metaphysics.[4]

Secondly, there is a growing body of philosophically informed research into Aristotle's conception of science whose cumulative effect has been to discredit the anti-scholastic myths and socio-politically charged caricatures of the early modern period in the eyes of many scholars.[5] Contrary to received opinion, Aristotle's natural science was not a purely deductive discipline that starts from supposed self-evident truths. In fact, Aristotle has been rehabilitated as one of the most significant pioneers of systematic empirical research. Moreover, the infamous errors of Aristotle's physics turn out to be tentative hypotheses entirely peripheral to his central commitments. It is possible to correct the shortcomings in Aristotle's science without simultaneously jettisoning the insights that are contained in his metaphysics.

Thirdly, the prolonged efforts of physicists and philosophers to make sense of the quantum revolution have forced reflective thinkers in both disciplines to begin digging more deeply into the metaphysical toolbox, recovering long-lost tools in order to cope with the problems raised, for instance, by the holism of quantum entanglement,[6] or the apparently irreversible nature of measurement,[7] or the ineliminable role of real potentialies,[8] and the coexistence of both commuting (classical) and non-commuting (quantal) properties.

Finally, instead of moving toward greater unification with 'fundamental physics', conceived in some monolithic or positivistic sense, the special sciences have achieved greater autonomy through the discovery of many levels of emergent phenomena.[9]

In an age of fragmentation and specialization in which modern philosophy has, at best, been permitted to play second fiddle to the sciences (or, at worst, been disregarded as irrelevant and outmoded), neo-Aristotelians are inspired by a revitalized conception of the philosopher as having an *essential* and *integrative* role to play in achieving knowledge of the natural order. Given the ambiguous and incomplete picture of reality painted by our best physical theories, the shift from physical reductionism in the philosophy of science toward a stronger appreciation for the integrity of the special sciences, and the resurgent interest in metaphysics in contemporary analytic philosophy, neo-Aristotelian philosophers today are invigorated by the prospect of achieving a unified *metaphysical* account of reality, enhanced by the insights of a rich philosophical tradition and informed by contemporary science. We think the essays we have gathered together in this collection convey something of that intellectual optimism and spirit of critical engagement.

Part 1

In what follows, we offer a brief summary of each of the papers included in this anthology. The first part of this collection is concerned with fundamental physics.

In Chapter 1, Xavi Lanao and Nicholas Teh defend Nancy Cartwright's conception of the natural world as 'dappled'. The current state of science points to the need for a plurality of theoretical frameworks, none of which has a claim to either universality or fundamentality. Such theoretical pluralism fits well with the Aristotelian conception of a natural world consisting of fundamental substances of a variety of kinds at a number of compositional levels. Lanao and Teh defend this pluralism by critically examining the strongest possible case for *fundamentalist unification*, which consists in setting aside the mysteries of quantum theory and arguing for the fundamentality of Classical Continuum Mechanics in relation to the remaining branches of physical theory, a case recently made by Sheldon Smith. Lanao and Teh show that even under these most favourable possible assumptions, the case for unification falls short. In order to secure the universality of continuum mechanics, the content of the theory must be reduced to nullity.

In Chapter 2, Edward Feser considers the challenge to traditional Aristotelian metaphysics posed by the 'block universe' of special relativity, as modeled by Minkowski spacetime. In this model, past, present, and future seem to be equally real and actual, in contrast to the Aristotelian conception of change, which consists in 'selectively conferring actuality on what are initially only potentialities'. Feser argues that the Aristotelian account of change is consistent with both the B Theory of time and with a wide variety of versions of the A Theory, in which case, the conflict between relativity and Aristotelian metaphysics would be only apparent. Even on the supposition that the Aristotelian account entailed the most extreme version of the A Theory, namely, Presentism, Feser contends that Aristotelianism would be consistent with the scientific core of special relativity. Feser outlines four possible approaches to reconciling special relativity with Presentism, each of which involves the postulation of a metaphysically privileged relation of simultaneity that is empirically inaccessible.

In Chapter 3, Robert Koons turns the focus toward quantum physics, more specifically, to the modern (Oxford) version of the Everett or many-worlds interpretation of quantum mechanics. The Everettian interpretation has four crucial advantages over its competitors: it postulates a unified, deterministic evolution of the quantum wave function, it does not involve the postulation of any additional, so-far-unverified collapse mechanism, it permits the consistent and unified description of the entire cosmos within the quantum framework, and it complements nicely recent work on the measurement problem (within the decoherence and consistent-histories approaches). However, Koons argues that the interpretation suffers from two critical flaws, both of which can be solved by the addition of an Aristotelian account of substantial form. First, the Oxford Everettians cannot apply Savage's account of subjective probabilities in order to make sense of the derivation of quantum probabilities via the Born rule, since Savage's axioms presuppose the *uniqueness* of outcomes. Second, the Oxford Everettians reliance on the strategy of functionalist reduction of the 'emergent' world of macroscopic objects leads to a radical proliferation of emergent realities, with one corresponding to every consistent or nearly consistent story. The solution to the second problem requires the addition to the theory of a fixed inventory of Aristotelian essences, each of which can be realized only in very narrowly specified ways by the underlying wave functions. The addition of such Aristotelian essences can also be used to solve the first problem, since the macroscopic substances realizing those essences can collectively select one world at each branch as the uniquely actual one, restoring the uniqueness of outcomes required by Savage's axioms. The result is a new interpretation of quantum mechanics, the *Traveling Forms interpretation*.

In Chapter 4, Alexander Pruss, offers a formal development of the Traveling Forms interpretation (TF) and defends it from various objections. Pruss also begins with the many-worlds interpretation, and, like Koons, he presses the objection that the MWI cannot make sense of quantum probabilities. Solving that problem leads naturally to the 'Many Minds' interpretation of Loewer and Albert, which can, in turn, be improved by moving to the 'Traveling Minds' interpretation of Squier and Barrett. However, both the Many Minds and Traveling Minds interpretations require an unreasonably extreme version of mind-body dualism. In contrast, the TF interpretation is consistent with mind-body supervenience and avoids the possibility of mindless zombies. Pruss develops a version of the TF interpretation that builds on the foundation of the modal interpretation of quantum mechanics: in particular, he relies on the indeterministic modal interpretation of Bacciagaluppi and Dickson. Pruss demonstrates that all substantial forms will naturally travel together through the branching structure of possible worlds.

In Chapter 5, William Simpson uses quantum physics to launch an attack on neo-Humean metaphysics (the main competitor to Aristotelianism in contemporary analytic metaphysics), in particular, a contemporary Humean account of the nature of causal powers. The phenomenon of quantum entanglement contradicts one of the two principal tenets of neo-Humeanism, as developed by David Lewis, namely, the principle of Humean Supervenience, according to which all facts supervene upon the distribution of intrinsic (local) qualities across ordinary spacetime. In a recent paper, Toby Handfield has attempted to rescue neo-Humeanism by jettisoning Humean Supervenience whilst retaining the other central tenet of Lewis's system, the Principle of Recombination. Simpson argues that the two theses are not as neatly separable as Handfield supposes. Any accommodation of quantum entanglement threatens to destroy the very possibility of localized powers and dispositions, resulting in a version of Spinoza's monistic necessitarianism. Quantum holism prevents processes (and their structures) from being isolated in the way that Handfield requires, with the result that the cosmos comprises just one gigantic and indivisible process. This lack of isolation makes the Recombination principle vacuous. Simpson also explores several alternative strategies for preserving neo-Humeanism, including ones involving Bohmian or objective-collapse mechanics. He argues that, whilst a modified form of Humean Supervenience can be reclaimed, such strategies end up with too impoverished a set of facts in the world's ultimate supervenience basis, excluding not only causal powers, secondary qualities and facts about consciousness, but also physical properties, like charge, mass, and spin, leaving only a world of featureless stuff. In fact, to rebuild a world of experiments and experimenters upon that slim foundation would require exactly the kind of unconstrained global functionalism that Koons criticized in Chapter 3.

In Chapter 6, Tuomas Tahko also considers the threat to localized causal powers posed by quantum holism. He challenges Jonathan Schaffer's recent proposal of *priority monism*. According to this account, whilst the cosmos is the only *fundamental* substance, it nonetheless has many *real* parts. Tahko points out that quantum entanglement, in the absence of an Aristotelian account of multiple substances, entails that the proper parts of the cosmos exist only as *appearances*, not in reality (whether fundamental or not). Tahko suggests that the distinctness of substances might be linked to incompatible properties in the quantum formalism.

Part 2

The second part of this anthology is focused on the special sciences of biology and psychology.

In Chapter 7, Christopher Austin and Anna Marmodoro propose a special kind of power possessed by certain composite entities—namely, a *dynamic structural power*, by which the entity sustains itself in existence over time. They turn to the contemporary program of systems biology to provide real-world examples of such powers. Organisms maintain themselves in existence through the manifestation of a specialized kind of causal loop, a loop that is able to persist despite the organism's mereological inconstancy. They argue that the ontological ground of this capacity for self-perpetuation is encoded in, but not reducible to, the features of the microscopic parts and their spatial and causal relations. The phenomenon of *phenotypic plasticity* entails that the stability that characterizes organic life is quite abstract: it is not the maintenance of a single, unchanging morphology, but rather, the unfolding of a morphological space, a landscape that is continuous and dynamically connected. An account of the unity of this space requires reference to a stable Aristotelian essence.

In Chapter 8, Christopher Austin further develops this picture of substantial diachronic unity by way of examining the 'evo-devo' (evolutionary-developmental) program in biology. Instead of focusing on whole organisms, Austin illuminates his account with reference to stable *developmental modules*, which persist across swaths of evolutionary history. Although such modules are not plausible candidates for the status of substances, they could be integral parts of such substances, with an identity and nature that is tied to the identity and nature of the organisms to which they belong. Just as Aristotle explains that the existence of a hand or eye is dependent on that of the whole animal, so the reality of developmental modules provides indirect evidence of the substantiality of organisms. Austin again appeals to the phenomenon of *phenotypic plasticity* and the existence of morphospaces with privileged regions of attraction as evidence of regulatory stability at the level of the organism.

In Chapter 9, David Oderberg responds to several objections to the existence of biological substances based on several kinds of borderline cases. Like Austin and Marmodoro, Oderberg argues that substantial form is needed as a principle of both synchronic and diachronic unity for organisms. He argues that, consequently, each substance can contain only one substantial form, and so substances cannot contain other substances as parts. In this chapter, he seeks to provide principled grounds for distinguishing biological substances (organisms) from non-substantial entities, such as autonomously existing organs and tissues, on the one hand, and multi-organism colonies and communities, on the other. Oderberg finds that the power of reproduction must play a central role in making these distinctions: organisms have the power to reproduce themselves, while mere organs and tissues do not. Although cells can reproduce themselves in special circumstances (as, for example, when we cultivate lines of stem cells), they are naturally suited to form tissues and organs that can only reproduce themselves via their contribution to the reproduction of an entire organism of the appropriate kind. Similarly, multi-species communities like lichens do not reproduce themselves. Instead, each species within the symbiotic community has its own independent mode of reproduction. In the case of natural communities of social animals, like colonies of ants, Oderberg relies on clear physiological analogies between individual ants and non-social organisms in closely related phylogenetic families, like beetles or roaches. The similarity between an individual ant and a member of a non-social insect species makes the inference to the substantial nature of the individual ant irresistible, with the consequence that the ant colony is a real but non-fundamental entity.

In Chapter 10, Janice Chik Breidenbach seeks to rehabilitate the idea of animals as agents. She rejects the accounts of those who seek a kind of causal exceptionalism for human or rational agents (like Descartes, Roderick Chisholm, or Donald Davidson). Instead, by building on the work of Helen Steward, she develops a fully naturalistic account of causation by agents. For Steward, animals are 'settlers' of matters of fact and are characterized by rudimentary intentional states. Following the lead of Elizabeth Anscombe, Steward and Breidenbach reject the identification of causation with the existence of strictly sufficient conditions. Breidenbach defends the possibility of real downward causation of parts by wholes in an indeterministic world by arguing that causation by substances is *more* fundamental than the event causation that has held center stage in philosophy since the time of David Hume. Like Marmodoro and Austin, Breidenbach emphasizes the importance of accounting for the diachronic unity of mereologically inconstant organisms, and she deploys the notion of substantial form as part of her response to Jaegwon Kim's challenge to the very possibility of synchronic downward causation.

In Chapter 11, William Jaworski argues for the superiority of an Aristotelian conception of psychology on the grounds that it enables us to dispense with the mind/body dichotomy. Instead of thinking of psychology as the study of a separate domain of mental phenomena, Jaworski proposes that we follow John Dewey and understand psychology and the social sciences as the sciences of persons and other organisms, understood as structured individuals. For Jaworski, the role of Aristotelian substantial form is played by *structure*, which (like Austin and Marmodoro) he defines as the substance's *power* to organize its material parts in such a way as to maintain itself and its characteristic activities in existence. Jaworski argues that our sciences also require structure (or form) as an account of the unity of certain biological and intentional *activities*. In order for structure to do the work of unifying powerful individuals and their activities, these structures cannot be identified with higher-order properties and relations of the organism's parts, as it is by functionalists and other reductionists. Instead, structure must be posited as a simple and irreducible form of causal power. Jaworski argues that we should embrace ontological naturalism, according to which we should take the practices of our best sciences as our guide to ontology. Given the recognition of structured activities of organisms (including human beings) in the biological and social sciences, such naturalism licenses a commitment to the real existence of powerful, structured individuals at the level of organisms.

In Chapter 12, Daniel De Haan suggests the possibility of some degree of rapport between Aristotelian hylomorphism and the New Mechanist Philosophy, a movement within the philosophy of biology and psychology. De Haan identifies a number of points of similarity between the two philosophical programs: both are committed to an ontological realism about organization and organized entities, both embrace pluralism about causation, explanation, and properties, and the two programs include similar accounts of the part-whole relation, with the causal powers of the parts partly grounded in features of the whole. Neither camp takes ontological emergence to be limited to the rational or psychological domain: instead, both see it as ubiquitous in biology. Even with respect to teleology, the differences lie within each camp rather than between them. In both cases, some prominent advocates defend a robust and non-perspectival realism about final causes.

Reflections

We (the editors) suggest that the two parts of this anthology may be seen to complement one another in the following way: we think the papers in Part 1 point to the need to recover the Aristotelian concept of *substantial form*, both as a source of unity and continuity (in relation to the amorphous, indeterminate character of the local, microscopic aspects of the quantum wave function) and as a source of distinction, separation, and locality (in relation to the cosmic holism of the wave function); the papers in Part 2, on the other hand, suggest that the life sciences provide us with *evidence* for the kind of substantial form that is required—one that grounds both the synchronic and diachronic unity of organisms in a way that is irreducible to the properties of the microscopic components.

In the quest for a metaphysical account of nature that is adequate to the richness of reality, we think it is important that the physical and life sciences should be considered together, not simply analysed in separation. In the absence of Part 1, we might legitimately question whether the evidence for the conceptual, explanatory, and methodological autonomy of the special sciences (discussed in Part 2) amounts to anything more than a 'polite form of microphysicalism', to paraphrase Bernard Williams.[10] In other words, we might still find ourselves wondering whether, when God made microphysical reality, he had to make anything extra in order to *complete* the world, to paraphrase Saul Kripke. However, Part 1, we suggest, does much to advance the case for the radical *incompleteness* of modern physics, at both the microscopic and cosmic scales, and the need for some form of ontological supplementation affecting the intermediate, macroscopic range. Conversely, in the absence of Part 2, we might worry that the paradoxes of contemporary physics are simply insoluble, an inescapable set of mysteries. However, the cumulative weight of Part 2 suggests we have good grounds for believing in substantial forms that unify macroscopic objects (namely, organisms and their functional parts), despite mereological and qualitative inconstancy. The biological sciences thereby provide us with ample evidence for the existence of causally powerful entities that are both fully *localized* and adequately *determinate*; that is, precisely those features that seem to be absent in our best current physical theories, as they have typically been understood.

Notes

1 See the suggestions for *Further Reading* at the end of this section.
2 Jonathan Schaffer, 'On What Grounds What', in David Manley, David J. Chalmers and Ryan Wasserman, eds., *Metametaphysics: New Essays on the Foundations of Ontology* (Oxford: Oxford University Press, 2009), 347–84.
3 P.F. Strawson, *Individuals* (London: Methuen, 1959); José A. Benardete, *Infinity: An Essay in Metaphysics* (Oxford: Oxford University Press, 1964); G.E.M. Anscombe, 'Causality and Determination', in Ernest Sosa, ed., *Causation and Conditionals* (Oxford: Oxford University Press, 1975), 63–81; Roderick M. Chisholm, *On Metaphysics* (Minneapolis: University of Minnesota Press, 1989); Peter van Inwagen, *Material Beings* (Ithaca, NY: Cornell University Press, 1990).
4 See Nancy Cartwright, 'Aristotelian Natures and the Modern Experimental Method', *Inference, Explanation and Other Philosophical Frustrations* 44–71(1992), and *The Dappled World: A Study of the Boundaries of Science* (Cambridge: Cambridge University Press, 1999).
5 See the section on 'Aristotle's science' in *Further Reading*.
6 Paul Humphreys, 'How Properties Emerge', *Philosophy of Science* 64(1997):1–17; Fred Kronz and Justin Tiehen, 'Emergence and Quantum Mechanics', *Philosophy of Science* 69(2002):324–47.
7 P. Coveney and R. Highfield, *The Arrow of Time: A Voyage Through Science to Solve Time's Greatest Mystery* (London: Allen, 1990); Ilya Prigogine, *Order Out of Chaos: Man's New Dialogue With Nature*, 1st ed. (London: William Heinemann Ltd., 1984).
8 The significance of this was first noted by Werner Heisenberg: W. Heisenberg, *Physics and Philosophy: The Revolution in Modern Science* (New York: Harper, 1958).
9 See the section on 'Emergence' in *Further reading*.
10 Bernard Williams, 'Hylomorphism', *Oxford Studies in Ancient Philosophy* 4(1986):189–99.

For Further Reading

For students or scholars who are interested in becoming better acquainted with the neo-Aristotelian revival in philosophy, we offer the following reading suggestions:

Ethics

Philippa Foot, *Natural Goodness* (Oxford: Clarendon Press, 2001).
Rosalind Hursthouse, *On Virtue Ethics* (Oxford: Oxford University Press, 1999).

Rosalind Hursthouse, Gavin Lawrence, and Warren Quinn (eds.), *Virtues and Reasons* (Oxford: Oxford University Press, 1995).
Alasdair MacIntyre, *Dependent Rational Animals* (Chicago: Open Court, 1999).
Martha C. Nussbaum, *The Fragility of Goodness* (Cambridge: Cambridge University Press, 1986).

Metaphysics

Kit Fine, "Essence and Modality: The Second Philosophical Perspectives Lecture," *Philosophical Perspectives* 8(1994):1–16.
Kit Fine, "Things and Their Parts," *Midwest Studies in Philosophy* 23(1999):61–74.
Ruth Groff and John Greco (eds.), *Powers and Capacities in Philosophy: The New Aristotelianism* (New York: Routledge, 2013).
John Hawthorne, *Metaphysical Essays* (Oxford: Clarendon Press, 2006).
Jonathan D. Jacobs (ed.), *Causal Powers* (Oxford: Oxford University Press, 2017).
Mark Johnston, "Hylomorphism," *The Journal of Philosophy* 103(2006):652–98.
Kathrin Koslicki, *The Structure of Objects* (New York: Oxford University Press, 2008).
Michael Loux, "Aristotle's Constituent Ontology," in Dean Zimmerman (ed.), *Oxford Studies in Metaphysics*, vol. 2 (Oxford: Clarendon Press, 2006), pp. 207–50.
E.J. Lowe, *The Four-Category Ontology: A Metaphysical Foundation for Natural Science* (Oxford: Clarendon Press, 2006).
Anna Marmodoro, "Aristotle's Hylomorphism, Without Reconditioning," *Philosophical Inquiry* 36(2013):5–22.
C.B. Martin, "Substance Substantiated," *Australasian Journal of Philosophy* 58(1980):3–10.
Trenton Merricks, *Objects and Persons* (Oxford: Oxford University Press, 2003).
George Molnar, *Powers: A Study in Metaphysics* (Oxford: Oxford University Press, 2003).
Daniel D. Novotny and Lukas Novak (eds.), *Neo-Aristotelian Perspectives in Metaphysics* (New York: Routledge, 2016).
David S. Oderberg, *Real Essentialism* (New York: Routledge, 2008).
Alexander R. Pruss, *Actuality, Possibility, and Worlds* (New York: Continuum, 2011).
Michael C. Rea, "Hylomorphism Reconditioned," *Philosophical Perspectives* 25(2011):341–58.
Tuomas Tahko (ed.), *Contemporary Aristotelian Metaphysics* (Cambridge: Cambridge University Press, 2013).
David Wiggins, *Sameness and Substance* (Cambridge, MA: Harvard University Press, 1980).

Philosophy of Science

Alexander Bird, *Nature's Metaphysics: Laws and Properties* (New York: Oxford University Press, 2010).
Nancy Cartwright, *How the Laws of Physics Lie* (Oxford: Oxford University Press, 1983).
Nancy Cartwright, *Nature's Capacities and Their Measurement* (Oxford: Oxford University Press, 1994).
Brian Ellis, *Scientific Essentialism* (New York: Cambridge University Press, 2001).

Rom Harré and E.H. Madden, *Causal Powers* (Oxford: Blackwell, 1977).

Stephen Mumford and Rani Lill Anjum, *Getting Causes From Powers* (Oxford: Oxford University Press, 2011).

Emergence

P.W. Anderson, "More Is Different," *Science* 177(1972):393–6.

Terence W. Deacon, "The Hierarchic Logic of Emergence: Untangling the Interdependence of Evolution and Self-organization," in B.H. Weber and D.J. Depew (eds.), *Evolution and Learning: The Baldwin Effect Reconsidered* (Cambridge MA: MIT Press, 2003), pp. 273–308.

Carl Gillett, "Samuel Alexander's Emergentism: Or, Higher Causation for Physicalists," *Synthese* 153(2006):261–96.

Carl Gillett, "On the Implications of Scientific Composition and Completeness: Or, the Troubles, and *Troubles*, of Non-Reductive Physicalism," in Antonella Corradini and Timothy O'Connor (eds.), *Emergence in Science and Philosophy* (New York: Routledge, 2010), pp. 25–45.

Robin F. Hendry, "Is There Downward Causation in Chemistry?" in D. Baird, L. McIntyre, and E.R. Scerri (eds.), *Philosophy of Chemistry: Synthesis of a New Discipline* (Dordrecht: Springer, 2006), pp. 173–89.

Kalevi Kull, Terrence Deacon, Claus Emmeche, Jesper Hoffmeyer, and Frederik Stjernfelt, "Theses on Biosemiotics: Prolegomena to a Theoretical Biology," *Biological Theory* 4/2(2009):167–73.

Robert Laughlin, *A Different Universe: Remaking Physics From the Bottom Down* (New York: Basic Books, 2005).

Michael Polanyi, "Life Transcending Physics and Chemistry," *Chemical Engineering News* 45(1967):54–66.

Michael Polanyi, "Life's Irreducible Structure," *Science* 160(1968):1308–12.

Mariam Thalos, *Without Hierarchy: The Scale Freedom of the Universe* (Oxford: Oxford University Press, 2013).

Aristotle's Science

David Charles, *Aristotle on Meaning and Essence* (Oxford: Clarendon Press, 2002).

David Charles (ed.), *Definition in Greek Philosophy* (Oxford: Oxford University Press, 2010).

Montgomery Furth, *Substance, Form and Psyche: An Aristotelian Metaphysics* (Cambridge: Cambridge University Press, 1988).

Allan Gotthelf, "Aristotle's Conception of Final Causality," *The Review of Metaphysics* 30/2(1976):226–54.

Allan Gotthelf and J.G. Lennox (eds.), *Philosophical Issues in Aristotle's Biology* (Cambridge: Cambridge University Press, 1987).

Monte Johnson, *Aristotle on Teleology* (Oxford: Oxford University Press, 2005).

J.G. Lennox, *Aristotle's Philosophy of Biology: Studies in the Origins of Life Science* (Cambridge: Cambridge University Press, 2001).

J.G. Lennox and Robert Bolton (eds.), *Being, Nature, and Life in Aristotle* (Cambridge: Cambridge University Press, 2010).

G.E.R. Lloyd, *Aristotelian Explorations* (Cambridge: Cambridge University Press, 1996).

Theodore Scaltsas, David Charles, and Mary Louise Gill (eds.), *Unity, Identity, and Explanation in Aristotle's Metaphysics* (Oxford: Oxford University Press, 1994).

Part 1

The Philosophy of Physics

1　Dodging the Fundamentalist Threat

Xavi Lanao and Nicholas J. Teh

1. Introduction

As demonstrated by the collection of essays in this volume (as well as Tahko (2012) and Novotny and Novak (2014)), Aristotelianism has recently been enjoying a revival in various subfields of philosophy, including metaphysics and the philosophy of science. To take an example that will play a motivating role in our discussion, the theme of "hylomorphism" has been widely discussed in analytic metaphysics (see Koons 2014) and references therein), as well as in the philosophy of quantum theory (Pruss 2017), the philosophy of mind (Jaworski 2017), and the philosophy of biology (Austin 2017).

Although the term "hylomorphism" is sometimes used very loosely, Neo-Aristotelians of a strict observance will want to reserve this term for a theory of substance that avoids collapsing into either dualism or materialism. In particular, in order to avoid the latter collapse, Neo-Aristotelian substances must exhibit a strong form of unity, i.e. a unity such that the powers of a substantial whole cannot be completely grounded in the powers of its parts. (On this score, see Koons (2014), who argues for the more specific thesis that there must be a mutual relationship of partial grounding between a substantial whole and its parts.) More generally, many Neo-Aristotelians are interested in the possibility of a kind of emergent causal efficacy (such as the powers of an organism) which cannot be reduced to forms of causal efficacy that are sometimes regarded as more "fundamental" (such as the powers of the electrons that comprise the organism).

While it is a truism that one cannot directly read metaphysics off of science, the above Neo-Aristotelian position can still be threatened by weaker claims about the relationship between science and metaphysics. Consider, for instance, the metaphysical thesis called "fundamentalism", which holds that the laws of a (hypothetical) unified physical theory exhaustively govern all of material reality: In this view, it is difficult to see how the powers of substances could be anything other than entirely grounded in the powers of the entities of the unified physical theory; hence, this metaphysical picture is inconsistent with Aristotelian hylomorphism.[1] But why should anyone believe in such a metaphysical thesis? It is at this juncture that many thinkers implicitly or explicitly invoke a connection with scientific practice:[2]

Fundamentalist Unification

The success of science (especially fundamental physics) at providing a unifying explanation for phenomena in disparate domains is good evidence for fundamentalism.

Thus, although Neo-Aristotelians are not directly threatened by anything in science, they will want to find ways of warding off this interpretation of the "unifying role" that physics plays with respect to disparate domains of phenomena.

The goal of this essay is to recommend a particular set of resources to Neo-Aristotelians for resisting Fundamentalist Unification and thus for resisting fundamentalism. The set of resources in question originates in the work of Nancy Cartwright, who has famously drawn on the details of scientific practice in order to launch an argument against fundamentalism.[3] We would like to urge two points in particular:

(i)   Anti-fundamentalism is a live option, because genuine arguments in favor of Fundamentalist Unification are hard to come by, and the best (and most fully worked-out) argument for it rests on assumptions that beg the question against Cartwright's epistemology of scientific models.
(ii) Neo-Aristotelians should find Cartwright's epistemology of scientific models appealing because it adopts the broadly Aristotelian approach of prioritizing the concrete over the abstract—call this approach "concretism".

Bringing these two points together, we submit that Cartwright's approach offers Neo-Aristotelians a distinctively concretist epistemology of scientific models that has the added benefit of making room for robust forms of Aristotelian metaphysical doctrines, such as hylomorphism. Nonetheless, we will also urge that Cartwright's approach is in many ways underdeveloped and that it needs to be more fully worked out if it is to be incorporated into a compelling Neo-Aristotelian picture of the relationship between metaphysics and science.

The plan of the essay is as follows. Section 2 discusses the topic of what scientific practice-based reasons one might marshall in favor of fundamentalism. As we see it (and as the question has been understood in the literature), any practice-based attempt to adjudicate this issue has to reckon with two considerations that seem to pull in opposite directions. On the one hand, even our most impressive scientific theories seem to only apply to domains of reality in a "patchwork" way, and on the other hand, it is undeniable that much scientific work consists in devising theoretical structures that are in *some sense* "unifying". Intuitively, the first consideration provides prima

facie evidence against fundamentalism, and the second consideration provides prima facie evidence in favor of it. Thus, a successful argument for Fundamentalist Unification needs to provide a convincing account of "theoretical unification" that explains away the appearance of theories only having a patchwork application to reality. One such account that claims to be "practice-based" is that of Smith (2001), and we will consider his argument in favor of Fundamentalist Unification in this section.

In Section 3, we will argue that Smith's account of theoretical unification turns on a specific and controversial epistemology of scientific models. We then highlight how Cartwright's concretist account of the epistemology of models explicitly rejects Smith's assumptions and leads to a different way of understanding substantive theoretical unification. Indeed, this rival epistemology forms the basis of Cartwright's famous argument against fundamentalism, which we then discuss.

Section 4 draws on our previous discussion to sketch some general guidelines for the project of developing a concretist epistemology of models. Although we take inspiration from Cartwright's anti-fundamentalist morals, we also highlight some ways in which the approach that we recommend diverges from hers.

2. Fundamentalism and Its Justification

Metaphysical fundamentalists believe that the universe is exhaustively governed by a limited set of principles, which are often called "fundamental laws of nature". The popularity of metaphysical fundamentalism presumably derives from the metaphysical hope—shared by various scientists and philosophers alike—that science will eventually discover the laws of nature that *exhaustively govern all of material reality*, from the causal agents implicated in quantum gravity to human and non-human organisms. These fundamental laws of nature are usually taken to be truths expressed in mathematical language, which accurately describe the behavior of all things in the world, at all times and places. However, they are typically not taken to be the actual laws of our most fundamental physical theories, but the laws of some future "Final Science" or "True Physics" that our current scientific efforts aim at (cf. Sklar 2003: Sec. 5; Hoefer 2010: 308).

In this paper, we will put aside armchair metaphysical speculation and instead focus on the following question: Based on the *practice of science*, what reasons might one have for accepting or rejecting metaphysical fundamentalism? By our lights, all parties to the debate will have to reckon with two practice-based facts that appear to be in tension with each other:

(1) Scientific theories have the appearance of having "patchwork" domains of application;
(2) Theories have been successful at providing "unifications" of such domains.

Regarding (1), both fundamentalists and anti-fundamentalists should agree that our explanatory and predictive practices in science suggest a much more "dappled" or "patchwork" picture of scientific activity than what the fundamentalist hopes for. For instance, biology and chemistry are scientific disciplines which seem to operate autonomously from physics: They often construct theories, give explanations, and make successful predictions without taking into account any fundamental laws of physics. Furthermore, fundamentalists should concede that even upon restricting scientific activity to "physics", it often appears to be the case that different domains of phenomena are described by different physical theories: To give an elementary example, point particle mechanics and fluid dynamics are physical theories that apply to relatively disjoint sets of classical phenomena.

With respect to (2), fundamentalists and anti-fundamentalists should likewise agree that various theories have had empirical and theoretical success in playing a "unifying role" with respect to phenomena in different domains. The issue at stake is how such "unifications" should be understood and what we are justified in inferring from them.

Let us briefly consider a fundamentalist narrative that emphasizes a particular understanding of (2) and uses this to explain away (1).[4] Suppose that fundamentalists and anti-fundamentalists agree that, at least within the confines of certain experimental scenarios, we have good reason to believe in the truth of mathematical laws describing the behavior of basic kinds of particles/fields and their interactions. One fundamentalist strategy for describing the unifying role of particle physics is to then elaborate on the narrative as follows: We also have good reason to believe that everything in the physical world is made up of these same basic kinds of particles. So, from the fact that everything is made up of the same basic particles and that we have reliable knowledge of the behavior of these particles under some experimental conditions, it is plausible to infer that the mathematical laws governing these basic kinds of particles within the restricted experimental settings also govern the particles everywhere else, thereby governing everything everywhere (Hoefer 2010: 317–18). A fundamentalist of this stripe would resist claims that the "patchwork" picture of science constitutes *prima facie* evidence in favor of anti-fundamentalism by denying that the patchwork picture carves at the joints of reality. Thus, for instance, Sklar claims that although explanations in biology and chemistry describe real phenomena in the world and are certainly useful for predictive purposes, they are not characterizing how things "really are" (Sklar 2003: Sec. 4).

From the anti-fundamentalist's perspective, the problem with this assertion is that it takes fundamentalism for granted, and then uses it to dismiss the *prima facie* evidence for anti-fundamentalism, viz. (1). If fundamentalists want to provide non-question-begging arguments for their position, they cannot assume that fundamentalism is true, since that would just be a statement of their (metaphysical) faith in fundamentalism. Instead, what

fundamentalists should do is to provide *evidence* in the form of an analysis of scientific practice that supports (2) and dispels the worries raised by (1). In other words, what is primarily at stake in the debate is whether Fundamentalist Unification is true.

Smith's (2001) essays provide just such an account, albeit one that is limited to the domain of classical phenomena, i.e. systems for which the units of action are large in comparison with Planck's constant, and whose speeds are small relative to the speed of light. He focuses on the question of whether classical phenomena are theoretically unified by a single theory, viz. Classical Continuum Mechanics (CCM). Although the focus on the case of classical mechanics might be seen as incurring a loss of generality, this is a particularly relevant case study for the debate about Fundamentalist Unification for two reasons. First, it focuses on well-established scientific theories that have been at the epicenter of this debate from the beginning (the origin of the debate can be traced back to Cartwright (1994)). And, second, this case does not involve inter-level relations or relations among different sciences, but focuses exclusively on a particular set of physical theories for which the length/energy scale is held fixed.[5] Accordingly, this is arguably the case in which it is *easiest* for the fundamentalist to demonstrate the unity of the domain of classical mechanics, because there are no complicating factors, such as complex inter-level relations, or a lack of well-established scientific theories that operate in that domain. Conversely, if we can show that the fundamentalist fails to even make it plausible that the domain of classical physics is unified, this would provide a strong case for the plausibility of anti-fundamentalism.

In constructing his argument, Smith draws on the work of Clifford Truesdell (1991), who sought to place Continuum Mechanics on a deductive foundation akin to that of Euclid's deductive axioms for geometry. Recall that in the theory of fluids and continuous media, "constitutive equations" are used to describe the response of a particular material to external stimuli—in other words, such equations provide a formal specification of some particular material. Truesdell aspired to provide a general theory of the constraints that should be satisfied by "physically reasonable" constitutive equations such as "determinism", "local action", and various invariance principles; let us call these the "Truesdell Constraints".[6] With that in mind, we are now in a position to lay out Smith's argument in favor of the following classical version of the Fundamentalist Unification thesis:

Classical Fundamentalist Unification (CFU)

The unificatory success of classical continuum mechanics demonstrates the truth of fundamentalism about the classical domain, viz. the laws of continuum mechanics govern the behavior of all phenomena within the classical domain.

Smith's Fundamentalist Argument for CFU

(F1) A theory only applies to a domain insofar as it provides a prin-cipled way of generating a set of models that are jointly able to describe all the phenomena in that domain.

(F2) The theory of CCM admits of an infinite number of constitutive equations (or *formal* models of materials).

(F3) The admissible constitutive equations of (F2) satisfy a small num-ber of simple laws, viz. the Truesdell Constraints, which ensure that they obey some set of "physically reasonable" axioms.

(F4) By (F2) and (F3), CCM has an infinite number of principled con-stitutive equations/models.

(F5) The constitutive equations of (F2) suffice to describe almost any classical phenomenon.

(F6) By (F1), (F4), and (F5), CCM applies to the whole classical domain.

(Note that although our reconstructed argument focuses only on the case of contact forces, i.e. forces acting on the surface of a body, such as the force of the wind, a completely analogous argument can be constructed to cover the case of body forces, i.e. forces that act throughout the body, such as gravitation.)

The argument's conclusion, i.e. (F6), is that CCM applies to the whole domain of classical mechanics, in the sense that for any possible phenom-enon in that domain, the theory has a "principled" model that accurately describes that phenomenon. Thus, the argument attempts to establish Fundamentalist Unification by providing an *interpretation* of the unifying power of continuum mechanics that allows one to deduce CFU.

Notice that (F1) defines the conditions under which we can consider a particular domain to be unified by a theory. At a schematic level, (F1) is a shared premise between fundamentalists and anti-fundamentalists; how-ever, fundamentalist and anti-fundamentalist may disagree on the determi-nate interpretation of what counts as a principled model, which will in turn inform their judgments about when a domain counts as theoretically uni-fied. In particular, (F2–4) assume a particular understanding of what it is for a theory to generate a model in a principled manner. For Smith, all that is required to generate such a principled model is to show the existence of a constitutive equation that satisfies the Truesdell Constraints (which define an infinite class of formal models for materials) (Smith 2001: 471). (F5) then asserts that such models suffice to provide a good description for "almost" any classical phenomena, thus effecting the desired "theoretical unification" of the classical domain. The next section elaborates on, and interrogates, the underlying assumptions of this particular account of theo-retical unification; we will argue that once these assumptions are uncovered, Smith's argument loses whatever initial plausibility it might have had as a non-question begging defense of CFU.

Section 3: Anti-Fundamentalism and Cartwright's "Concretist" Epistemology of Models

In Section 3.1, we will first show that Smith's argument in favor of CFU assumes a particular (and controversial) epistemology of models. We then sketch a rival epistemology of models that is due to Cartwright (1999)—we call this "concretism". Section 3.2 then explains how concretism is deployed by Cartwright in her argument against CFU. The moral of this section, then, is that CFU (and Fundamentalist Unification, more generally) is extremely sensitive to one's preferred epistemology of models. Moreover, the Neo-Aristotelian has little reason to accept anything close to the epistemology that Smith proposes.

Section 3.1: Aristotelian Concretism About Models

As noted above, the main point of contention between Smith and Cartwright is their interpretation of a "theoretical unification". Consider that by invoking the notion of a "principled" model, premise (F1) of Smith's argument is referring to a set of criteria that need to be met in order for a theory to count as "unifying" a particular domain: Theories only apply to a domain *D* insofar as they can provide principled models for phenomena in *D*; thus, a *unified* theory of *D* needs to be able to provide principled models for *all* phenomena in *D*. In its schematic form, (F1) is shared by fundamentalists and anti-fundamentalists alike, because both parties agree that not just any model will count as principled! However, disagreements arise once we get into the specifics of the criteria according to which a model will count as principled, which will in turn determine what a "theoretical unification" consists in.

We will now explore Smith's understanding of what it takes for a model to be principled. Recall that in his Fundamentalist Argument, premises (F2) and (F3) jointly describe the existence of an infinite set *S* of models that satisfy physically reasonable assumptions that we called the Truesdell Constraints; however, these premises do not yet claim that such models count as "principled" ways in which the theory can be applied to describe classical phenomena. Indeed, as far as these premises go, such a characterization might be consistent with the models being purely *ad hoc* or phenomenological (relative to the theory). The additional claim that the models in *S* are principled is made in (F4), and the claim that any classical phenomenon will be described by a model in *S* is made in (F5)—these two premises, then, are the controversial ones that we will now interrogate.

Why should one think that (F4) is true? A plausible way of reconstructing Smith's reasoning here is as follows: Evidently, we have a general method for demonstrating the existence of constitutive equations which satisfy the Truesdell Constraints; indeed, we know that there is an infinite set *S* of such models (cf. Smith 2001: 471–2). Now, several of these models happen to be *explicitly constructible* and have been applied with great empirical success; the success of these models can then be taken to justify other models in *S* (Smith 2001: 474).

Based on this line of thought, let us spell out more precisely what Smith means by a "principled" model. According to Smith, a model is principled relative to a particular theory if it (i) fulfils the formal constraints that the theory sets for the generation of models (i.e. the Truesdell Constraints), (ii) these constraints represent some relevant physical aspects of the phenomena in the domain (i.e. general physical properties of materials), and (iii) some models that satisfy the constraints have been successfully used to describe, predict, and explain phenomena in that domain. This account of "principled" is not a purely formal one, because *some* models of *S* need to successfully predict empirical phenomena for this account to get off the ground; however, it is still highly formal in the sense that once this minimal empirical requirement has been met, the warrant of those models is transferred to *S* as a whole by very formal means, i.e. simply in virtue of the models in *S* satisfying the Truesdell constraints; the models in *S* are then judged to be principled. We shall call this general approach to the epistemology of models an "abstractionist" epistemology, since it is willing to count models as principled on the basis of highly formal criteria.

It seems to us that this "abstractionist" epistemology of models should raise the suspicions of Neo-Aristotelians, as well as others who philosophers who share the intuition that warrant does not transfer in virtue of the kind of formal relationship that Smith wishes to lean on. As a toy example, consider the relationship between the following two models A and B in some set *S* of models which satisfies "physically reasonable constraints": Let A be a model that has been extremely fruitful and empirically well-confirmed, such as the simple harmonic oscillator, and let model B simply be some other element of *S* whose existence we can perhaps only guarantee abstractly. Why should the empirical confirmation that attaches to A (in virtue of which it intuitively counts as "principled"), or some part thereof, transfer to B simply in virtue of their both being elements of *S*? Plausibly, the fact that they both satisfy "physically reasonable constraints" cannot be made to bear the weight of such a transfer of warrant, and, furthermore, it is at least conceivable that many models in *S* might simply fail to apply to anything in reality (or even in possible worlds that are "close" to the actual world). By contrast, A and B usually have a much tighter relationship in cases where we think the warrant has some chance of transferring, e.g., a case in which A is the simple harmonic oscillator, and B is a modification of A to include a damping term. Returning now to the case of CCM, one might reasonably expect that the grounds on which a model is justified should include knowledge of how that model is related to some actual set of (real) materials and experimental settings, i.e. knowledge that goes beyond the idea that all "reasonable" constitutive equations should satisfy the Truesdell Constraints. Thus, from the concretist point of view, all that

Smith has shown is that CCM has the resources to articulate some very schematic degrees of freedom (i.e. the abstract notion of a constitutive equation satisfying the Truesdell constraints) which can be "filled in" in many different ways. Upon being filled in, they may or may not be applicable to real phenomena, and even if they are applicable, the extent to which such an application counts as "principled" should be taken to be an open question.

Let us now shift our focus to (F5): Why should we think that the set of constitutive equations compatible with the Truesdell Constraints is able to describe *all* real materials within the classical domain? According to Smith, the fact that there is an almost infinite number of constitutive equations that satisfy the Truesdell Constraints is good reason to believe that each phenomena of the domain will be described by some constitutive equation (Smith 2001: Sec. 4.2). However, just pointing out that the set of models that satisfy the Truesdell Constraints is infinite does not show that this set will contain all the models needed to describe each phenomenon in the domain. After all, it is reasonable to think an infinity of potential models might not be able to account for the diversity of real materials that one encounters in nature. Thus, an argument is needed here in order to show that any real material can be modeled (in a principled manner) by an element of S. We now consider an argument to this effect that is implicit in Smith's remarks.

Smith thinks there are reasons to be optimistic about not just the number but also the variety of models contained in S—they are so various, he thinks that it is almost unthinkable that they could fail to model a real material in a principled way. When discussing the potential infinity and variety of contact forces, he presents as an example that he calls "Lindsay's function", which is a function introduced in Lindsay's classic textbook on CCM. Lindsay's function is an exponential "ansatz" that can account, via Fourier series theorems, for almost any function, subject to a small set of constraints. And since almost any force can be modeled by such a function, or so Smith's story goes, Lindsay's function provides principled models for such forces. Smith then wishes to run an analogous argument in the case of CCM, where the analog of the models generated by Lindsay's function is S, i.e. the set of constitutive equations satisfying the Truesdell constraints (Smith 2001: Sec. 4.1).

The problem with this strategy for generating an infinite—and infinitely diverse—set of principled models that will account for any real material is that it severely vitiates the plausibility of Smith's understanding of "principled". If we allow almost any function to fulfill the theoretical constraints and to define a principled model (as Smith suggests is the case in his example regarding Lindsay's function), the relationship between empirically successful principled models and the merely potential principled models turns out to be very loose indeed. Why should we think that we can transfer the epistemic warrant from empirically successful models in the set S to all the other models of S if we allow almost any function to define a principled model in the set? It seems that the reasons Smith gives to believe that (F5) is justified also severely undermine the justification for (F4), thereby threatening the ability of Smith's own definition of "principled" to set any substantive constraints on what counts as a principled model.

As we have just seen in detail, the disagreement between fundamentalists and anti-fundamentalists boils down to a disagreement about how to understand principled models. The Neo-Aristotelian philosopher may acknowledge that CCM can generate a set of models potentially capable of describing phenomena involving fluids and inter-body gravitational forces; however, the Neo-Aristotelian would take issue with understanding the relationship between the theory, the model, and the model's application to material phenomena in the highly formal terms that Smith relies on. Rather, philosophers of a Neo-Aristotelian inclination tend to think that experimental and material knowledge play a far more substantive role in the generation and justification of principled models. We now turn to a way of spelling out these Neo-Aristotelian thoughts that has been championed by Nancy Cartwright.

Nancy Cartwright has developed an in-depth analysis of scientific practice defending what we call a "concretist" epistemology of models. The immediate consequence of her epistemology for the case at hand is that in order for a model to be principled, it is not enough that it accords to some general formal principles; we also need to have the relevant material and experimental knowledge that allow us to determine and justify whether some specific mathematical model fits the phenomenon at hand.

Cartwright does not offer an explicit definition of what it is for a model to be principled; however, she does offer heuristics concerning how principled models are generated in scientific practice.[7] As we understand Cartwright, a model M of a phenomenon X in experimental context Y is principled only if it has been generated from theory T by means of a process that satisfies the following conditions:

(i)   The construction of the model takes into account information regarding the material phenomenon X and experimental context Y (e.g., what causal factors of the phenomenon X are relevant, how to properly shield the experimental context Y in order to isolate X, etc.),

(ii) The construction of the model takes into account information regarding how to construct a computationally tractable model (e.g., how to generate a mathematical model that can be used for calculations, simulations, etc.), and

(iii) The construction of the model takes into account knowledge about how to bridge the gap between theoretical and experimental knowledge (e.g., what can be simplified or idealized, what parameters should be defined, etc.).

Cartwright captures these conditions for the derivation of principled models in scientific practice by referring to a set of rules that she calls "bridge principles". Bridge principles are the primary guidelines for determining how we can generate principled models from theory; indeed, they also determine the scope of the theory's application by constraining the range of models that can be derived from a theory in a principled way. Note, however, that Cartwright emphasizes that bridge principles do not generate models solely on the basis of theoretical input; one must also have in mind the target of the model and its intended experimental context. In this sense, bridge principles should provide a way of determining whether the model matches the target phenomenon that is to some degree independent of the theory.

In Section 4, we will elaborate on how to more precisely understand a concretist epistemology of models more generally. For now, let us just emphasize once more that Smith's argument requires us to accept an abstractionist epistemology of models in order for premises (F4) and (F5) to be justified. As we have noted, this abstractionist epistemology relies on allowing very permissive constraints on what counts as a principled model that strongly contrasts with Neo-Aristotelian methodological inclinations regarding the epistemic relevance of material and experimental knowledge. Fortunately for the Neo-Aristotelian, an abstractionist epistemology is not the only epistemology of models available. As we have seen, Cartwright presents a concretist epistemology of models that undermines key premises of Smith's argument. Furthermore, as we show in Section 3.2 below, Cartwright's concretist epistemology can be put into work in order to construct a counter-argument against Fundamentalist Unification.

Section 3.2: Cartwright's Anti-Fundamentalist Argument

The general underlying principle behind the above discussion is that, in order to understand what the standards are for a good theory-model relationship (i.e. how to produce and recognize principled models), we need to go beyond the theory. In addition to the theory, we also need to take into account theoretical know-how regarding what is computationally tractable, experimental know-how regarding the material conditions to implement the model, and practical know-how regarding how to bridge from theory to experiment.

On the basis of her analysis of scientific practice, Cartwright famously argues that science does not offer us a fundamentalist picture of the world. Instead, it offers us an image of the world as "dappled", i.e. the world is not governed by a limited set of universal laws that apply to all domains, but is instead divided into small pockets of restricted regularities (Cartwright 1994). Here is the structure of her argument, as applied to the classical domain:

Cartwright's Anti-Fundamentalist Argument

    (F1)   Theories only apply to a domain insofar as there is a principled way of generating a set of models that are jointly able to describe all the phenomena in that domain.

(AF2)   Classical mechanics has a limited set principled models, so it only applies to a limited number of sub-domains.

(AF3)   The limited sub-domains of AF2 do not exhaust the entire classical domain.

(AF4)   From (F1), (AF2), and (AF3), the domain of classical mechanics is not universal, but dappled.

As we mentioned above, (F1) is a shared premise between fundamentalists and anti-fundamentalists. The differences arise in how they respectively understand what it takes for a model to be principled; Cartwright's view on this is captured by premises (AF2) and (AF3) of the anti-fundamentalist argument.

Cartwright justifies (AF2) by pointing out that we only have the knowhow to reliably apply a handful of bridge principles that cover only a limited set of sub-domains. This is because most of our material and experimental knowledge is limited to situations we can control by shielding the target system from external interferences. Given these limitations on our material and experimental knowledge, we cannot generate principled models that describe phenomena outside these shielded sub-domains. Accordingly, Cartwright's strategy to justify (AF3) is just to point out that the classical domain is much larger that the conjunction of all the sub-domains defined by those shielded situations for which we have principled models that describe them.

The main insight driving Cartwright's argument for the disunity of classical mechanics is precisely that for many phenomena within the domain of classical mechanics, we lack the material and experimental knowledge necessary to generate and apply models in a principled way. We may well have a formal mechanism that allows us to show the existence of a model, as suggested by Smith's analysis, but this is insufficient to provide the right kind of theoretical unification.

Cartwright takes her concretist epistemology as a starting point for her analysis of scientific practice, which leads to the conclusion that the domain of classical mechanics is not universal, but dappled (i.e. AF4). Accordingly, Cartwright's argument is an attempt to undermine CFU (and thereby Fundamentalist Unification) by providing an analysis of the relation between theories, models, and phenomena that highlights the limitations of classical mechanics to unify its whole domain.

Section 4: Towards a Concretist Epistemology of Models

One of the limitations of Cartwright's concretist epistemology is that, as presented, the normative principles that she recommends are rather vague, and it is difficult to see how they generalize to cover different kinds of cases. The root cause of these limitations is what we might call Cartwright's extreme "particularism" about the philosophy of science. Her writing sometimes suggests that she does not think one can give any general guidelines and/or constraints for how to identify principled models and bridge principles, as well as the role that they play; all we can do is investigate the details of how they are used in scientific practice. For instance, she remarks the following with respect to interpretative models:

> And what kinds of interpretative models do we have? In answering this, I urge, we must adopt the scientific attitude: we must look to see what kinds of models our theories have and how they function, particularly how they function when our theories are most successful and we have most reason to believe in them. In this book I look at a number of cases which are exemplary of what I see when I study this question. It is primarily on the basis of studies like these that I conclude that even our best theories are severely limited in their scope.
>
> (Cartwright 1999: 9)

So, for Cartwright, the amount of details and precision that a general account of principled models (or most scientific concepts) is very limited precisely because of the richness and variety that we find in scientific practice. At the end of the day, if we want to know the scope of a particular theory or whether a specific model is principled, we have to do the hard work of looking into the details of that theory and that model and how are they used in scientific practice.

In this section, we attempt to depart from Cartwright's specific strain of particularism in order to attempt to define some general guidelines for a concretist epistemology of models. Section 4.1 presents Cartwright's main case study in favor of her concretist epistemology of models, what she calls the "Neurath's Bill" case. We use this case as an illustration of some of the central desiderata for a general concretist epistemology. The next sub-section, Section 4.2, will use the lessons learned in 4.1 to define the main desiderata for a concretist epistemology of models emphasizing the differences with Smith's abstractionist epistemology. Finally, Section 4.3 addresses the potential worry of whether and how a concretist epistemology of models can accommodate the role of theory and theoretical unification in scientific practice.

4.1 Some Lessons From the Neurath's Bill Example

In keeping with her belief that the philosophy of science cannot be seriously conducted except by attending to the details of scientific practice, Cartwright's arguments are generally framed within the context of a physical example. For instance, her "Neurath's Bill" example is the setting for trying to demonstrate "dappledness" within the domain of classical mechanics. In this example, a $1,000 bill is dropped from the top of St. Stephen's Square, travels erratically over the air while being swept away by the wind, and finally lands somewhere in the square. Cartwright is skeptical that any theory will provide a "good description" for such cases, which has aspects best described by classical fluid mechanics, on the one hand, and aspects best modeled by point particle mechanics, on the other hand. It is worth quoting her conclusions from this example in full:

> Let us set our problem of the 1000 dollar bill in St. Stephen's Square to an expert in fluid dynamics. The expert should immediately complain that the problem is ill defined. What exactly is the bill like: is it folded or flat? straight down the middle, or . . . ? is it crisp or crumpled? how long versus wide? and so forth and so forth and so forth. I do not doubt that when answers can be supplied, fluid dynamics can provide a practicable model. But I do doubt that for every real case, or even for the majority, fluid dynamics has enough of the 'right questions' to ask to allow it to model the full set of causes, or even the dominant ones. I am equally sceptical that the models that work will do so by legitimately bringing Newton's laws (or Lagrange's for that matter) into play. How then do airplanes stay afloat? Two observations are important. First, we do not need to maintain that no laws obtain where mechanics runs out. Fluid dynamics may have loose overlaps and intertwinings with mechanics. But it is in no way a subdiscipline of basic physics; it is a discipline on its own. Its laws can direct the 1000 dollar bill as well as can those of Newton or Lagrange. Second, the 1000 dollar bill comes as it comes, and we have to hunt a model for it. Just the reverse is true of the plane. We build it to fit the models we know work. Indeed, that is how we manage to get so much into the domain of the laws we know.
>
> Many will continue to feel that the wind and other exogenous factors must produce a force. The wind after all is composed of millions of little particles which must exert all the usual forces on the bill, both at a distance and via collisions. That view begs the question. When we have a good-fitting molecular model for the wind, and we have in our theory (either by composition from old principles or by the admission of new principles) systematic rules that assign force functions to the models, and the force functions assigned predict exactly the right motions, then we will have good scientific reason to maintain that the wind operates via a force. Otherwise the assumption is another expression of fundamentalist faith.
>
> (Cartwright 1999: 284–5)

Let us highlight some important points in the quoted passage that bear on Cartwright's concretist epistemology of models. First, it is important to note that Cartwright does not present this example as a challenge to the possibility of giving a scientific explanation of the trajectory of the bill, but as a challenge to the fundamentalist claim that all "classical phenomena" fall under the domain of the same theoretical principles or laws. The core disagreement between fundamentalists and anti-fundamentalists is not about whether there are actual or possible scientific explanations for all classical phenomena, but about to what extent all these (actual and possible) explanations that science provides fall under a unified set of theoretical principles or laws.

Second, the target of Cartwright's skepticism is not the possibility of finding a model for the 1000 dollar bill; rather, her skepticism is about the possibility of finding a principled model for that situation, that is, a model that is derived from the theory through systematic rules (i.e. bridge principles). Accordingly, the reason why the Neurath's Bill example is a challenge to fundamentalists is not only because it is a complex situation, but because it is a situation sufficiently different from any situation that we have good models for and we have no systematic rules that tell us how to generate models for situations like this one.

There is a rather delicate but important issue that arises in the context of Neurath's Bill (and other examples with a similar structure), and we believe that it is one about which Cartwright is not sufficiently careful. This can be seen from the fact that Cartwright slides from "skepticism about whether any theory will provide a good *description* for some phenomena" to "skepticism about whether any theory will provide a principled *predictive* model for Neurath's Bill". This move is only licensed if one identifies "description" with "prediction", which is not one of Cartwright's explicit commitments. Furthermore, notice that any system that displays extreme sensitivity to initial conditions (such as Neurath's Bill) will not admit of predictive models in practice, and so if "description" were to be identified with "prediction", principled models in such cases would be ruled out by *fiat*. This seems unfair to Cartwright's interlocutors and furthermore threatens to make Cartwright's philosophy of models collapse into some form of operationalism. We thus conclude that the relevant question about cases such as Neurath's Bill is not whether they admit of principled predictive models, but rather, whether they admit of principled models that adequately *describe* the empirical scenario.

Finally, note that Neurath's Bill is supposed to be just one out of a cornucopia of scenarios within this domain which are not described by a unified set of laws and theoretical principles. Cartwright's more general point is that classical mechanics is not unified because there are no laws or theoretical principles that apply in all the domain of classical mechanics. There are some domains of classical mechanics that we have a very good grasp of, such as point particle mechanics and fluid dynamics; however, there are many other domains where we do not have a good grasp of how to understand (explain and predict) the phenomena occurring there. So, according to Cartwright, classical mechanics is not unified because its domain is divided in different sub-domains, not all of which can be accounted by a single set of theoretical principles or laws.

How can we build on the above morals to sketch some general desiderata for a concretist epistemology of models? To be clear, what we are trying to do here is sketch guidelines for an approach that accepts the primacy of studying how particular theories/models work in order to make philosophical progress, while still avoiding the particularist extreme that Cartwright sometimes inclines towards (on which no general rules can be articulated). Based on the previous discussion, it seems to us that in the concretist's view, a principled model needs to at least possess the following features:

(i)   It is predictively accurate when applied to the sub-domain D' of the relevant domain D;
(ii) It is descriptively accurate when applied to the relevant domain D;
(iii) It is formally adequate with respect to the relevant theory;
(iv) It is materially adequate with respect to the relevant experimental context.

(i) is an uncontroversial criterion that both concretist and abstractionist epistemologies will agree on. On the other hand, disagreements will begin to crop up with respect to (ii): For abstractionists such as Smith, the fact that *some* elements in a set S of models (where S is perhaps picked out by the requirement that its models satisfy some generic physical assumptions) satisfy (i) can be used to secure (ii) for other, non-predictive, elements of S—in other words, (i) is used to bootstrap to (ii) for D much larger than D'. On the other hand, Cartwright does not seem particularly interested to consider cases in which D differs from D', i.e. she is generally only interested in models which are predictively accurate. We recommend that the concretist explore a *via media* between the Smith and Cartwright positions: In particular, it seems plausible that some models can be descriptively accurate without being predictively accurate, perhaps in virtue of the specific (not merely formal) relationship that they bear to predictively accurate models.

(iii) is again an area in which one might expect some common ground between concretists and abstractionists (although there is quite a bit of room for disagreement about how tight these "formal relationships" have to be). And finally, (iv) will be another point of contention between concretists and abstractionists: as we saw in Section 2, Smith is not particularly concerned with the material adequacy of models, thus leading to his highly permissive conception of principled models.

4.2 *Accommodating Abstraction Into Scientific Practice*

Neo-Aristotelians emphasize the inherent complexity of the natural world, and show a healthy skepticism regarding the possibility of capturing such complexity with overly general and abstract methods. However, this skepticism should not translate into a rejection of formalization and abstraction as theoretical tools which play a role (albeit a limited role) in understanding the natural world. Fortunately, the advocate of a concretist epistemology does not have to reject abstraction altogether as a theoretical tool; rather, concretism grants abstraction a legitimate theoretical role in scientific practice while keeping in check the philosophical excesses that stem from a blind faith in the power of abstraction.

In particular, abstraction plays a crucial role in a kind of unification that we call **Schematic Unification (SU)**. We have a case of SU when one tries to unify sub-theories by formulating a more abstract theory such that some laws/concepts/quantities of the sub-theories turn out to be instances of some laws/concepts/quantities of the abstract theory. Notice that, in itself, SU implies nothing about the prospects for theoretical unification, in the sense that finding a mathematical abstraction of a successful set of models (which, in turn, allows us to generate many new models) in no way guarantees that such an extension will describe a real domain. Indeed, radically generalizing a theory so that its range of models becomes infinite might in fact make it more difficult for the theory to have an empirical grip on reality.

Accordingly, one way of re-formulating Cartwright's criticism of Smith's account of theoretical unification is to say that Smith overemphasizes the role that SU has in theoretical unification. However, this is not to deny that SU has no legitimate role in theoretical unification. SU is a crucial step in setting up a common formal framework that allow us to understand and manipulate different theories and models under the same formalism. Physics provides various cases of SU forming part of a such a strategy to secure theoretical unification. For instance, take the case of Newtonian point-particle mechanics (PM). Let our sub-theories be (i) PM equipped with Universal Gravitation as a special force law (Gravitational PM); and (ii) PM equipped with Coulomb's Law as a special force law (Coulomb PM). We then introduce a unifying theory "Super PM", which is PM equipped with a purely schematic force function F. Evidently, SU is achieved by subsuming the different concrete forces under a single abstract force concept. Notice, however, that there is also a non-schematic, supplemental element that accompanies SU in this case: We are told how to combine different concrete forces, i.e. by vector addition. In other words, the schematic character of force is not only used to accomplish SU, but also provides a theoretical framework for describing interactions between Gravitational PM and Coulomb PM.

Accordingly, *under appropriate conditions*, SU can yield a genuine strategy for obtaining a larger theory that can unify two domains of phenomena. From a concretist perspective, these "appropriate conditions" are those in which we have enough information about the material systems and the experimental context that we can supplement the schematic character of the concepts of the unifying theory with specific principles that bridge the gap between the mathematical abstraction and the concrete phenomena at hand.

Smith's emphasis on SU exemplifies many of the general worries that motivate neo-Aristotelians' skepticism with respect to the use (and abuse) of formal methods in the sciences. By distinguishing SU from theoretical unification and analyzing their relation and epistemic differences, a concretist epistemology of models allows for a more careful and empirically grounded understanding of SU in theoretical unification. This thus provides an example of how a Neo-Aristotelian framework can incorporate the use of formal methods as legitimate without compromising the relevance of materiality and concreteness.

Section 5: Conclusion

Our main aim in this paper was to provide the Neo-Aristotelian philosopher with the conceptual tools to resist the fundamentalist challenge. We have argued that the best argument in favor of fundamentalism (Smith's Fundamentalist Argument) is based on an implausible abstractionist epistemology of models that begs the question against the anti-fundamentalist and violates the Neo-Aristotelian ethos of emphasizing materiality and concreteness over formalization and abstraction.

Fortunately for the Neo-Aristotelian philosopher, Smith's abstractionist epistemology of models is not the only viable option. Cartwright's concretist approach to the epistemology of models allows the Neo-Aristotelian philosopher to, first, resist Smith's arguments for the theoretical unification of the classical domain; and, second, to implement the conceptual resources of Cartwright's concretist epistemology to construct a positive argument for anti-fundamentalism, thereby shifting the burden of the proof in the debate.

Cartwright's concretist epistemology is limited by her particularism, so it is difficult to generalize from her specific, detailed examples to a more general concretist account of the epistemology of models. In order to overcome this difficulty, we analyzed Cartwright's concretist approach and tried to extend it by sketching the form of a set of constraints that need to be satisfied in order for a model to be "principled". These constraints include both criteria of formal and material adequacy. The resulting framework provides an excellent example to Neo-Aristotelian philosophers of how one can be critical of the excessive use of formalization and abstraction while remaining faithful to actual scientific practice.

We started the paper by emphasizing the threat that fundamentalism may pose to Neo-Aristotelian ontologies. Let us conclude by drawing some general morals that may give hope to ontological projects of a Neo-Aristotelian flavor. First, the arguments against fundamentalism presented in this paper give Neo-Aristotelians ammunition to resist the main threat from fundamentalism: The accusation that Neo-Aristotelian ontologies do not square well with contemporary scientific theories or the scientific consensus.

Second, it provides a prima facie reason to explore alternative ontologies to traditional fundamentalist ones. To the extent that fundamentalism is one of the central motivations for proposing fundamentalist ontologies of the world, the undermining of fundamentalism—through embracing a concretist epistemology of models—also undermines the motivation for exclusively focusing on fundamentalist ontologies. Thus, anti-fundamentalist creates room to entertain a plethora of alternative ontologies.

Finally, the dappled view of science resulting from this concretist episte-mology fits better with power ontologies than with traditional law ontolo-gies (as noted by Cartwright 1994). Furthermore, a dappled view of the world leaves much more space for non-reductive accounts of causation and causal powers, thereby opening many more ontological possibilities to the Neo-Aristotelian, including different varieties of substance ontologies and hylomorphism.

Notes

1 For the introduction of the term "fundamentalism" into the philosophical lexi-con, see Cartwright (1994).
2 See, for instance, Sklar (2003), Hoefer (2010), and Smith (2001).
3 Cartwright introduces anti-fundamentalism (or, as she calls it, the "patchwork" view of science) in Cartwright (1994) and further develops her anti-fundamentalist views in Cartwright (1999). For some responses to Cartwright's work, see Bovens et. al. 2010.
4 For examples of this strategy, see Hoefer (2010) and Sklar (2003).
5 Many discussions about scientific fundamentalism tend to assume that theoreti-cal unification comes about in a "vertical" way, i.e. through the reduction of less fundamental theories to a more fundamental theory. For instance, it has sometimes been thought that statistical mechanics presents evidence that the entities discussed in thermodynamics are ultimately or really nothing more than the entities which form the subject matter of some fundamental kinetic theory. But Cartwright observes, first, that theoretical unifications that involve reduc-tion are not the only kinds of unifications that are relevant to scientific practice (many unifications are instead "horizontal" in the sense that the unified theory is supposed to apply at the same "scale" as the theories that are being unified), and, second, that the key question for fundamentalism is whether the unified theory really expands the domain of application of the theories that are being unified. We shall thus focus on this "domain-expanding" aspect of theoretical unification.
6 These constraints are introduced by Smith (2001: Sec. 4.1), who cites Truesdell (1991: 200–4) as the source.
7 Most of the central elements of her concretist approach to the epistemology of models can be found in Cartwright (1999: ch. 2 and ch. 8).

Bibliography

Austin, C. (2017) 'A Biologically Informed Hylomorphism', in W. Simpson, R. Koons and N. Teh (eds.) *Neo-Aristotelian Perspectives on Modern Science* (New York: Routledge).
Bovens, L., Hartmann, S. and Hoefer, C. (2010) *Nancy Cartwright's Philosophy of Science* (Oxford: Taylor & Francis Group).
Cartwright, N. (1994) 'Fundamentalism vs. the Patchwork of Laws', *Proceedings of the Aristotelian Society*, New Series, 94(January): 279–92.
——— (1999) *The Dappled World: A Study of the Boundaries of Science* (Cam-bridge: Cambridge University Press).
Hoefer, C. (2010) 'For Fundamentalism', in Luc Bovens, Stephan Hartmann and Carl Hoefer (eds.) *Nancy Cartwright's Philosophy of Science* (Oxford: Taylor & Francis Group).

Jaworski, W. (2017) 'Psychology Without a Mental-Physical Dichotomy', in W. Simpson, R. Koons and N. Teh (eds.) *Neo-Aristotelian Perspectives on Modern Science* (New York: Routledge).

Koons, R. (2014) 'Staunch vs Faint-Hearted Hylomorphism: Toward an Aristotelian Account of Composition', *Res Philosophica* 91(2): 151–77.

Novotny, D. and Novak, L. (eds.). (2014) *Neo-Aristotelian Perspectives in Metaphysics* (New York: Routledge).

Pruss, A. (2017) 'A Traveling Forms Interpretation of Quantum Mechanics', in W. Simpson, R. Koons and N. Teh (eds.) *Neo-Aristotelian Perspectives on Modern Science* (New York: Routledge).

Sklar, L. (2003) 'Dappled Theories in a Uniform World', *Philosophy of Science* 70(2): 424–41.

Smith, S. (2001) "Models and the Unity of Classical Physics: Nancy Cartwright's Dappled World." *Philosophy of Science*, 68(4): 456–75.

Tahko, T. (ed.). (2012) *Contemporary Aristotelian Metaphysics* (Cambridge: Cambridge University Press).

Truesdell, C. A. (1991) *A First Course in Rational Continuum Mechanics* (Boston: Academic Press, Inc).

2 Actuality, Potentiality, and Relativity's Block Universe

Edward Feser

I. Introduction

The distinction between actuality and potentiality (or, in the more traditional jargon, act and potency) is fundamental to Aristotelian metaphysics, especially as it has been developed in the Thomistic tradition. It is the heart of Aristotle's account of how, contra Parmenides and Zeno, change is possible. The Aristotelian hylomorphic analysis of physical substances as composites of form and matter is an application of the distinction, with form corresponding to actuality and matter to potentiality. The Thomistic theory of the real distinction between essence and existence is another application, with essence corresponding to potentiality and existence to actuality. The Aristotelian-Thomistic account of God as Unmoved Mover (or "purely actual actualizer") of the world crucially depends on the distinction. And so forth.[1]

It might seem that the distinction has been rendered obsolete by Einstein, and in particular by the Minkowski space-time interpretation of the Special Theory of Relativity (STR). Michael Lockwood sums up a common view:

> To take the space-time view seriously is indeed to regard everything that ever exists, or ever happens, at any time or place, as being just as real as the contents of the here and now. And this rules out any conception of free will that pictures human agents, through their choices, as selectively conferring actuality on what are initially only potentialities. Contrary to this common-sense conception, the world according to Minkowski is, at all times and places, actuality through and through: a four-dimensional *block universe*.[2]

Leave aside the question of free will, which is not my concern here. What is relevant is that Lockwood's remarks suggest that there is, just as Aristotelians hold, an essential connection between time, change, and the actualization of potential, so that to deny one is to deny the others. The idea is that the block universe concept rules out the reality of time, or at least of temporal passage. But temporal passage follows upon change, so that if there is no temporal passage there can be no change either. Change, however (Lockwood's implicit argument continues), is just the actualization of potentiality, so that if there is no change, there is no actualization of potentiality. Hence, on the Minkowskian interpretation of STR, there is in the natural order no actualization of potentiality; everything in the world, whether "past," "present," or "future," is all "already" actual, as it were. Thus, as Karl Popper noted, does Einstein recapitulate Parmenides.[3]

So, the entire edifice of Aristotelian metaphysics might (given the centrality to it of the theory of actuality and potentiality) appear to have been toppled by space-time physics. But as space-time physicists like to emphasize, appearances can be deceiving. When the relevant issues are carefully disentangled and analyzed, it turns out that relativity by no means renders inapplicable the distinction between actuality and potentiality. At most it affects *how* we apply this distinction, but not *whether* we need to apply it. That, in any event, is what I will argue in this paper.

The plan of the paper is as follows. I will argue first that, even prescinding from the details of the physics of relativity, we have good *a priori* reason to believe that it could not undermine the theory of actuality and potentiality. For on the one hand (and as I argue in section II) we can know on independent metaphysical grounds—and in particular from the metaphysical presuppositions of any possible physics—that change, temporal passage, and thus the actualization of potential must be real features of the world. And, on the other hand (and as I argue in section III), we also have good *a priori* reason to believe that the appearance of a timeless and changeless world presented by relativity is *merely* an appearance, an artifact of the *method* by which physics studies physical reality rather than a discovery about the nature of physical reality itself. In particular, it is a byproduct of the mathematization of nature that is the hallmark of modern physics. While this method may capture important aspects of physical reality, it does not capture the whole of it, so that the absence of a feature from the mathematical picture of nature is not evidence of that feature's absence from nature itself. This idea might be developed in either an instrumentalist or a structural realist direction, and the latter is my own preferred approach. Structural realism is itself susceptible of alternative interpretations, and section IV spells out the sense in which the position defended in this paper is a structural realist one.

Of course, all of this still leaves us with the question of exactly *how* change, temporal passage, and the actualization of potential relate to the picture of the world presented by relativity if they are not actually at odds with that picture. This is a question usefully approached by way of consideration of the various theories of temporal passage on offer in contemporary philosophy of time and how their defenders have reconciled them with relativity—the topic of section V. To affirm that temporal passage, change, and the actualization of potential are real features of the world is to commit oneself to an A-theory or tensed theory of time. Reconciling any version of the A-theory with relativity would suffice to reconcile the theory of actuality and potentiality with relativity, though Aristotelianism is traditionally associated with presentism, the version of the A-theory which might seem most difficult to reconcile with relativity. But reconciliation is in fact defensible, even in the case of presentism.

Consideration is also given in section V to the B-theory or tenseless theory of time, which is commonly thought to be the view most in harmony with relativity. Naturally, since temporal passage and thus (the Aristotelian argues) real change would be absent from the world as described by the B-theory, that theory cannot, to that extent, be reconciled with Aristotelianism. However, the *actualization of potential* need not be absent from such a world.[4] To that extent, even an utterly static, four-dimensional Minkowskian block universe would be reconcilable with at least the *core* notion of Aristotelian-Thomistic metaphysics.

I do not say much, in what follows, about the details of the physics of relativity, or for that matter about the details of Aristotelian metaphysics. The reason is that in this case, the devil is *not* in the details, but rather in the big picture. As I will argue, the apparent conflict between relativity and the theory of actuality and potentiality has less to do with either theory as such than it has to do with general issues in the metaphysics of physics.

II. Time and Change

In Aristotelian metaphysics, the theory of actuality and potentiality is introduced as necessary in order to make sense of the reality of change and temporal passage (though it turns out to have applications beyond this). The way that relativity is supposed to pose a challenge to that theory is not by showing that it is mistaken as an analysis of the preconditions of change and temporal passage, but rather by showing that there are no such things as change and temporal passage in the first place. Accordingly, the focus in what follows will be on the question of whether change and temporal passage are real, not on whether the theory of actuality and potentiality is necessary in order to make sense of them.[5]

On an Aristotelian analysis, a real change (as opposed to a mere Cambridge change) involves the gain or loss of some attribute, but also the persistence of that which gains or loses the attribute.[6] For example, when a banana goes from being green to being yellow, the greenness is lost and the yellowness is gained, but the banana itself persists. If there were no such persistence, we would not have a *change* to the banana, but rather the annihilation of a green banana and the creation of a new, yellow one in its place.

Time, on the Aristotelian analysis, is just the measure of change thus understood. When we say that it took a certain banana four days to go from being green to fully yellow and then another eight days to turn brown, what this temporal description captures is the rate at which the events in question succeeded each other. Absent such change, there would be nothing to measure, and thus no time. The Aristotelian thus takes a "relational" view of time rather than the "absolutist" view associated with Newton. (That is not to say that the Aristotelian regards *simultaneity* as relative, but that is an issue I will address later on.)

Now, this understanding of time is, of course, controversial. For example, Sydney Shoemaker suggests that we can conceive of a world in which the inhabitants of three regions A, B, and C each occasionally observe the other regions go into a "frozen" or unchanging state for a period of a year.[7] Given that each region is observed by the inhabitants of the others to do so according to a regular pattern (every three, four, and five years, respectively) they would have reason to conclude that every 60 years, the regions must all be "freezing" together, and thus that in their world, no change is occurring for a year's time. Hence (it is argued), it is possible for time to exist without change.

But even leaving aside the tendentious assumption that we can deduce what is really possible from what is conceivable, this argument is problematic in ways noted by E. J. Lowe.[8] For one thing, it reasons from the claim that the inhabitants of Shoemaker's imagined world would have *evidence* for time without change to the conclusion that it is *possible* that there could be time without change. But this gets things the wrong way around, for we first have to know that something really is possible before we can be confident that there could be evidence for it. For another thing, if we suppose that A, B, and C really are all frozen or unchanging, then it becomes utterly mysterious what causes Shoemaker's world ever to become unfrozen, or unfrozen after exactly a year's time, rather than at some earlier or later time.

In any event, for present purposes, what matters is whether time is real in *at least* the sense the Aristotelian claims it is (again, as the measure of change), not whether time is real in *at most* that sense. And it is difficult to see how the reality of change, and thus of time in at least that sense, can coherently be denied. In particular, it is difficult to see how it could coherently be denied *in the name of science*, given the presuppositions of the very practice of science. Science involves *perceptual* activities like observation and experiment, and *cognitive* activities like entertaining concepts, formulating theories, inferring consequences from those theories, weighing evidence, and so forth. All of these activities entail the existence of change and time in the relevant senses.[9]

Hence, consider even the simplest observational or experimental situation, such as watching for the movement of a needle on a dial. When the needle moves from its rest position, it loses one attribute and gains another (namely a particular spatial location), and it is one and the same needle that loses and gains these attributes and one and the same dial of which the needle is a component. If there were no gain or loss of attributes, or if the needle or dial were not the same, the observation would be completely useless. For example, if what you are doing is testing a prediction about whether the needle which is at its rest position at t_1 will be at a different position at t_2, it would be completely irrelevant to such a test if the needle you observed at t_2

was a *different* needle from the one you observed at t_1, and there would be nothing to watch for if it were not possible for this same one needle to gain or lose an attribute.

Naturally, this presupposes the realist assumption that the needle and dial are mind-independent objects, but the basic point would hold even on a phenomenalist interpretation of science. Hence, suppose that all you are really observing when you read the dial are certain sense data rather than any mind-independent objects, and suppose we interpreted scientific theories as mere descriptions of the relationships between sense data. Observational and experimental situations like the one we are considering would somehow have to be interpreted in a way consistent with this. But however that would go, you would still have to suppose that the person who has the initial sense data at t_1 (namely you) is the *same* person who has the different sense data at t_2, and that this same one person is capable of gaining and losing attributes (namely the sense data in question). The observation would be completely irrelevant if the person who has the sense data at t_2 was a *different* person from the one who had them at t_1, or if that same one person was incapable of gaining of losing attributes like sense data.

Consider also even the simplest cognitive activity involved in science, such as reasoning from a premise to a conclusion via the inference rule *modus ponens*. When you reason from the premises *If p, then q* and *p* to the conclusion *q*, you lose one attribute (namely, the attribute of having the conscious thought that *If p then q, and p*) and gain another (namely, having the conscious thought that *q*). Moreover, it is one and the same person (you) who both loses the one attribute and gains another. If the person who had the second thought were not the same as the person who had the first one, there would not be any *reasoning* going on, any more than there would be if (say) Donald Trump had had the conscious thought that *If p, then q, and p*, and Hillary Clinton had, a moment later by sheer coincidence, the conscious thought that *q*. Nor would there be any reasoning going on if the same one person were incapable of losing one attribute (having the conscious that that *If p, then q, and p*) and gaining another (having the conscious thought that *q*).

A skeptic might object that the changes apparently involved in perception and cognition could be merely *illusory*. But the trouble is that, for all the skeptic has shown, this skeptical scenario *itself* presupposes change. The skeptic initially thinks that he has perceptual experiences and cognitive processes that manifest change; then, he entertains arguments to the effect that this may all be illusory; then, he concludes that such changes don't really occur after all. But all of *that* evidently involved changes of various sorts (for example, the skeptic first having one belief and then giving it up and coming to have another).[10]

Could the skeptic plausibly accuse such a response of merely begging the question against him? No, because the response is not a matter of simply dogmatically appealing to a premise that the skeptic denies (to the effect that change exists) and then pretending to refute him on that basis. Rather, it is a matter of pointing out that the skeptic *himself* in fact seems implicitly to *accept* the premise in question, even in the very act of denying it. Hence, the only reply open to the skeptic is to show that he is *not* implicitly committed to the premise. That is to say, the skeptic needs to give some account of how it is possible for him to so much as entertain his skepticism given that change does not exist. In the absence of such an account, it is the *skeptic*, and not his critic, who is being dogmatic. Yet, no such account is forthcoming.

Anyone who claims that science has shown that change and temporal passage are illusory thus faces a dilemma. If he acknowledges the existence of the cognitive and perceptual states of scientists themselves, then he is implicitly committed to there being at least *some* change and temporal passage after all, namely the change and temporal passage that exist within these thinking and experiencing conscious subjects. This not only merely relocates rather than eliminates change and temporal passage, but opens up a Cartesian divide between the conscious subject and the rest of reality, with all of its attendant problems.[11] (How does a temporal and changeable conscious subject arise within an atemporal and changeless natural order? Could there be causal interaction between two realities so radically unlike? Are we left with epiphenomenalism?) If instead, the skeptic about change and temporal passage takes an eliminativist line and *denies* the existence of the cognitive and perceptual states of scientists, then he will be throwing out the evidential basis of the scientific theory that led him to deny the reality of change and temporal passage in the first place. Into the bargain, he will face the further incoherence problems that notoriously afflict eliminativism about cognitive states, even apart from the problem of undermining the evidential basis of science.[12]

This argument for the incoherence of denying the reality of change is very simple, but, I think, utterly decisive. Yet, I suspect that its simplicity and decisiveness are, paradoxically, precisely why its force is not more widely seen. The thesis that change is an illusion seems to many (quite wrongly, but still, it *seems* to them) to have the full weight of modern physics behind it. Hence, the suspicion among such people (so I would wager) is that precisely because the argument about incoherence appears to be so obvious and decisive, there simply *must* be something wrong with it. Surely it *can't* be *that* easy to refute what (they suppose) physics is telling us! Surely only an argument of great technical sophistication and complexity could do that! Thus does the argument about incoherence get put to one side without anyone having actually answered it. The problem of how to fit the temporal and changeable world of the conscious subject into the wider timeless and changeless world purportedly revealed by physics is treated as a mere puzzle to be solved somehow, someday, by some future science.

But it is simply unreasonable either to suppose that the success of science justifies ignoring the problem or to express glib confidence that science will eventually solve the problem, because *science is precisely what creates the problem*. Or rather, what creates the problem is the mistaken assumption that the scientific description of nature is an *exhaustive* description of nature.

III. Mathematics and Method

In 1962, philosopher Max Black wrote:

> But this picture of a 'block universe,' composed of a timeless web of 'world-lines' in a four-dimensional space, however strongly suggested by the theory of relativity, is *a piece of gratuitous metaphysics*. Since the concept of change, of something happening, is an inseparable component of the common-sense concept of time and a necessary component of the scientist's view of reality, it is quite out of the question that theoretical physics should require us to hold the Eleatic view that nothing happens in 'the objective world.' Here, as so often in the philosophy of science, *a useful limitation in the form of representation* is mistaken for a deficiency of the universe.[13]

Black's characterization of change as "a necessary component of the scientist's view of reality" essentially sums up the point of the previous section. His characterization of relativity's picture of a changeless world as "a useful limitation in the form of representation" but also "a piece of gratuitous metaphysics" if taken to reflect anything *more* than a limitation in science's form of representation, sums up the argument of the present section.

There are two key points here. The first is that the absence of change and temporal passage from relativity's "representation" of the universe is plausibly regarded as an artifact of the *methods* of physics, rather than necessarily reflective of the reality studied *by means of* those methods. That is to say, change and temporal passage could be there in nature even if they don't show up in the physicist's picture of the world, because the methods the physicist uses to paint that picture wouldn't capture them even if they were there. The second point is that, for that reason, physics *by itself* cannot tell you one way or the other whether change and temporal passage are real. Physics could tell you that only if conjoined with an *independent, philosophical* argument to the effect that the picture painted using the methods of physics is adequate as a *metaphysics* of nature. Let's develop these points in order.

Suppose an artist produced images of a certain building using only black and white materials (pen and ink, wash, and so forth). Suppose the images were so vivid, detailed, and skillfully rendered that one could accurately predict from them what one would see when he entered the building and went down a certain hallway or walked into a certain room, could use them as a basis for designing furniture that would fit in the rooms, and so forth.

Obviously, there is nothing in this situation that would warrant the conclusion that since red, green, blue, yellow, and other colors are not captured in the artist's drawings, they must not really be there anywhere in the building itself either. In particular, the elegant simplicity, predictive success, and technological utility of the drawings would not warrant it. Nor would they warrant the conclusion that everything one sees in the images is actually there in the building. For example, it would obviously be a mistake to think that the walls, doors, tables, chairs, and other objects in the building must all have black lines around them, on the basis of the fact that the contours of the objects in the drawings are rendered by lines drawn in black ink. The absence of colors and the presence of black lines are byproducts of the *method of representing* the building and have nothing to do with the nature of the building itself.

That a similar phenomenon exists with *mathematical* representations of the world is obvious from everyday life. Consider, for example, the engineer concerned with designing a commercial aircraft, who wants to know how many passengers could be flown in it given that the engines have a certain thrust capacity, etc. He might determine that a passenger of average weight would be 152.5 pounds (or whatever), and he would ignore considerations such as the ethnicity of the passengers, their meal and entertainment preferences, and the like. Suppose that the airplane he designs works very well, that the design exhibits an elegant simplicity compared to those of other aircraft, etc. Obviously, that would not warrant the conclusion that actual airline passengers have no ethnicity or meal and entertainment preferences, or even that any of them weigh 152.5 pounds (since it could turn out that no actual passenger has exactly that weight). The absence from the engineer's description of the passengers of any reference to ethnicity, meal preferences, etc. and the presence of passengers of average weight are artifacts of the *method of representing* them and do not necessarily reflect the actual passengers themselves.

Now, the mathematical representation of nature is the hallmark of modern physics, and what is true of the humble examples just given is true also of the sophisticated mathematical models developed by the physicist. They too are bound to exclude features that exist in reality and add features that do not exist in reality, and their elegance and predictive and technological success do not show otherwise. As Jeffrey Koperski writes:

> Continuum mechanics, for example, treats matter as if it were smoothed out and continuous across a region rather than atomic. Aerodynamics treats the airflow over a wing the same way, and these are perfectly good idealizations for the scale at which we normally deal with materials, especially fluids and gases. Spacetime theorists make this same move by ignoring [the] midscale structure [of the universe].[14]

It is important to put aside immediately an objection that might seem obvious but in fact has no force. The objection is that, unlike the representations produced by the artist or the aircraft engineer, which capture only part of physical reality, the representation of nature produced by physics, especially space-time physics, captures *all* of physical reality. The reason this is a bad objection is that it blatantly begs the question. Whether the representation of physical reality produced by physics captures all of physical reality is precisely what is in question. The Aristotelian metaphysician—and not only him, as we will see—denies precisely that that is the case. Notice that I said "*physical* reality." The point is not just that there might be aspects of reality beyond the physical which physics does not capture. The point is that from the Aristotelian point of view, physics does not capture everything there is, *even in physical reality*.

The point, as we will see, is also by no means limited to considerations about change and temporal passage, but let's begin with those, since they are our main concern here. John Bigelow has noted the role that mathematics and formal logic have played in modern physics in suggesting the picture of a timeless and changeless world, even apart from relativity.[15] The Newtonian analysis of motion represents the speed of a body in terms of the slope of a curve on a graph which measures time on one axis and space on another. Different times and different places are thereby represented in the same way, viz. as points on the graph, all of which exist at once, as it were. Then there is the use of calculus to characterize speed in terms of the limits of infinite sequences. To formulate statements about such sequences in modern predicate logic requires quantifying over past and future bodies and events no less than present ones. Again, this suggests a picture on which all times and places exist at once, as timeless and changeless Platonic mathematical objects.

Physicist Lee Smolin has also emphasized that a timeless picture of physical reality is a byproduct of the physicist's mathematical manner of representing reality, noting that "the process of recording a *motion*, which takes place in time, results in a *record*, which is frozen in time—a record that can be represented by a curve in a graph, which is also frozen in time."[16] Noting that the concept of four-dimensional spacetime is the result of a thoroughgoing application of this method of representing nature, Smolin warns against too quickly drawing metaphysical conclusions from the successes of the method:

> Some philosophers and physicists see this [method] as a profound insight into the nature of reality. Some argue to the contrary—that mathematics is only a tool, whose usefulness does not require us to see the world as essentially mathematical . . .
>
> [They] will insist that the mathematical representation of a motion as a curve does not imply that the motion is in any way identical to the representation. The very fact that the motion takes place in time whereas its mathematical representation is timeless means that they aren't the same thing . . .

By succumbing to the temptation to conflate the representation with the reality and identify the graph of the records of the motion with the motion itself, [some] scientists have taken a big step toward the expulsion of time from our conception of nature.

The confusion worsens when we represent time as an axis on a graph . . . This can be called spatializing time.

And the mathematical conjunction of the representations of space and time, with each having its own axis, can be called *spacetime* . . . If we confuse spacetime with reality, we are committing a fallacy, which can be called the fallacy of the spatialization of time. It is a consequence of forgetting the distinction between recording motion in time and time itself.[17]

His proposed label for it notwithstanding, the error Smolin identifies here is not that of reducing time to a kind of spatial dimension—that may or may not make sense at the end of the day, but that is a separate issue— but rather the fallacious inference involved in supposing that the usefulness of the mathematical representation *all by itself justifies* such a metaphysical reduction. That conclusion simply does not follow, any more than the parallel conclusions follow in the examples of the black-and-white drawing and the aircraft design. Or, to take another example from philosopher Craig Bourne:

[J]ust because something is represented spatially, we cannot draw the conclusion that it is a spatial dimension or that it is in anyway [sic] analogous to a spatial dimension. For consider . . . a three-dimensional colour space which illustrates the possible ways in which things can match in colour . . . [I]t would be misconceived to draw the conclusion that brightness, hue, and saturation were each spatial dimensions, just because they were represented spatially. And to go on to conclude that each of these dimensions must be alike just because they comprise the different dimensions of colour space would be equally fallacious, since they're not. We should, then, be equally wary of drawing conclusions from Minkowski space-time diagrams.[18]

Propositions of mathematics and formal logic of their nature have a "timeless" Platonic character. Hence, representing something in purely mathematical terms or in the language of formal logic is bound to give it the appearance of something devoid of change or temporal passage. It doesn't follow that the thing represented really is itself devoid of change or temporal passage, for that appearance may instead be a byproduct of the mode of representation. The representation might thereby be *adding* something to the picture that isn't there in reality, just as the black-and-white drawing adds in black outlines around the objects that aren't really there in the objects themselves. Since the representation will also be unable to capture anything not susceptible of mathematical or formal representation, it might also be *leaving out* aspects that are there in reality. The mathematical representation will necessarily *abstract from* those features even if they are really there.

Though Smolin's remarks are very recent, his basic point is not a new one.[19] It was often made during the first half or so of the twentieth century by thinkers of diverse interests and theoretical commitments. Process philosopher Alfred North Whitehead gave the label "The Fallacy of Misplaced Concreteness" to the tendency to confuse an abstract mathematical representation of reality with the concrete reality represented, and the error identified by Smolin is a special case of this fallacy.[20] Phenomenologist Edmund Husserl emphasized that the "mathematization" of nature results in an "idealization" which does not capture the whole of reality.[21] Henri Bergson complained that the mathematician's conception of nature leads us wrongly to think of time as a series of frozen moments, like the still photographs that make up a film strip.[22] Historian of science E. A. Burtt lamented the tendency of the modern scientist to "make a metaphysics out of his method."[23] As we have already seen, analytic philosopher Max Black made a similar point. Unsurprisingly, Aristotelian-Thomistic philosophers raised such criticisms as well. Jacques Maritain argued that mathematics captures only one of three "degrees of abstraction" from concrete physical reality,[24] and other Aristotelian-Thomistic philosophers endorsed Whitehead's "Fallacy of Misplaced Concreteness" objection.[25] But as these various examples illustrate, one hardly need have an Aristotelian ax to grind to suppose that there might be more to physical reality than is captured in the mathematical representation developed by modern physics.

So, we could be justified in deducing from that mathematical representation the conclusion that change and temporal passage are illusory only if we had a good independent, metaphysical argument to the effect that what can be found in that representation *exhausts* physical reality. (As Laurence Sklar has said, "one can extract only so much metaphysics from a physical theory as one puts in."[26]) There also have to be no *stronger* independent metaphysical arguments for the contrary conclusion than what can be found in that representation *does not* exhaust physical reality. Are there good arguments of the former sort? Are there any of the latter sort?

Einstein had at least an *implicit* metaphysical argument for the conclusion that change and temporal passage must be illusory if they don't show up in physical theory. Unfortunately, that argument was essentially an appeal to verificationism, whose influence on Einstein's formulation of STR is well known. Needless to say, though influential at the time the theory was formulated, verificationism has fallen on hard times, and for good reason. It is difficult to formulate verificationism in a way that is not self-defeating; it does not fit all actual scientific practice (not even that of Einstein, who later abandoned it); and it is incompatible with scientific realism, for which there are powerful arguments. What about that last redoubt of the verificationist manqué, the appeal to Ockham's razor? The idea here would be that while there might (contra positivist verificationism) in *principle* be something more to physical reality than what physics reveals, in *practice* we should not suppose that there is, because we simply do not need to postulate anything more in order to explain what needs to be explained.

The problem here is that we plausibly *do* need to postulate more, and we have already seen one reason why we do, in the previous section. Again, we cannot make sense even of the scientist's own cognitive and perceptual states, on which the evidence for physical theory rests, unless we affirm the reality of change and temporal passage. If physics' mathematical representation of the world doesn't capture the latter phenomena, the very practice of physics nevertheless presupposes that they are real and thus presupposes that the mathematical representation doesn't capture the whole of physical reality.

But more can be said. As Bertrand Russell (no Aristotelian or Thomist, and from the start an important expositor and advocate of relativity[27]) came increasingly to emphasize in his later work, the knowledge that mathematical physics gives us is essentially knowledge of *structure* rather than of the intrinsic nature of the concrete reality which has the structure (about which "physics is silent").[28] Indeed, it was precisely developments in physics like Einstein's relativity that led him to this position. Yet, there must *be* such an intrinsic nature to physical reality, for structure cannot exist without it, any more than the Cheshire cat's grin can exist without the cat. (More on this below.) Since physics' mathematical representation of nature does not give us that, we know it doesn't tell us everything even about physical reality.

There is still more to be said in response to any facile appeal to Ockham's razor. The early Russell had questioned the reality of causation, on the grounds that the physicist's equations make no reference to it.[29] Now, that this is a fallacious inference should be evident from what has already been said. That some extra-mathematical feature is absent from the physicist's mathematical representation of nature simply does not entail that it is absent from nature itself, any more than the absence of color from a black-and-white drawing entails that color is absent from the objects represented by the drawing. But what matters for the present point is that, as the later Russell himself came to emphasize, the notion of causation is in fact crucial to the epistemology of physics. For the only reason we have for supposing that observation and experiment provide evidence for physical theory is that our perceptual experiences are *causally related* to actual physical reality. If they were not so related, then experience would float free of physical reality and thus provide no evidence of its nature. But if the epistemology of physics requires causation, then that shows us yet again that physics' mathematical representation of nature simply does not capture all there is to physical reality.

One further consideration. An immediate consequence of the mathematization of nature inaugurated by Galileo, Descartes, and the other early moderns was the introduction of the primary versus secondary attribute distinction. Color, sound, heat, cold, and the like as common sense understands them, since they are qualitative rather than quantitative, were treated as features merely of the mind's perceptual representation of physical reality rather than of physical reality itself. This is the origin of the "qualia problem," the problem of fitting the qualitative features of conscious experience into the physical world. For if color, sound, etc. as common sense understands them do not really exist in matter, then it seems they do not exist in the brain, since the brain is one material thing among others. And yet, they do exist in the *mind*. So, it at least *seems* to follow in turn that the mind, and in particular consciousness, is not material. Accordingly, and as Thomas Nagel has been emphasizing for 40 years, it is modern science's own conception of the physical that opens up this apparent gap and makes the mind-body problem so intractable.[30] Erwin Schrödinger made a similar point:

> We are thus facing the following strange situation. While all building stones for the [modern scientific] world-picture are furnished by the senses qua organs of the mind, while the world picture itself is and remains for everyone a construct of his mind and apart from it has no demonstrable existence, the mind itself remains a stranger in this picture, it has no place in it, it can nowhere be found in it.[31]

Now, since conscious experience provides the observational evidence on which physics rests, physics presupposes the reality of conscious experience. Hence, if conscious experience is left out of physics' mathematical representation of nature, then the epistemology of physics once again entails that there must be more to reality than is captured by that mathematical representation.

Obviously, the materialist will have much to say in response to all of this, but the point isn't to take sides on the mind-body problem. The point is to offer a further consideration against any facile appeal to Ockham's razor. Given its reliance on the distinction between primary and secondary attributes, and the qualia problem that this seems to open up, the mathematization of nature *creates* at least as many difficulties as it solves. Hence, a concern for parsimony would hardly justify us in supposing it to provide an exhaustive description of reality.

IV. Structural Realism

Let's pursue in a little more depth the reason *why* there must be more to physical reality than can be captured by mathematics. For someone might agree that physics captures only the mathematical structure of the natural world, but then go on to deny that there is anything more to that world *to* capture. To hold that physics captures the mathematical structure of physical reality but that there is more to physical reality than that is to adopt a version of *epistemic structural realism*. To hold that physics captures that structure and that there is *not* more to physical reality than that is to adopt a version of *ontic structural realism*. A critic could accept the basic thrust of the idea that physics gives us only mathematical structure but still attempt to block the overall argument of this paper by opting for ontic structural realism rather than epistemic structural realism.[32] So why prefer the latter to the former?

To be sure, the previous two sections have already indicated why there must be more to *reality* than mathematical structure. For I have argued that the existence of the experiences and cognitive processes of the conscious subject entails the existence of change and temporal passage; that there must be causal relations between the experiences of this conscious subject on the one hand and physical reality on the other if the former are to provide evidence about the latter; that the mathematical models of physics don't capture this change, temporal passage, or causation; etc. But someone might accept all of this consistent with ontic structural realism. He might opt for a Cartesian picture according to which mathematical structure exhausts *physical* reality, but there is nevertheless a non-physical conscious subject distinct from that reality which is (somehow) causally related to it. Needless to say, contemporary naturalist philosophers wouldn't be attracted to such a picture, but why rule it out?

A standard objection to ontic structural realism is that knowledge of the mathematical structure of the world is essentially knowledge of *relations* and that there cannot be relations without *relata*. Hence, it cannot be that the mathematical structure exhausts reality. There must be relata which are related by the mathematical relations described by physics. I think this is a good objection, though Anjan Chakravartty (who has put forward the objection himself) suggests that ontic structural realists might regard it as question-begging, on the grounds that their point is precisely that we need to revise our concept of a relation in such a way that it can exist without relata.[33] But by itself, this is hardly a powerful response. If I assert that there could be round squares and offer nothing more in response to the charge that this is incoherent than the bare suggestion that we need to revise our concepts of roundness and squareness, I have hardy made my assertion more plausible. I need to give some positive account of exactly *how* such a revision might be accomplished.

James Ladyman has suggested that there are at least two ways in which we might be able to make sense of the idea of relations without relata.[34] To understand the first, consider a universal like the relation *being larger than*. We could grasp this universal even if we were to deny that there are any things that instantiate it, any two things such that one is larger than the other (just as we grasp the universal *unicorn* even though we know that it is not instantiated). Hence, we have a sense in which we might conceive of a relation without relata. An obvious problem with this suggestion, though, is that when talking about the natural world it is precisely instantiations and *not* universals we are interested in. Even if we regarded the natural world as a single, four-dimensional object, we could distinguish between the world itself as a concrete particular and the universal that it instantiates—a universal which, unlike the natural world itself and qua universal, is abstract rather than concrete, in principle multiply instantiable, causally inert, and so on. And the problem is that there is no way to make sense of concrete *instances* of relations without relata. Nor will it do for the ontic structural realist to try to dodge this problem by suggesting that the natural world really just *is* a kind of universal or Platonic object. Among other problems, this would make it utterly mysterious how physics is or could be an *empirical* science any more than mathematics is and would thereby threaten to undermine the very evidential basis of physics.

The second way Ladyman suggests we might be able to make sense of relations without relata is by thinking of the purported relata as *themselves* analyzable in terms of further relations, and those in terms of yet further relations, all the way down, as it were. But it is hard to see how this solves the problem. Either relations require relata or they do not. If, as ontic structuralists claim, relations do not require relata, then what is the point of positing an infinite regress of relations to serve as relata? Why not just stop with the top level set of relations and be done with it? But if relations do require relata, then how does positing an infinite regress of relations serve to *identity* those relata (as opposed to endlessly *deferring* an identification)? What non-question-begging reason could the ontic structural realist give for claiming that an infinite regress of relations is any less problematic than relations without relata?

It seems, then, that the "no relations without relata" objection stands, and ontic structural realism fails.[35]

To summarize the argument of this and the previous section: First, given its purely mathematical character, physics' representation of nature is bound to leave out change and temporal passage even if they are really there in nature itself. Hence their absence from that representation has, by itself, no metaphysical implications. It could have metaphysical implications only if we had good independent reason to think that that representation is an *exhaustive* picture of nature, and no good independent reason to think that it is *not* exhaustive. Second, in fact there are no good reasons to think that the representation is exhaustive, and several independent reasons (the ineliminability of change and temporal passage from the cognitive and experiential processes of the conscious subject; the argument from the need for intrinsic or non-structural qualities of some sort or other and the difficulties facing the ontic structural realist's attempt to avoid them; the argument from causation; the argument from secondary qualities) to think that it is not exhaustive. So, we can conclude, *even apart from* a consideration of the details of relativity, that relativity qua purely mathematical description of nature cannot give us good reason to doubt the reality of change and temporal passage. And the argument of section II shows that change and temporal passage *are* in fact real.

The skeptical reader will, of course, nevertheless want to know exactly *how* the reality of change and temporal passage fit into the picture of the world afforded by relativity, given that there at least *seems* to be a conflict. That is a perfectly reasonable question, but it is important to emphasize that answering it is first and foremost a problem *for the physicist*, not for Aristotelians and other defenders of the reality of change and temporal passage. The scientist who pretends that it is the *latter* who have the primary burden here is like the party guest who trashes the house and then demands of the host: "So how do you propose to clean this mess up?" For as we have seen, the physicist himself no less than anyone else ultimately has to affirm the reality of change and temporal passage. If he puts forward a theory that at least appears to deny them, then, he is the one who has some explaining to do. The predictive and technological successes of relativity are undeniable, but that must not blind us to the fact that *metaphysically*, it is something of a mess, and that the physicists are the ones who made the mess.

V. Relativity and A-Theories of Time

On the other hand, a metaphysician is better placed to clean up a metaphysical mess, whoever made it. And there is a sizable philosophical literature on the issue of how relativity might be reconciled with an A-theory of time, i.e. a theory that affirms the objective reality of change and temporal passage. Now, presentism is the version of the A-theory which holds that the present alone exists, and thus that past things and events no longer exist and future things and events do not yet exist. This is the version of the A-theory which might seem most obviously at odds with relativity. For if, as STR holds, simultaneity is relative to frames of reference and there is no privileged frame, then whether or not a moment is present would be relative to a frame of reference, and there would be no frame by reference to which we could define an absolute present. Moreover, if the universe is a four-dimensional block, then all things and events (past and future no less than present) would seem to be equally real. So, if presentism can for all that nevertheless be reconciled with relativity, then it seems any A-theory could be. Let's begin with presentism, then.

There are essentially four general approaches to reconciling presentism with relativity. They differ in what they take the physics of relativity to tell us about objective reality. The first and most obvious approach to reconciliation would be to back away from even a structuralist brand of realism and hold that the physics of relativity does not really tell us *anything* about objective physical reality in the first place, but should be given a purely instrumentalist or other anti-realist interpretation. If a model of the universe on which there is no absolute present is merely a useful fiction, then naturally there is no incompatibility with the metaphysical claim that there nevertheless is in reality an absolute present moment. Nor can anti-realism be easily dismissed in this context given relativity's historical and conceptual connections with positivist verificationism, which is itself a kind of anti-realism. As Sklar writes:

Certainly the original arguments in favor of the relativistic viewpoint are rife with verificationist presuppositions about meaning, etc. And despite Einstein's later disavowal of the verificationist point of view, no one to my knowledge has provided an adequate account of the foundations of relativity which isn't verificationist in essence.[36]

One way to develop an anti-realist approach would be along the lines of Arthur Prior's suggestion that relativity has merely *epistemological* significance and in particular shows only that we cannot *know*, of some events, whether they are absolutely simultaneous with one another, but not that there is no fact of the matter about whether they are.[37] Another way is proposed by David Woodruff, who suggests that relativity seems inconsistent with presentism only if we think of Minkowski's four-dimensional space-time manifold as a concrete *substance* in its own right.[38] Thus understood, its future and past components naturally seem no less existent than its present ones. But we can deny the existence of any such substance and hold instead that what exist are only present objects and events of the more ordinary sort. The space-time manifold can then be thought of as a "geometric representation" of "what will happen to accelerated bodies and how it will affect measurements of time and space" as well as "what causal interactions are possible."[39] Since this makes of the manifold a tool for making predictions about the behavior of things, Woodruff allows that his position might be seen as a kind of instrumentalism (though with an important qualification to be noted presently).

An anti-realist solution has the merit of being simple and straightforward, though of course it inherits all the usual difficulties with anti-realism. But if this approach is rejected, there are still three remaining, more or less realist ways of attempting a reconciliation. This brings us to the second approach to reconciling presentism and relativity, which would be to affirm that the physics of relativity really does capture objective physical reality, but to maintain that Einstein and Minkowski simply got that physics wrong. This is the approach of William Lane Craig, who proposes a neo-Lorentzian relativity theory.[40] Lorentz's theory, which is empirically equivalent to Einstein's, affirms absolute simultaneity and thus allows for a privileged present moment. A famous difficulty with it is that it has to posit an empirically undetectable aether by reference to which to define a privileged frame of reference. But there are, Craig argues, various alternatives to the aether as Lorentz understood it (for example, the cosmic microwave background radiation, or quantum non-locality).

A variation on this approach would be to criticize anti-presentist appeals to STR on the grounds that STR is even from an Einsteinian point of view not strictly correct in the first place, but merely an approximation to the general theory of relativity (GTR).[41] Moreover (this line of argument continues) even GTR is not the last word, but is itself merely an approximation to whatever the correct theory of quantum gravity turns out to be.[42]

Unlike Craig, who holds that Einstein got the physics of relativity wrong, this line of argument holds that Einsteinian relativity is not wrong so much as incomplete, that there is a larger framework of still developing physical theory in the context of which relativity must be interpreted, and that this larger framework might turn out to be more favorable to presentism.[43]

Of course, such proposals are highly controversial, and a completed physics may turn out *not* to favor absolute simultaneity any more than STR does. But this brings us to the third approach to reconciling presentism and relativity, which suggests that STR as it stands may in fact already be reconcilable with presentism, or at least with a modified presentism. Theodore Sider suggests three ways this idea might be developed (though he does not endorse any of them and is not himself a presentist).[44] They all involve the presentist affirming the existence of a part of the Minkowskian manifold. The first affirms the existence of some point in the manifold together with everything in its past light cone; the second affirms the existence of the point and everything in its future light cone; and the third affirms the existence of the point and everything spacelike separated from it. Because each of these options preserves at least part of the manifold, it is to that extent realist. Because each denies other parts, it preserves, to that extent, the presentist idea that all points of time are not equally real. Sider quite rightly notes that the first two options nevertheless depart considerably from presentism as usually understood, since the first allows (to use Sider's examples) that dinosaurs are still part of reality, and the second that future Martian outposts are part of reality. The third option, he suggests, most closely preserves the basic thrust of presentism.

Sider not implausibly objects to these sorts of proposals that they privilege a particular point in space-time in a way that has no justification in relativity physics itself.[45] They also privilege such a point in a way *presentists* typically would not. For presentism, as usually understood, it is a *class* of points (or things and events) that exist in the present. Fitting this idea into relativity would require positing, within the Minkowskian manifold, a privileged hyperplane of simultaneity relative to some frame of reference, and taking that alone to be real. The problem, Sider argues, is that there is nothing in the geometry of Minkowskian space-time to justify taking any hyperplane to be privileged in this way.[46]

That brings us to the fourth approach to reconciling presentism and relativity. Suppose nothing in the geometry of space-time either as understood within STR or even within the correct theory of quantum gravity reveals any point, region, or privileged slicing of the manifold with which the present might be identified. One could still argue (as, for example, Dean Zimmerman does) that physics, while correctly describing objective physical reality as far as it goes, nevertheless does not provide an *exhaustive* description and needs to be supplemented by metaphysics.[47] Philosophical arguments for presentism, on this view, provide us with independent evidence that there must be a privileged slicing of the manifold described by physics, even if physics itself cannot tell us what it is. Zimmerman takes the Minkowskian manifold to represent "the set of locations at which events could happen," with the present amounting to a "wave of becoming" that moves through the manifold, as it were.[48] The presentist can admit the existence of non-present points in the manifold and simply deny that there are any objects and events that occupy them.[49] The presently occupied points constitute a privileged slicing of the manifold, and while Sider rejects such a slicing as too "scientifically revisionary," Zimmerman responds that his position does not *revise* the physics of relativity—he is not claiming that it is wrong as far as it goes—but simply supplements it.[50]

Insofar as Zimmerman's position allows for some kind of reality even to non-occupied points of the Minkowskian manifold, it is plausibly realist. But it is worth noting that even Woodruff's view can, as he notes, be read as a kind of realism rather than instrumentalism.[51] For one thing, the idea that the Minkowskian four-dimensional space-time manifold exists as a kind of concrete substance is not, Woodruff argues, actually part of the physics of relativity or strictly implied by the physics. Rather, it is a metaphysical interpretation one may (or may not) wish to give the physics. Hence, to deny its existence as a concrete substance is not to take an anti-realist interpretation of the actual *science*. Second, in taking the manifold to represent what will happen to accelerated bodies, how they can interact causally, etc., he regards relativity as telling us something about *the real features of actual things themselves*, not merely about how reality appears to us or how we represent it. As Jeffrey Koperski suggests, we can think of Minkowski space-time as a kind of *phase space* representing the possible states of the evolving universe.[52] And to affirm even that is to go beyond a purely instrumentalist positon, at least as instrumentalism is usually understood. Accordingly, Zimmerman's position too would arguably remain at least minimally realist even if he discarded the notion of the reality of non-occupied points of space-time.

In summary, then, the four general approaches to reconciling presentism and relativity are, first, to deny that the physics of relativity really describes objective physical reality in the first place; second, to affirm that it does describe objective physical reality, but argue that existing models are mistaken or incomplete and that a correct and completed physics will end up vindicating rather than undermining presentism; third, to affirm that relativity does describe objective physical reality and to argue that existing physics in fact already contains sufficient resources to vindicate presentism; and fourth, to affirm that physics does describe objective physical reality and allow that neither current physics nor a future, completed physics suffices to vindicate presentism, but that physics nevertheless can and ought to be supplemented from outside by metaphysics in a way that will vindicate presentism. (These are, of course, idealized summaries of the four general approaches, and a presentist philosopher might combine elements from more than one of them.[53])

But an A-theorist could also depart from presentism and mitigate the apparent conflict with relativity by conceding a kind of reality to things and events other than present ones. One way of doing this is by opting for a "growing block" theory on which past things and events are as real as present ones, with the present constituting the growing edge of a four-dimensional universe.[54] Another way is to go even further in a realist direction and concede the existence even of future events no less than past ones, while maintaining that present events are unique in being illuminated by the "moving spotlight" of the "now."[55] Aspects of some of the four approaches to reconciling presentism with relativity (anti-realism, appeal to a completed physics, appeal to STR as it stands, or supplementing physics with metaphysics) might then be adapted to a reconciliation of either the "growing block" or "moving spotlight" versions of the A-theory with relativity.

Of course, each of these approaches has generated a large literature and the various arguments for and against them need to be carefully weighed. The point, however, is not to adjudicate the contemporary debate or to opt for any particular version of the A-theory. The point is rather simply to note that a wide variety of approaches to defending the reality of change and temporal passage in the face of relativity have been developed in contemporary philosophy, *entirely independently of any concern with upholding the Aristotelian theory of actuality and potentiality*. They are already sitting there "on the shelf" for deployment by any neo-Aristotelian who wants to reconcile that theory to relativity. By no means, then, is such a reconciliation necessarily *ad hoc* or an exercise in anachronism.

Now, suppose it turned out that no variation on the A-theory was ultimately defensible. Suppose it turned out that an eternalist or B-theory of time is correct, and that the universe really is a static four-dimensional block from which real temporal passage is entirely absent. Even in this case, the Aristotelian theory of actuality and potentiality would not be refuted, though the range of its applicability would be severely restricted. The reason is that the existence of such a block universe would still be *contingent*. There would be nothing about its nature that requires that a block universe of precisely that sort, or any block universe at all for that matter, exists. It would in that sense be of itself *potential* and in need of *actualization*.

The way this idea would be spelled out in the Thomistic branch of the Aristotelian tradition is that a four-dimensional block universe, considered as one big substance, would have an *essence* distinct from its *existence*, where the essence of a thing is, considered by itself, a kind of potentiality and its existence is a kind of actuality.[56] A natural way for the Thomist to extend this line of thought is to argue that what actualizes the potential existence of the universe is a divine First Cause the essence of which *just is* existence—an uncaused cause which is (as it is often put) Subsistent Being Itself. But whether one wants to go in this specific direction is irrelevant to the present point. One could instead opt for (say) an infinite regress of actualizers of potential, with the four-dimensional block universe being actualized by some cause C, C in turn actualized by some other cause, which is in turn actualized by another, and so on *ad infinitum*. Whether this alternative, non-theological scenario is ultimately defensible can be put to one side for present purposes.

Whether interpreted in theological terms, in terms of an infinite causal regress, or in some yet different terms, it is important to stress that the Thomistic analysis is concerned with an *atemporal* or *timeless* kind of causation, and thus with a kind that doesn't involve change in the strict sense.[57] If the position expressed in the passage from Lockwood quoted earlier is correct, there is no actualization of potentiality *within* the universe. The Thomist point is that the universe *itself*, *as a whole*, would still require actualization. Nor would the basic situation be changed if we thought of the universe as arising out of some other universe or as a part of some multiverse. For it would, in that case, nevertheless be the case (the Thomist argues) that this *larger set* of universes requires actualization. Naturally, this claim is controversial, but even if it is judged to be mistaken, there is nothing in *relativity theory* or in the notion of a four-dimensional block universe that shows that it is mistaken.

So, even if relativity were to have the worst implications for Aristotelianism it is thought by some to have, it still would not refute the theory of actuality and potentiality. An atemporal, timeless, or eternal actualization of potentiality would still be defensible, as a kind of limit case of the actualization of potentiality more familiar to common sense as everyday change. To be sure, the applicability of the theory would be *very* greatly reduced. It could no longer ground a hylomorphic analysis of everyday material substances. Perhaps its only application would be within natural theology.

But that the distinction would nevertheless survive even the worst case scenario reinforces the point that one ought to take with a grain of salt glib assertions to the effect that relativity has put paid to Aristotelian metaphysics, or any other metaphysics for that matter. I have in this paper left many questions unsettled, such as exactly how the mathematical models of physics relate to concrete physical reality, which of the various contemporary theories of time to adopt, and how relativity ought to get interpreted in light of such a theory. No doubt some readers will think this a deficiency of my argument, but, in fact, it is a strength. What it reflects is not any weakness in the Aristotelian position, but rather *how little the physics of relativity by itself actually tells us about metaphysics and how wide open is the range of possible interpretations*. Given this fact, and given also the incoherence of any attempt entirely to expunge change and temporal passage from our picture of reality, the question of how to work out the metaphysics of relativity is not the *Aristotelian's* problem. It is *everyone's* problem.[58]

Notes

1 For exposition and defense of the theory of actuality and potentiality, and an account of its application to hylomorphism, to the real distinction between essence and existence, and to other aspects of Aristotelian-Thomistic metaphysics,

see Feser (2014). For exposition and defense of the theory's application within natural theology, see Feser (2017), especially chapter 1.

2 Lockwood (2005: 68–9).

3 Popper (1998). Julian Barbour is one contemporary physicist happy to follow out and embrace the radically Parmenidean implications he sees in modern physics. Cf. Barbour (1999: 1).

4 As we will see, for Thomist Aristotelians, though change entails the actualization of potential, the actualization of potential does not necessarily entail change.

5 Again, for a defense of the theory, see Feser (2014), especially chapter 1. A currently popular approach to analyzing change that does not make use of the distinction between actuality and potentiality is *temporal parts theory*. (Cf. Quine 1961; Lewis 1986; Heller 1990; and Sider 2001). From an Aristotelian point of view, however, what the theory leaves us with is not genuine change at all. For either temporal parts amount to a series of ephemeral things which are successively created and annihilated, and which do not add up to a single thing that persists through change; or they amount to an eccentric kind of spatial parts, the difference between which no more entails change over time than the fact that an object can at the same moment be green on one side and red on another entails a change of color. For criticism of temporal parts theory from an Aristotelian point of view, see Feser (2014: 201–8), Oderberg (1993: ch 4), and Oderberg (2004).

6 Contemporary philosophers typically speak of change in terms of the gain or loss of a "property," but the term is inappropriate here given the narrow technical sense in which it is used by Aristotelian and Thomistic philosophers. In Aristotelian-Thomistic metaphysics, a "property" is a "proper accident" or attribute which flows from a thing's essence, as opposed to a merely "contingent accident." Naturally, change is not limited to the gain or loss of "properties" in this sense.

7 Shoemaker (1969).

8 Lowe (2002: 247–9).

9 Cf. Healey (2002).

10 Cf. Dummett (1960) and Zwart (1975).

11 Cf. Mundle (1967).

12 Cf. Baker (1987: ch 7).

13 Black (1962: 181–2). Emphasis added.

14 Koperski (2015: 136).

15 Bigelow (2013). See especially pp. 155–8.

16 Smolin (2013: 34).

17 Ibid.: 34–5.

18 Bourne (2006: 158). Bourne attributes the point to Mellor (2005).

19 Nor is he alone among prominent living physicists in making it. In a July 22, 2014 online interview with *Scientific American*'s John Horgan, George Ellis urges his fellow physicists to "consider . . . what lie[s] outside the limits of mathematically based efforts to encapsulate aspects of the nature of what exists" and emphasizes that "mathematical equations only represent part of reality, and should not be confused with reality" (Horgan 2014).

20 Whitehead (1967: 51, 58).

21 Husserl (1970).

22 Bergson famously raised such objections against Einstein beginning in the 1920s, in Bergson (1998) and other works. Jimena Canales (2015) has recently revisited the dispute between Bergson and Einstein.

23 Burtt (1952: 229).

24 Maritain (1959).

25 Cf. Van Melsen (1954: 83–4) and Weisheipl (1955: 109).

26 Sklar (1985: 292). Cf. Van Melsen (1954: 187).
27 Russell (2009), first published in 1925.
28 Russell (1985: 13). Russell treated the theme at length in Russell (1927).
29 Russell (2003). I criticize the argument of this article at greater length in Feser (2014: 114–18).
30 Cf. Nagel (1979).
31 Schrödinger (1956: 216).
32 This is essentially the position defended in Tegmark (2014).
33 Chakravartty (2007: 77).
34 Ladyman (2014).
35 Yet, does epistemic structural realism not have problems of its own? A long-standing objection was raised by M. H. A. Newman (1928) against Russell's version of the view. Newman claimed that Russell's position implies that the knowledge physics gives us is entirely trivial—that physics, as understood by Russell, imposes nothing more than a cardinality constraint on the natural world, viz. a mere requirement that there be a certain minimal number of entities related by the structure described by physics. But this is implausible; surely physics tells us more than *that*. However, as Russell noted in reply to Newman, it isn't quite right to say that epistemic structural realism entails that we know from physics *absolutely nothing* but the mathematical structure of the world. We can also know, in Russell's view, of the isomorphism between a perceptual state and the extra-mental physical state that causes it. (Cf. the discussion of the correspondence between Newman and Russell in Landini (2011: 331–3.) Similarly, structural realist John Worrall (2007) notes that there is observational as well as theoretical content to a physical theory and that this puts stricter constraints on what the natural world must be like than a mere cardinality constraint.
36 Sklar (1985: 303). For discussion of Einstein's verificationism and its role in the development of relativity, see Brown (1991: ch 5) and Craig (2001: ch 7).
37 Prior (1970) and (1996).
38 Woodruff (2011). Unlike other presentists, Woodruff does not think it essential to hold on to absolute simultaneity, though that aspect of his position (which seems to me implausible) appears to be detachable from the thesis I'm summarizing here.
39 Ibid.: 118.
40 Cf. Craig's brief presentation of this position in Craig (2008) and his book-length presentation in Craig (2001).
41 Dean Zimmerman (2011b) emphasizes this point, though his own position is best understood as a variation on the fourth approach I will describe presently.
42 Cf. Monton (2006).
43 Some of the developments potentially useful to presentists are summarized in Koperski (2015: 129–34).
44 Sider (2001: 48–52).
45 An even more extreme privileging along these lines is represented by a fourth proposal considered by Sider, according to which a certain space-time point—by itself, without its future light cone, past light cone, or spacelike separated points—is all that exists (2001: 45–7). The idea is entertained, without being endorsed, in Sklar (1985) and in Hinchliff (2000).
46 Sider (2001: 47–8). Another way of trying to marry presentism to STR as it stands is to allow that what is present is relative to the observer. (The idea is discussed without being endorsed in Stein (1968). As Jeffrey Koperski points out, though (2015: 119), this would seem to make of presentness a mere subjective or observer-dependent secondary quality, which is hardly what the presentist intends.

47 Zimmerman (2011b). Cf. also Zimmerman (2008) and (2011a).
48 Zimmerman (2011a: 140).
49 Zimmerman (2008: 219).
50 Another variation of the approach of supplementing physics with metaphysics might be to argue that relativity theory simply isn't using the word "time" in the same sense as the presentist metaphysician is. Cf. Quentin Smith's distinction between "Metaphysical Time" and "Special-Theory-of-Relativity-Time" in Smith (1993: 229–31). (Though Smith's own application of this distinction is not to supplement STR with presentist metaphysics. Rather, he defends a variation of the neo-Lorentzian view that Einstein got the physics wrong. Cf. Smith (2008).
51 Woodruff (2011: 120–4).
52 Koperski (2015: 129).
53 One of the editors of this volume suggests that some readers, noting that all four of the approaches just described are committed to there being a metaphysically privileged present moment whether or not this is empirically detectable, might wonder why these are to be regarded as *different* approaches rather than variations on the same theme. The answer is that the difference between the approaches lies, not in their *own* metaphysical implications, but rather in what they take the metaphysical implications of the *physics of relativity* to be.
54 Influential defenses by philosophers include Broad (1923) and Tooley (1997). A prominent defender of the view among physicists is George Ellis (2014).
55 The "moving spotlight" theory has recently been defended in Cameron (2015) and discussed sympathetically in Skow (2015).
56 Cf. Feser (2014: 241–56).
57 In particular, it doesn't involve a persisting substrate that is present both before and after the actualization, as there in the case of melting ice, ripening bananas, and other everyday examples of change. Rather, the kind of actualizing in question here is a matter of timelessly making it to be the case that there exists anything at all, including any substrate that might underlie change.
58 I thank the editors of this volume for helpful comments on an earlier version of this paper.

Bibliography

Baker, L. R. (1987) *Saving Belief* (Princeton: Princeton University Press).
Barbour, J. (1999) *The End of Time* (Oxford: Oxford University Press).
Bergson, H. (1998) *Creative Evolution* (Mineola, NY: Dover).
Bigelow, J. (2013) 'The Emergence of a New Family of Theories of Time', in H. Dyke and A. Bardon (eds.) *A Companion to the Philosophy of Time* (Oxford: Wiley-Blackwell): 151–66.
Black, M. (1962) 'Review of G. J. Whitrow's "The Natural Philosophy of Time"', *Scientific American* CCVI: 181–2.
Bourne, C. (2006) *A Future for Presentism* (Oxford: Clarendon Press).
Broad, C. D. (1923) *Scientific Thought* (London: Routledge and Kegan Paul).
Brown, J. R. (1991) *The Laboratory of the Mind: Thought Experiments in the Natural Sciences* (London: Routledge).
Burtt, E. A. (1952) *The Metaphysical Foundations of Modern Physical Science*, Revised edition (Atlantic Highlands, NJ: Humanities Press).
Cameron, R. P. (2015) *The Moving Spotlight: An Essay on Time and Ontology* (Oxford: Oxford University Press).
Canales, J. (2015) *The Physicist and the Philosopher: Einstein, Bergson, and the Debate That Changed Our Understanding of Time* (Princeton: Princeton University Press).

Chakravartty, A. (2007) *A Metaphysics for Scientific Realism* (Cambridge: Cambridge University Press).

Craig, W. L. (2001) *Time and the Metaphysics of Relativity* (Dordrecht: Kluwer Academic Publishers).

—— (2008) 'The Metaphysics of Special Relativity: Three Views', in W. L. Craig and Q. Smith (eds.) *Einstein, Relativity, and Absolute Simultaneity* (London: Routledge):11–49.

Dummett, M. (1960) 'A Defense of McTaggart's Proof of the Unreality of Time', *Philosophical Review* 69: 497–504.

Ellis, G. (2014) 'Time Really Exists! The Evolving Block Universe', *Euresis Journal* 7: 11–26.

Feser, E. (2014) *Scholastic Metaphysics: A Contemporary Introduction* (Heusenstamm/ Piscataway, NJ: Editiones Scholasticae/Transaction Books).

—— (2017) *Five Proofs of the Existence of God* (San Francisco: Ignatius Press).

Healey, R. (2002) 'Can Physics Coherently Deny the Reality of Time?', in C. Callender (ed.) *Time, Reality, and Experience* (Cambridge: Cambridge University Press): 293–316.

Heller, M. (1990) *The Ontology of Physical Objects* (Cambridge: Cambridge University Press).

Hinchliff, M. (2000) 'A Defense of Presentism in a Relativistic Setting', *Philosophy of Science* 67: S585–76.

Horgan, J. (2014) 'Physicist George Ellis Knocks Physicists for Knocking Philosophy, Falsification, Free Will', *ScientificAmerican.com*, at https://blogs.scientificamerican.com/cross-check/physicist-george-ellis-knocks-physicists-for-knocking-philosophy-falsification-free-will/ [last accessed 23.4.17].

Husserl, E. (1970) *The Crisis of European Sciences and Transcendental Phenomenology*, trans. by D. Carr (Evanston, IL: Northwestern University Press).

Koperski, J. (2015) *The Physics of Theism* (Oxford: Wiley Blackwell).

Ladyman, J. (2014) 'Structural Realism', *Stanford Encyclopedia of Philosophy*, at https://plato.stanford.edu/entries/structural-realism/ [last accessed 23.4.17].

Landini, G. (2011) *Russell* (London: Routledge).

Lewis, D. (1986) *On the Plurality of Worlds* (Oxford: Blackwell).

Lockwood, M. (2005) *The Labyrinth of Time* (Oxford: Oxford University Press).

Lowe, E. J. (2002) *A Survey of Metaphysics* (Oxford: Oxford University Press).

Maritain, J. (1959) *Distinguish to Unite: or, The Degrees of Knowledge* (New York: Charles Scribner's Sons).

Mellor, D. H. (2005) 'Time', in F. Jackson and M. Smith (eds.) *The Oxford Handbook of Contemporary Philosophy* (Oxford: Oxford University Press): 615–35.

Monton, B. (2006) 'Presentism and Quantum Gravity', in D. Dieks (ed.) *The Ontology of Spacetime* (Amsterdam: Elvesier): 263–80.

Mundle, C. W. K. (1967) 'The Space-Time World', *Mind* 76: 264–9.

Nagel, T. (1979) 'What Is It Like to Be a Bat?', in T. Nagel (ed.) *Mortal Questions* (Cambridge: Cambridge University Press): 165–80.

Newman, M. H. A. (1928) 'Mr. Russell's Causal Theory of Perception', *Mind* 37: 137–48.

Oderberg, D. S. (1993) *The Metaphysics of Identity Over Time* (London: Macmillan).

—— (2004) 'Temporal Parts and the Possibility of Change', *Philosophy and Phenomenological Research* 69: 686–708.

Popper, K. (1998) 'Beyond the Search for Invariants', in K. Popper (ed.) *The World of Parmenides* (London: Routledge): 146–222.

Prior, A. (1970) 'The Notion of the Present', *Studium Generale* 33: 245–8.

—— (1996) 'Some Free Thinking About Time', in B. J. Copeland (ed.) *Logic and Reality: Essays on the Legacy of Arthur Prior* (Oxford: Oxford University Press): 47–52.

Quine, W. V. (1961) 'Identity, Ostension, and Hypostasis', in W. V. Quine (ed.) *From a Logical Point of View* (New York: Harper and Row): 65–79.

Russell, B. (1927) *The Analysis of Matter* (London: Kegan Paul).

—— (1985) *My Philosophical Development* (London: Unwin Paperbacks).

—— (2003) 'On the Notion of Cause', in Stephen Mumford (ed.) *Russell on Metaphysics* (London: Routledge): 163–82.

—— (2009) *ABC of Relativity* (London: Routledge).

Schrödinger, E. (1956) 'On the Peculiarity of the Scientific World-View', in E. Schrödinger (ed.) *What Is Life? And Other Scientific Essays* (New York: Doubleday): 178–228.

Shoemaker, S. (1969) 'Time Without Change', *Journal of Philosophy* 66: 363–81.

Sider, T. (2001) *Four-Dimensionalism* (Oxford: Clarendon Press).

Sklar, L. (1985) 'Time, Reality, and Relativity', in L. Sklar (ed.) *Philosophy and Spacetime Physics* (Berkeley and Los Angeles: University of California Press): 289–304.

Skow, B. (2015) *Objective Becoming* (Oxford: Oxford University Press).

Smith, Q. (1993) *Language and Time* (Oxford: Oxford University Press).

—— (2008) 'A Radical Rethinking of Quantum Gravity: Rejecting Einstein's Relativity and Unifying Bohmian Quantum Mechanics With a Bell-neo-Lorentzian Absolute Time, Space and Gravity', in W. L. Craig and Q. Smith (eds.) *Einstein, Relativity, and Absolute Simultaneity* (London: Routledge): 73–124.

Smolin, L. (2013) *Time Reborn* (New York: Houghton Mifflin Harcourt).

Stein, H. (1968) 'On Einstein-Minkowski Space-time', *Journal of Philosophy* 65: 5–23.

Tegmark, M. (2014) *Our Mathematical Universe* (New York: Alfred A. Knopf).

Tooley, M. (1997) *Time, Tense, and Causation* (Oxford: Clarendon Press).

Van Melsen, A. G. (1954) *The Philosophy of Nature*, 2nd edition (Pittsburgh: Duquesne University).

Weisheipl, J. A. (1955) *Nature and Gravitation* (River Forest, IL: Albertus Magnus Lyceum).

Whitehead, A. N. (1967) *Science and the Modern World* (New York: The Free Press).

Woodruff, D. M. (2011) 'Presentism and the Problem of Special Relativity', in W. Hasker, T. J. Oord and D. Zimmerman (eds.) *God in an Open Universe: Science, Metaphysics, and Open Theism* (Eugene, OR: Pickwick Publications): 94–124.

Worrall, J. (2007) 'Miracles and Models: Why Reports of the Death of Structural Realism May be Exaggerated', in A. O'Hear (ed.) *Philosophy of Science* (Cambridge: Cambridge University Press): 125–54.

Zimmerman, D. (2008) 'The Privileged Present: Defending an "A-Theory" of Time', in T. Sider, J. Hawthorne and D. W. Zimmerman (eds.) *Contemporary Debates in Metaphysics* (Oxford: Blackwell): 211–25.

—— (2011a) 'Open Theism and the Metaphysics of the Space-Time Manifold', in W. Hasker, T. J. Oord and D. Zimmerman (eds.) *God in an Open Universe: Science, Metaphysics, and Open Theism* (Eugene, OR: Pickwick Publications): 125–57.

—— (2011b) 'Presentism and the Space-Time Manifold', in C. Callender (ed.) *The Oxford Handbook of Philosophy of Time* (Oxford: Oxford University Press): 163–244.

Zwart, P. J. (1975) *About Time* (Amsterdam: North-Holland).

3 The Many Worlds Interpretation of QM

A Hylomorphic Critique and Alternative[1]

Robert C. Koons

1 Introduction

The so-called Many Worlds Interpretation of quantum mechanics has been extant now for nearly 60 years, beginning as H. Everett III's doctoral dissertation in 1956 [Everett 1956], with further contributions by B. DeWitt and N. Graham in their 1973 book, *The Many Worlds Interpretation of Quantum Mechanics* [DeWitt and Graham 1973]. The Everett approach takes quantum mechanics both realistically and as a stand-alone, autonomous theory of the world, not in need of a separate theory of measurement to bridge the apparent gap between the deterministic evolution of the wavefunction in a highly abstract, probabilistic space, and empirically observable statistics in the laboratory. Instead, Everett proposed that all of the apparently contradictory macroscopic results assigned some finite probability by the theory are equally real, co-existing in distinct sets of *relative states*. DeWitt and others later identified these clusters of mutually consistent relative states with distinct and co-existing *worlds* or *branches* of the world.

These early versions of the interpretation faced a huge problem: there were no *worlds* or *branches*, describable in macroscopic terms, to be found in the formalism of quantum mechanics itself. We can find within the formalism something called *superpositions*, which are states that seem to attribute to particular systems (like particles) a plurality of mutually inconsistent properties, each with a certain amplitude, but there seems to be no way to recover macroscopic instruments and determinate measurement relations from these isolated superpositions. This problem is often described as the problem of finding a *preferred basis*, since the decomposition of the world into discrete branches can only take place relative to a selection of a certain set of orthogonal parameters. Any selection of such a basis seemed arbitrary and unprincipled, and so the objectivity of the co-existing branches was thrown into doubt. In addition, there is nothing in the wavefunction that corresponds to the *persistence* or *splitting* of branches. Probabilities of various states simply fluctuate over time: there is no way to *trace* where the probability that once belonged to a given state has moved (either as a unified packet or through fission).

In the 1970s, 80s, and 90s, a great deal of theoretical work commenced on the problem of giving a fully quantum-theoretic account of measurement.

This work comprises the programs of *decoherence* of W. Zurek 1982 and H. D. Zeh 1973 and the *consistent histories* approach of Griffiths 1984, Omnès1988, and Gell-Mann and Hartle 1990, 1993. The decoherence results show that under favorable circumstances a stable, approximately classical domain can be expected to emerge from the quantum-mechanical descriptions of a measuring system, its object, and the surrounding environment (Wallace 2011). What decoherence left unsolved was why we see the emergence of just *one* such quasi-classical domain when interacting with quantum superpositions. A marriage of decoherence with the Everett interpretation was inevitable, with the Everett interpretation explaining the apparent uniqueness of result as a product of the relativity of our perspective in this or that branch, and the decoherence providing the missing preferred basis and explaining how to extract persistent and apparently "splitting" quasiclassical domains from quantum descriptions.

The consistent histories approach was even closer to the spirit of the Everett interpretation, since it sought to extract approximately classical domains from the quantum function for the entire cosmos, rather than looking at particular instrument-object-environment arrangements. Here again, the two approaches seemed designed to resolve each other's deficiencies: with consistent histories providing the preferred basis, and the Everett interpretation dissolving the worry about what to do about certain regions or phases of the cosmic history in which there are no consistent histories at all. On the Everett interpretation, only the quantum wavefunction describes fundamental reality, so only it can be expected to have universal validity. Consistent histories simply describe the approximate emergence of quasiclassical branches under favorable circumstances, including, presumably our own.

In recent years, the Many Worlds Interpretation has found a new home in Oxford, among both physicists and philosophers of science, including David Deutsch, Simon Saunders, David Wallace, Christopher Timpson, and Harvey Brown. The Oxford group has developed the idea of using decoherence and consistent histories approaches to solve the preferred basis problem, explaining the emergence of approximately classical "domains" from the wavefunction. They have also, building on seminal work by Deutsch 1999, attempted to solve the other central problem of the interpretation, which is that of making sense of the precise probabilities ascribed to different outcomes by applying Born's rule to the wavefunction.

In the next two sections, I will raise two objections to the new, Oxford-style Everettian interpretation. First, in section 2, I will argue that Deutsch's strategy cannot make sense of the probabilities that play such a central role in quantum mechanics. The Many Worlds Interpretation cannot explain the rational necessity of one of the crucial axioms (Savage's Sure Thing Principle) upon which the modern theory of subjective probability depends. Then, in a much longer section 3, I will argue that the Oxford Everettians attempt to use the philosophical framework of *functionalism* to elucidate the relation between the manifest world of scientific experiment and observation and the underlying, fundamental quantum reality ends in failure. Specifically, I will idenify four failures of this account:

1. I will use Putnam's paradox to demonstrate a radical indeterminacy of content that would afflict all of our scientific theories.
2. I will demonstrate that any consistent story of the world (no matter how fantastic) would count as equally *real*.
3. As a consequence, it would be impossible for any of our scientific theories to be wrong, making it equally impossible for them to be empirically confirmed.
4. This failure of empirical testability would deprive us of any reason for believing in quantum mechanics in the first place.

In section 4, I will critically examine seven possible strategies that Oxford Everettians might rely on to solve the Putnamesque problems identified in section 3. These strategies are (1) appealing to the concept of *emergence*, (2) appealing to "our" actual language and theories, (3) relying on causal constraints to fix the interpretation, (4) appealing to *natural* or *eligible* properties, (5) using realism about spacetime, (6) appealing to simplicity, and (7) relying upon decoherence to solve the problem. I will argue that all seven strategies fail, although there is a version of the appeal to simplicity that might turn out to provide at least a partial solution. However, even this appeal to simplicity (if it were ultimately successful) would leave us without an adequate account of the *reality* of the manifest, macroscopic world, and it would still have the consequence that none of our theories in the special science (theories of the emergent, macroscopic world) can ever be false, with all of the epistemological catastrophe that such a result would bring

I turn in section 5 to sketching a neo-Aristotelian alternative to the Oxford Everettian interpretation. This new interpretation adds the additional metaphysical constraints needed to solve the Putnamesque paradoxes in the form of a set of *essences* of macroscopic substances. This also enables us to use the actualization of these essences as a way of distinguishing the one *actual* branch from all the *merely possible* ones. Alex Pruss and I call the resulting interpretation *the traveling branches* interpretation. This interpretation builds on both the realism about the quantum wavefunction and the results of decoherence theory, in exactly the same way as these are treated by the Oxford Everettians, and yet it ends up in a metaphysical and semantic position that is much more defensible.

2 Probability and the Oxford-Style Everett Interpretation

The basic problem can be stated quite simply: since all possible outcomes will in fact occur with probability *one*, what meaning can be assigned to the varying strengths of probability assigned by Born's rule to different outcomes? On the Oxford interpretation, it makes no sense to count branches, since the very existence of branches is only a non-fundamental and inherently vague phenomenon, resisting any perfect precisification. Deutsch's answer, developed further by Saunders and Wallace (Saundersa and Wallace 2008, Wallace 2007, Wallace 2011), is to use the pragmatic approach to subjective probabilities developed in the early 20th century by Frank Ramsey, Leonard Savage, John von Neumann, and others in order to argue that perfectly rational agents must, given certain constraints including perfect knowledge of the quantum state, *act as if* they assigned the appropriate probabilities to the various branches.

In a paper entitled "Truth and Probability,"[Ramsey 1988] Frank P. Ramsey sought to provide an operational or behavioral definition of the notion of *degree of belief* or *degree of confidence of truth*, as well as the correlative notion of *desirability* or *utility* or *subjective value*. Ramsey imagined an idealized experimental setup in which both the degrees of belief in various propositions of the experimental subject (i.e., the subject's *subjective probability function*) and the degrees of desirability that the subject attaches to the states of affairs represented by those propositions (i.e., the subject's *utility function*) may be measured. Ramsey imagines that the subject is confronted by what he (the subject) believes to be an omnipotent and totally trustworthy Bookie, who offers the subject a series of choices between two options. Some of these options come in the form of simple bets: e.g., an option that might be offered to the subject could take the form: α if p is true; otherwise, β. If the subject's choices conform to certain principles of mutual coherency, then there exists a unique representation of the subject's state of mind in terms of a probability function taking as its values real numbers in the interval from 0 to 1 (inclusive) and a utility function taking real numbers as values.[2]

In 1954, Leonard Savage [Savage 1988] provided a slightly different axiomatization, from which he was able to prove a representation theorem of the appropriate kind. Savage's setup included the following elements:

1. A set of *states* of the world, S, with elements $s, s', s'', \ldots$ and subsets (the *events*) $E, E', E'', \ldots$.
2. A set of consequences or outcomes C, with elements $c, c', c'', \ldots$.
3. A set of acts $\mathcal{A}$, with elements $A, A', A'', \ldots$.
4. An assignment of a consequence from C to every act-state pair (A, s), designated $A(s)$.
5. A binary relation $\succeq$ between pairs of acts that is interpreted to mean *is preferred or equal to*.

His axiom system included the following axioms:

> **Axiom 1** *The relation $\succeq$ is a weak ordering of the acts: the relation is transitive, and any two acts are comparable (either $A \succeq A'$ or $A' \succeq A$, or both).*

Definition 1 $A \succeq_E A'$ if and only if: if acts A and A' are modified so that their consequences are the same for every state not included in E, and they are not modified for any states in E, the resulting modification of A is preferred or equal to the modification of A'. The propriety of this definition depends on the next axiom, Savage's Sure Thing Principle.

Axiom 2 The Sure Thing Principle *If acts A, A', B, B', and event E are such that (i) A and B agree on all states outside of E, (ii) A' and B' also agree on all states outside of E, (iii) A and A' agree on all states within E, (iv) B and B' agree on all states within E, then: $B \succeq A \leftrightarrow B' \succeq A'$.*

Savage's Sure Thing Principle asserts the irrelevancy of non-discriminating possibilities. That is, suppose that two actions A and B are known to have exactly the same consequences (as far as things are concerned that matter to the agent) on the assumption of E: then, if B is preferred to A, then B would still be preferred to A regardless of what those irrelevant consequences would be (regardless of what the common consequences are that would follow from either A or B on the supposition of E). The Sure Thing Principle (in conjunction with the other axioms) has a further consequence: the agent will also prefer B to A regardless of whether the agent knows E to be true or false or is left in some state of ignorance about E. The probability of E, given its irrelevance to the comparison of A and B, must be irrelevant to the preferability of B over A.

On the basis of his axioms, Savage was able to prove the following representation theorem, demonstrating that any reasonable agent must act as if guided by both a utility and probability function (with maximizing expected utility as the decision criterion):

Theorem 1 Representation Theorem

There exists a unique real-valued function P defined for the set of events, and a unique (up to positive linear transformations) real-valued function u defined over the set of consequences such that:

1. $P(E) \geq 0$ for all E.
2. $P(S) = 1$.

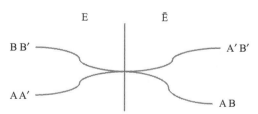

Figure 3.1

3.   If E and E' are disjoint, then $P(E \cup E') = P(E) + P(E')$.
4. E is not more probable than E' if and only if $P(E) \leq P(E')$.
5.   If the E_i's are a finite partition of S, and A is an act with conse-
     quence c_i on E_i, and if the E_i''s are another finite partition on S, and
     A' is an act with consequence c_i' on E_i', then $A \succeq A'$ if and only if:

$$\sum_{i=1}^{n} u(c_i)P(E_i) \geq \sum_{i=1}^{m} u(c_i')P(E_i')$$

2.1 The Failure of Savage's Principle in a Many-Worlds Setting

Deutsch [Deutsch 1999] and Wallace [Wallace 2010] argue that the formal
results of Ramsey and Savage can enable the Everettians to make sense of
the varying probabilities of the various branches of the wave function. If we
assume that our ideally rational agent must satisfy Savage's axioms (with
states now identified with branches) and must also respect certain physically
symmetries, then we can prove a representation theorem similar to Savage's,
but with the added feature that the probabilities assigned to the branches
must mimic the probabilities assigned to those branches by Born's rule. In
particular, Wallace incorporates Savage's Sure Thing principle into his condition of *diachronic consistency*.

I share the doubts about the cogency of the argument expressed by Huw
Price [Price 2010]. Price points out that by adding multiple worlds to our
inventory of reality, we give rational agents *new things to care about*, things
that can't be captured by any probability-weighted averaging of worldbound utilities. For example, a rational agent might assign a certain finite
amount of utility to the degree of *equality of outcome* enjoyed or suffered
by persons across the span of branches. Wallace [Wallace 2010, pp. 256–
7] argues that we can capture such cross-world utility considerations by
tinkering with the various world-bound measures, but it is easy to prove
that giving a positive weight to transworld equality will necessarily violate Wallace's diachronic consistency condition (by violating Savage's Sure
Thing Principle), despite the obvious rationality of that principle in a one-
world setting.

A core principle of Savage's axiomatization of decision theory is his *Sure
Thing Principle*, which is incorporated into Wallace's axiom system (in the
form of his *diachronic consistency condition*). As we have seen, the Sure
Thing Principle asserts the irrelevancy of non-discriminating possibilities.
That is, suppose that two actions A and B are known to have exactly the
same consequences (as far as things are concerned that matter to the agent)
on the assumption of E: if B is then preferred to A, then B would still be
preferred to A regardless of whether the agent knows E to be true or false.
This makes sense in a normal, one-world decision setting: if I know that
the world will end up either in E or in E' (exclusively), and I am indifferent

between A and B on the assumption of E (because I know that A and B would produce exactly the same results in that case), then my preference for A over B must be concerned exclusively with what would happen on the assumption of E'.

However, in a many-worlds setting, I have to imagine that both E and E' will become actual (albeit in different branches), and I might care about certain trans-branch facts. We can no longer assume that my utility function is simply a weighted sum of intra-branch values. For example, suppose that there are two people affected by my action, person 1 and person 2. In the E' branches, person 1 and person 2 will each receive zero units of value, regardless of whether I perform A or B. However, in the E branches, persons 1 and 2 get one unit of value if I do A, but if I do B instead, then person 1 gets 0.4 units of value and person 2 gets 2 units. Now suppose that I value outcomes on the basis of the total utility (across all branches) minus the average deviation from the mean (representing my dislike for interpersonal inequality). My desire to avoid overall inequality applies equally well to inter-branch inequality as to intra-branch inequality. Here is the resulting table:

	E	E'	Expected value
A	$\langle 1,1 \rangle$	$\langle 0,0 \rangle$	$2-0.5 = 1.5$
B	$\langle 0.4,2 \rangle$	$\langle 0,0 \rangle$	$2.4-2.8/4 = 1.7$

Note that if we eliminate E' from consideration, then I would definitely prefer action A, since it provides perfect equality and that outweighs in this case the somewhat greater total utility of action B. In contrast, however, if we take branch E' into account (as a genuine part of reality), the total value of action B now outweighs its somewhat lower average deviation from the mean, and B is now preferable to A. This is all perfectly reasonable in a many-worlds setting, in which all of the outcomes are equally real (enjoyed or suffered in reality in one branch or the other), but it clearly violates Savage's Sure Thing Principle (and, consequently, also Wallace's even stronger diachronic consistency condition). No amount of tinkering with utility functions can overcome this difficulty.

We can see that, in the Many Worlds setting, we can turn this example into a straightforward violation of Savage's Sure Thing Principle (and thereby, also, a violation of Wallace's diachronic consistency axiom). Let's add two new actions A' and B' to the setting. A agrees perfectly with A' on condition E, and B agrees with B' on that same condition. The two actions A' and B' produce exactly the same result under condition E'. Therefore, the Sure Thing Principle requires that the agent prefer A to B just in case she prefers A' to B'. However, considerations of equality that are germane in the Many Worlds setting lead the agent quite reasonably to prefer B to

A but also to prefer *A'* to *B'*. The Savage-style justification of the Born rule collapses.

	E	E'	Expected value
A	⟨1,1⟩	⟨0,0⟩	2–0.5 = 1.5
A'	⟨1,1⟩	⟨1,1⟩	4
B	⟨0.4,2⟩	⟨0,0⟩	2.4–2.8/4 = 1.7
B'	⟨0.4,2⟩	⟨1,1⟩	4.4–1.8/4 =3.95

Why doesn't this possibility of violating the Sure Thing Principle apply with equal force to those who reject the Many Worlds Interpretation? Because it is irrational to let what might have happened but did not actually happen influence one's evaluation of a particular possible outcome. We could express this idea by means of a slogan: *Value Supervenes on Being.* The value of an outcome is a function of what does or does not happen if that outcome were actual. It cannot depend on what happens in other, mutually incompatible outcomes. In the Many Worlds Interpretation, all of the branches are equally real, equally partaking of existence. Hence, Deutsch and Wallace have no grounds for excluding trans-branch values and so no way to validate Savage's axioms.

Ironically, the technical results of Deutsch and Wallace do provide strong grounds for accepting a One World Interpretation of quantum mechanics. Since any One World interpretation will satisfy Savage's axioms, the symmetry considerations cited by Deutsch and Wallace make it reasonable to suppose that any rational agent in a one-world setting must attribute objective chances to events in a way that corresponds to the Born rule, given certain knowledge of the actual quantum function. Thus, Deutsch and Wallace's hard work was not wasted: they simply deployed it on behalf of the wrong interpretation!

2.2 A Second Failure: Retrospective Probability and Anti-Darwinian Branches

There is a second critical gap in the Deutsch-Saunders-Wallace program: namely, its inability to justify purely *retrospective* uses of probability, including the grounds for our conviction that we do not inhabit any of the low-probability anti-Darwinian branches (branches in which highly maladapted populations have managed to survive despite that maladaptiveness). An anti-Darwinian branch is one in which the typical populations of organisms are *very* badly adapted to their environment, a world in which Empedocles' half-ox-half-man and other unfit monstrosities predominate. There is good reason to think both (i) that there are very many such anti-Darwinian branches lurking in the world's quantum-wave function, under some suitable interpretation function and (ii) that all such anti-Darwinian branches have extremely low quantum amplitudes.

One-world theorists can hold that all branches with extremely low amplitudes are *approximately impossible*—that is, close enough to being impossible as to be practically negligible. However, such a notion of "approximately impossible" is unavailable to many-world theorists. Each branch is, for a many-worlds theorist, just as real as any other. They are all *equally far* from being impossible, since all actual events are equally possible. It makes no sense to say that some are so close to being impossible as to be practically indistinguishable from it.

Many-worlds theorists (Deutsch, Saunders, Wallace) have focused their attention on two issues concerning probability: (i) arguing that it could be a constraint on rational action that the rational weight expected utilities of future branches proportionally to the square of the branches' amplitudes, and (ii) arguing that it could be rational for an investigator to treat sequences of observations that match those expected on relatively high-amplitude branches to count as confirming a statistical theory. Both of these issues concern agent-bounded uses of probability, in the sense that it is only the probabilities of events actually observed by the agent, either in his or her past or expected in the future, that are relevant.

However, when we use natural-selection arguments to exclude the possibility that we and other extant organisms are terribly maladapted to our environment, we are using probabilities in a way that extends far beyond our own past or present experiences. The arguments about decision theory and about statistical inference are irrelevant. We would be faced, on the Many Worlds Interpretation, with a problem of pure self-location, and there would seem to be no grounds for rationally assuming that we must be *ab initio* located in one sort of branch rather than other. I cannot say that it is virtually impossible for me to be located in a low-probability branch when there are in fact many counterparts of mine who are in fact located there. For example, Richard Dawkins has often stated, quite reasonably, that he is certain a priori that natural selection will have shaped the evolution of living organisms on any planet in the cosmos, no matter how remote. To have any validity, such a priori appeals to natural selection must embrace the whole of reality, or none of it. The alternative branches of the Many Worlds Interpretation are simply remote regions of reality, and yet there is no doubt that such a priori confidence in the validity of natural selection in all such branches would be profoundly misplaced. But charity begins at home: if we cannot apply the principle of selection to remote branches a priori, we would be unjustified to do so for our own world.

What cost do we pay if we forego such a priori appeals to natural selection? A very profound cost to the foundations of epistemology. Our natural reliance on our senses and memories is undermined if we cannot count on natural selection to ensure their reliability. Such undercutting of confidence in empirical knowledge will, in turn, deprive scientific theory (including the theory of quantum mechanics) of objective warrant.

3 Recovering the Manifest Image Through Ramseyfication

Leaving probability aside for the moment, the central problem for the Many Worlds Interpretation is that of bridging the gap between the scientific image of the quantum wavefunction and the "manifest image" (to use Sellars's phrase from Sellars 1962) of our approximately classical, macroscopic world, the world occupied by all experimenters, their instruments, and the results of their experiments. David Wallace's solution is an admirably simple one: all features and entities of the "manifest image" (macroscopic objects, organisms, sensible properties) are to be reduced to functional roles realized in some way by the quantum wavefunction—where these functional roles can either be identified with second-order, functional properties, or with the quantum-mechanical role-fillers corresponding to those properties (in style of David Lewis 1966, 1972, and 1980).

The functional properties can be identified with the result of Ramseyfying ([Ramsey 1929]) our ordinary folk ontology and our special sciences (including, perhaps, classical mechanics) in the language of pure quantum-mechanics (infinite Hilbert space, unitary Schrödinger evolution). There are three historical precursors to the kind of functionalization of the manifest image that Wallace has in mind: the phenomenalistic project, as typified by John Stuart Mill and the early Carnap, Bertrand Russell's functional-structural account of physics in *The Analysis of Matter* [Russell 1927], and the late-20th-century, behaviorism-inspired accounts of the mind, especially the Analytic Functionalism of David Lewis [Lewis 1966, 1972, 1980]. The ideal formal machinery for each attempted functional reduction is F. P. Ramsey's account of scientific theories [Ramsey 1929], well explained by Lewis in Lewis 1972.

In both the phenomenalist project and in Russell's 1927 structuralism, there was a significant epistemological dimension, based on the idea that we have privileged and certain access only to our own conscious states (a kind of Cartesian starting point for knowledge). That epistemological element is much reduced in Lewis's functionalism and entirely absent from Wallace's project, so I will, at the risk of some anachronism, present all four programs as if they were concerned solely with ontological issues, that is, with identifying the correct truthmakers or truth grounds for the reduced theory.

In this paper, I will use a model-theoretic version of Ramseyfication, in which, instead of introducing second-order variables and quantifiers for the predicates, we simply extend the interpretation function of a given model in order to turn a model of the original, base language into a model of the emergent theory in an appropriately expanded language.

We must also make use of a set of *possible worlds*, because all versions of functionalism require that we make some reference to the *dispositions* of things to respond or behavior in specified ways, even if the things never actualize these dispositions. The simplest formal semantics for such dispositions makes use of the subjunctive conditional: if P *were* true, Q *would be* true. We can represent the truth of such a subjunctive conditional at the actual world w^* by supposing that w is surrounded by a system of spheres of worlds, representing degrees of closeness or similarity of those worlds to w. The subjunctive conditional $(P \square\rightarrow Q)$ is true at w^* if the material conditional $(\neg P \vee Q)$ is true in all of the worlds contained by some P-permitting sphere.

For the sake of simplicity, I will assume that each of the individuals exists in only one world. Our interpretation function must assign to each constant (proper name) an individual in each world, and to each n-ary predicate, a set of n-tuples of individuals from that world. This will enable us to assign truth-values at each world to all logically complex formulas, using the usual clauses of Tarski's truth definition. That is, a negation $\neg\phi$ is true in a world w just in case ϕ is not true there, and a disjunction $\phi \vee \psi$ is true in w just in case either ϕ or ψ is true there. We can interpret the existential and universal quantifiers in the usual way, using at each world the domain of individuals that exists there.

1. A model frame F consists of a set of worlds W, a designated actual world $w^* \in W$ and, as in David Lewis's semantics [Lewis 1973], a system of spheres S, consisting of nested subsets of W, centered on w^*. I will assume that the sphere-system S is dense (between any two concentric spheres there is always a third) and that the number of worlds in every sphere-membership equivalence class is equal to the number of sets of atomic formulas of the language.

2. A model M consists of a model frame F plus a domain of "worldbound" individuals D (each existing in just one world) and an interpretation function I, which is used to interpret the predicates, function symbols, and simple singular terms (names or constants) of the language.

3. The set of individuals D is partitioned into disjoint cells, one for each world in W. We can think of D as a function from W into a set of disjoint sets, with $D(w)$ designating the worldbound individuals of world w.

4. For any n-ary predicate F, $I(|F|)$ is a function whose domain is W, and for each world $w \in W$, $I(|F|)(w)$ is a set of n-tuples of the members of $D(w)$.

5. For any constant c, $I(|c|)$ is a function whose domain is W, and for each world $w \in W$, $I(|c|)(w)$ is a member of $D(w)$.

6. For any atomic sentence $F(c_1, c_2,..., c_n)$, $I(|F(c_1, c_2,..., c_n)|)$ is a set of worlds in W, where each world w belongs to $I(|F(c_1, c_2,..., c_n)|)$ if and only if the n-tuple $\langle I(c_1)(w), I(c_2)(w),..., I(c_n)(w)\rangle$ belongs to $I(|F|)(w)$.

7. $I(|(\phi \& \psi)|) = I(|\phi|) \cap I(|\psi|)$, and similarly for the other sentential connectives.

8. $I(|\exists x\ \phi\ (x)|)$ = the infinite union of the sets $I\ (|\phi(c)|)$, for each constant c in the language L. (We'll assume that the language L has been enriched with enough constants to provide a witness for every existential generalization true in M.)
9. $I\ (|(\phi\Box\!\!\rightarrow\psi)|)$ = W if there is an $I\ (|\phi|)$-permitting sphere s in S such that every world in $I(|\phi|)\cap s$ is also in $I\ (|\psi|)$). Otherwise $I\ (|(\phi\Box\!\!\rightarrow\psi)|)$ = $\emptyset$.

A model $M = \langle F, D, I\rangle$ is a model of a theory T just in case, relative to I and D, the actual world w^* belongs to $I(|T|)$, where $I(T)$ is the intersection of the sets $I(|\phi|)$, for each formula ϕ in T. As usual, a theory is defined as a set of formulas closed under logical implication.

Let's suppose that we start with a model $M_{base} = \langle F, D, I\rangle$, defined for our base language L_{base}, which represents the fundamental level of reality. Now suppose that we extend the language L_{base} to a language $L_{base+emergent}$, by adding constants, function symbols, and predicates that signify an emergent, non-fundamental level of reality. A theory $T_{emergent}$ of this emergent world is realized in our base model M just in case the interpretation function I can be extended to a new function $I_{realizer}$, defined for $L_{base+emergent}$, such that the model $M_{extended} = \langle F, D, I_{realizer}\rangle$ is a model of $T_{emergent}$. In such a case, we can say that the function $I_{realizer}$ is a *realization* of the emergent theory $T_{emergent}$ in the original base model M_{base}. This model-theoretic version is a generalization of Ramsey's original idea, since it applies even to theories that are not finitely axiomatizable. Instead of taking a single formula that axiomatizes the emergent theory and replacing all the emergent terms and predicates with first- and second-order variables, we extend the interpretation function of the original model in order to provide extensions to all the terms and predicates of the emergent theory. In cases in which a theory can be axiomatized by a single formula, the two methods are exactly equivalent: the base model will verify the second-order Ramsey formula if and only if the model's interpretation function can be extended to produce a model of the corresponding theory.

3.1 Classical Phenomenalism and Russell's Structuralism

Using this model-theoretic approach to realization, classical phenomenalism could be seen as postulating that all truths about the existence and characteristics of physical objects are realized by truths about the private and subjective sense-experience that human observers have or would have under specified, counterfactual conditionals. As Mill put it, physical objects are "mere permanent possibilities of perception." So, we start with a base language P_{phen}, which includes terms for subjects of experience, terms for sense-data, and predicates that define sense experiences in terms of the locations of sense data in the egocentric spaces of subjects (with properties like *up* and *down*, *left* and *right*, *forward* and *back*), at times in private, egocentric time lines. The language will also include the subjunctive conditional. We will then consider a class of models for this language, consisting of a set of worlds W, a designated actual world w^*, an interpretation function I for evaluating atomic sentences in each world, and a system of concentric spheres S for the interpretation of subjunctive conditionals. For simplicity's sake, I will treat all sense-data and subjects as worldbound individuals (in Lewis's sense). We can then select the model $M_{true-phen}$ that incorporates all the actual truths about actual and counterfactual experiences. The set of formulas true in $M_{true-phen}$ is the set $TRUE_{phen}$, the set of all truths expressible in the vocabulary of L_{phen}.

Throughout this paper, I'm going to assume that the structure of the true model of fundamental reality is rich enough that there is a homomorphism from any canonical model for Lewis's subjunctive conditionals into that model, which, in the case of phenomenalism, we'll call $M_{true-phen}$.[3] If this were not the case, we would have little reason to believe that any of our theories of the emergent world have even approximate models in the true model of fundamental reality. Furthermore, we have good reason to think that the model of fundamental reality is very rich representationally, with a very large number of worlds, with a very rich set of relations of comparative similarity. There is every reason to think that the canonical model for any language using the subjunctive conditional can be mapped into such a rich model.

We now enrich the language by adding terms referring to physical objects, which will now be assigned locations and trajectories in a single three-dimensional (public) space, indexed by universal time. Since we still retain the subjunctive conditionals, we can now express conditional relationships between sentences expressed in purely phenomenal terms and sentences expressed in purely physical terms, and between pairs of sentences both of which are purely physical in form, as well as between sentences that mix both vocabularies.

- Call the resulting language $L_{phen+phys}$.
- Consider each theory expressible in $L_{phen+phys}$ that is consistent with the set of phenomenal truths, $TRUE_{phen}$.
- Let T_0 be one such a theory.
- Since T_0 is consistent with the set of phenomenal truths, we can extend the interpretation function I_{phen} to a function $I_{phen+phys}$ in such a way that theory T_0 is true in the model $M_{true-phen}$ relative to $I_{phen+phys}$.

The extended interpretation accomplishes exactly the same thing as would be accomplished by Ramseyfying the physical vocabulary in a finite axiomatization of T_0, if there is a such a thing. That is, $I_{phen+phys}$ assigns some property-intension or individual-concept-intension in $M_{true-phen}$ to every predicate and individual constant in the physical vocabulary of T_0 in such a way as to verify T_0. The model-theoretic approach that I've sketched is actually more general than Ramseyfication, since it will apply to any consistent theory, whether or not that theory can be finitely axiomatized. In addition, it means that we can keep everything in first-order logic. Each interpretation function I relative to which T_0 is true in $M_{true-phen}$ constitutes a distinct *realization* of T_0.

Russell's structuralist program in *The Analysis of Matter* is exactly iso-morphic to the classical phenomenalist program. The only difference is that Russell does not use subjunctive conditionals, as Mill did, but instead speaks of *causal* relations, both in the phenomenal and in the physical world. However, he does not offer a substantive account in the 1927 book of what causation consists in, so this is a difference we can, at least for the moment, set aside. In addition, of course, instead of speaking about phenomenal sense-data, Russell in 1927 speaks instead about *perceptions*, which he takes to be events in the brain with which we are immediately acquainted.

In general, there will be many realizations of any theory T_0 in the model $M_{true-phen}$, and there will be many other theories in the enriched language besides T_0 that are consistent with the set of all phenomenal truths (and which therefore have realizations in $M_{true-phen}$). In order to cut down the number of theories and realizations, we need some further constraints both on our theory T_0 and on the permissible realizations of that theory. We can accomplish both of these at once simply by restricting the interpretation function. We can then hope to pick out the one true theory of physics that has a unique permissible realization in the model $M_{true-phen}$.

In the case of both the phenomenalist and Russellian-structuralist pro-gram, these constraints consist in the *laws of perspective* that link geometri-cal properties described in terms of public, four-dimensional spacetime with properties described in terms of egocentric phenomenal space and time. We can put a constraint an any acceptable interpretation function, requiring that when it identifies a physical object in a world with a set of sense data associ-ated with subjects in that world, the interpretation function must assign a shape and size to the physical object that corresponds to the shape and size of each of the corresponding sense-data, with the correspondence relation fixed by the laws of perspective as applied to the physical location assigned to the relevant subject of experience. That is, the physical primary qualities of bodies must correspond to sense-data and subject-locations in such a way that each sense-datum accurately records the shape of the body, as it would appear to a subject at the location to which the subject is assigned.

This is quite a severe constraint—in fact, too severe, since it fails to take into account the existence of illusions and hallucinations. It is reasonable to suppose that only one theory-interpretation pair will maximize the degree of fit between the bodies and the corresponding sense-data, and we can take this pair to give us both the set of truths about the physical world and the corresponding truthmaker in the phenomenal world for each truth.

3.2 Analytic Functionalism About the Mind

David Lewis's version of Analytical Functionalism is exactly isomorphic to the phenomenalist or structuralist model sketched in the preceding subsec-tion. The differences are these: first, the base model with which we begin is a model of something like classical physics and chemistry, including facts about overt behavior, sensory-organ stimulations, and neural structures and patterns of firing. The true model of the world $M_{true-phys}$ yields a set of physicalistically acceptable truths, $TRUE_{phys}$ in a language of purely physical (and chemical, biological, and neurological) vocabulary L_{phys}. We want to extend this language to a language $L_{phys+psy}$ that includes the vocabulary of psychology, with predicates that assign beliefs, desires, and sensory experiences to a class of sentient and rational bodies (the human beings). Lewis assumes that we are already given not only the vocabulary of $L_{phys+psy}$, but also a fairly rich theory of *folk psychology* T_{folk} that specifies a large number of connections between psychological and physical states. This will include facts about the sensory experiences resulting from sensory-organ stimulations, coordinated in such a way that experiences are veridical under normal conditions. It will also include connections between belief-desire pairs and overt behavior, and certain kinds of overt behavior that result directly from certain experiences or desires, like wincing from pain.

- Let's assume that T_{folk} is consistent with the set of physical truths, $TRUE_{phys}$.
- If so, we can find an interpretation function $I_{phys+psy}$, relative to which T_{folk} is true in the true model of the physical world, $M_{true-phys}$.
- If there is such a function, it will be a *realization* (in Ramsey's sense) of the folk theory of psychology.
- If there is a unique such function, then we can use it to define the set of *all* psychological and psychophysical truths by simply identifying it with the set of sentences $TRUE_{phys+psy}$ in the language $L_{phys+psy}$ that are verified by the model $M_{true-phys}$ as extended by the interpretation function $I_{phys+psy}$.

Lewis is entitled to help himself to the psychophysical language $L_{phys+psy}$ and the folk theory T_{folk} in that theory, since the facts about what language humans speak and what sentences in that language they assert can be recovered with a high degree of determinacy from the physicalistically and behavioristically acceptable set of facts, simply by consulting users' overt verbal behavior (including their counterfactual behavior under all possible circumstances). This is the sort of task that Donald Davidson described as *radical interpretation*. [Davidson 1973] In any case, overt linguistic behavior (as described in $TRUE_{phys}$) would place very severe constraints on acceptable candidates for the language $L_{phys+psy}$ and the folk theory T_{folk}.

In addition to or as an alternative to reliance on the folk theory T_{folk}, we could rely, as Donald Davidson recommended, on a Principle of Charity, which could serve as a constraint on acceptable interpretation functions. We could require that the interpretation of sentences that attribute the belief with content ϕ be assigned intensions in which ϕ is also verified, at least, to as great an extent as possible. We could also apply a similar Principle of Charity to the assignment of sensory and mnemonic contents to human subjects, along with a Principle of Humanity or Reasonableness that requires that beliefs be reasonable, given a subject's sensory and mnemonic information.

3.3 Wallacian Functionalism

In a sense, Wallace's functionalism, inspired by Daniel Dennett's *Real Patterns* [Dennett 1991], is a combination of phenomenalism and analytical functionalism, with mental properties reduced to macroscopic (and chemical and biological) properties in something the form of Analytical Functionalism, and macroscopic properties reduced to states of the quantum wavefunction, in something like Russell's structuralism. The difficulty with this strategy, as we'll see, is that this leaves us trying to lift ourselves by our own bootstraps, with too little basis for constraining the kinds of emergent domains that can emerge.

It's reasonably clear what the reducing or fundamental model is supposed to be. We can take the language of pure quantum mechanics (with its description of the cosmic wavefunction and its deterministic Schrödinger evolution) and supplement it with a counterfactual or subjunctive conditional. This will require a model that contains a domain of worlds, each of which consists of a single quantum wavefunction evolved through time, one world designated as actual (which picks out the world's actual wavefunction), and a system of spheres *S* for the evaluation of subjunctive conditionals (the *worlds* of these models will not be Everettian branches, but different versions of the underlying quantum wavefunction). The system of spheres could be based, as in David Lewis's semantics [Lewis 1973], on a relation of comparative similarity between quantum worlds. This would require something beyond pure quantum theory, and in that sense, Wallacian functionalism does, like other interpretations of quantum mechanics, require some substantial supplementation to the theory. However, we might hope that the mathematics of the Hilbert space would provide us with a unique, natural measure of distance between possible wavefunctions, or at least a fairly small family of such measures. We would still have to decide whether to follow David Lewis's proposal, in which we must include among the possible worlds those with small, localized "miracles," to be preferred in closeness to worlds that verify the antecedent of the conditional that are non-miraculous but otherwise quite far from the actual world. Deciding how many such miraculous worlds to include and how to weigh their comparative similarity will introduce a large measure of subjectivity or conventionality to the project. However, for the sake of argument, I will waive objections along these lines.

In Wallace's proposal, the only constraint on the Ramsey realization of the emergent theory is this: all fillers of functional roles in the emergent theories must be entities and sets of entities to be found in the correct model of the formal language of pure quantum mechanics. In particular, there are no constraints on the extended interpretation function that can be expressed in terms of *causal connections* between emergent and quantum-mechanical entities or *pure semantic conditions* (such as metaphysically correct reference or truth-conditions for the emergent language) or *metaphysical priority* (no degrees of *naturalness* or *eligibility* that apply to sets of n-tuples quantum-mechanical entities), as I will argue in sections 3.4 and 4.1 below.

There is, however, another difficult choice for the Wallacian functionalist to make: is the high-dimensional space ($3N$, where N is the number of particle-systems) of the quantum wavefunction the whole of fundamental reality, or is there in addition a four-dimensional spacetime upon which the higher-dimensional space is defined? Wallace and Timpson [Wallace and Timpson 2010] prefer the latter, but this seems ad hoc and artificial, given their wavefunction Puritanism. Why posit the four-dimensional manifold, if there are no fundamental entities located there? The Schrödinger dynamics doesn't depend in any way on the familiar four-dimensional structure. In addition, this position seems inconsistent with the attempt to use decoherence to explain all of classical physics: see Halliwell's attempt to generate three-dimensional space as an *emergent* by-product of quantum cosmology.

In fact, by moving from a pure quantum wavefunction (in its $3N$-dimensional state space) to the wavefunction plus a four-dimensional manifold, Wallace and Timpson are moving in exactly the right direction. I will simply argue that they should move still further in that direction, admitting still more to the fundamental ontology of the world. By embracing spacetime realism, Wallace and Timpson are admitting that there is at least one entity, spacetime, with fundamental existence and a real essence that stands over and above the austere mathematics of pure quantum theory. Since this is a move in the right direction, I will allow the inclusion of spacetime in the base model $M_{true-QM}$.

So, let's turn now to the emergent domain. Our first problem is a very basic one: what language do we use, and what theory in that language? In the case of phenomenalism, we had the common vocabulary of geometry and the necessary laws of perspective to constrain the language and theory of the emergent domain of physical objects. In the case of Analytical Functionalism, we had a folk theory of psychology and psychophysics that could be recovered from, or at least powerfully constrained by, the overt verbal behavior of human beings, all of which was contained within the base model of fundamental things. In addition, the beliefs and sensory states attributed by the emergent theory have contents that match the vocabulary of the base theory. Now, we have only the language and theory of pure quantum mechanics to begin with, which by itself tells us nothing about the languages and beliefs of the denizens of an emergent world, and which lacks the direct access to our beliefs and concepts of the physical environment, as was available for the phenomenalist.

So, it seems that we must use *every* possible language and *every* possible theory. There are no languages or theories and no language users or believers explicit at the level of quantum reality. Any constraints we place on these theories (besides their sheer interpretability in the model of quantum mechanics) are going to be constraints of internal coherency. That is, we might reasonably demand of any theory $T_{emergent}$ of the emergent world that, according to $T_{emergent}$ itself, the human beings speak the language of $T_{emergent}$ and have beliefs and sensory and mnemonic experiences that mostly accord with $T_{emergent}$. We can also require that $T_{emergent}$ have the theoretical virtues valued by most people (as depicted in $T_{emergent}$), and that $T_{emergent}$ be well-confirmed, according to itself. Call the theories that meet these constraints the *internally ideal* or *coherent* theories.

3.4 Putnam's Permutation Argument for Semantic Indeterminacy

I will argue, in a way inspired by Putnam's argument for metaphysical anti-realism [Putnam 1978, 1980, 1981, Lewis 1984], that there is a radical indeterminacy of meaning and intension for all the names, predicates, and function symbols of the languages of our emergent theories. This isn't surprising, since all of the entities and properties posited by such theories are, from the point of view of Wallacian functionalism, mere useful fictions. In Wallace's picture, all that matters is that we find an interpretation of those theories in the true model of quantum mechanics that makes all of the formulas of that theory come out true (or at least approximately true) under that interpretation. The meaning of the emergent theories, the theories of the world's manifest image, is utterly holistic in character.

Suppose that $T_{emergent}$ is a theory of a world that is emergent relative to the model $M_{true-QM} = \langle F, D, I \rangle$. That means that there is an interpretation function, call it $I_{intended}$ that extends I to the language of $T_{emergent}$, resulting in a new model $M_{QM+emergent} = \langle F, D, I_{intended} \rangle$, with the theory $T_{emergent}$ true in $M_{QM+emergent}$. It is immediately obvious that there are an infinite number of alternative extensions of I that will also produce an extension of $M_{true-QM}$ relative to which $T_{emergent}$ is true. Take any permutation $\pi(w)$ for any world $w \in W$ of the objects in $D(w)$. Now apply the permutation $\pi(w)$ to the interpretation $I_{intended}$ with respect to the interpretation of all constants and predicate symbols at w. The resulting interpretation $I_{intended-\pi(w)}$ will also be a realization of $T_{emergent}$. Apply similar permutations to every world in W, resulting in the thoroughly scrambled interpretation $I_{bizarro}$. The extension of $M_{true-QM}$ by $I_{bizarro}$ will also be a model of $T_{emergent}$, and so $I_{bizarro}$ will be a realization of $T_{emergent}$ in $M_{true-QM}$.

So, for example, it is completely indeterminate what a predicate like 'is human' or 'is conscious' is true of or realized by. In the interpretation function $I_{bizarro}$, the intension of *human beings* might be the intension of *kumquats* in $I_{intended}$, and the interpretation of *is conscious* might be *contains vitamin B*. In fact, as Alexander Pruss [Pruss 2015] has pointed out, all the predicates that apply truthfully to the emergent world as it exists today (including mental-property predicates) could be interpreted in such a way that they apply truthfully only to the cosmos as it was 12 billion years ago.

> Any two worlds that are isomorphic under an isomorphism of the quantum structure (i.e., of the Hilbert spaces and the operator algebras) have the same functional properties. Now consider two worlds w_1 and w_2. Both are short-lived worlds: the temporal sequence of each is only a billion years long. Each world is an exact duplicate of a temporal portion of our world. Thus, w_1 is an exact duplicate of the temporal portion of our world from 13 billion years ago to 12 billion years ago, while w_2 is an exact duplicate of the temporal portion of our world from a billion years ago to the present. Then w_2 has the same kind of mental properties that obtained in our world over the last billion years. And w_1 has the same kind of mental properties that obtained in our world from 13 to 12 billion years ago. But there is a quantum-structure preserving isomorphism from w_1 to w_2. This isomorphism is simply given by the time-evolution operator U_{12} (where we measure time in billions of years). This operator is an isomorphism of the quantum structure. Hence w_1 and w_2 are exactly alike with respect to mental properties. Hence our world had exactly the same mental properties in the early 13-to-12 billion-years-ago period as in the last billion years. That's absurd. (For one, it makes us question how we could possibly know that the world is as old as we think it is.) [Pruss 2015]

Here is the key difference between Wallacian functionalism and the phenomenalistic functionalism of a Mill or Carnap, or the behavioristic functionalism of David Lewis. In the case of a phenomenalistic functionalism, the fundamental or base theory is a theory of our phenomenological experience, and the target or reduced theory is one of the "external" world. In this case, there is arguably some constraint on the content of the reduced theory that is non-holistic. For example, in the case of the primary qualities, we could insist that the geometrical properties assigned to physical objects in our external theory resemble the geometrical properties of the corresponding inner phenomena. So, if the external theory asserts the existence of something tetrahedral in shape, we could insist that the corresponding model of the phenomenal world include something that at least appears tetrahedral (perhaps, a two-dimensional projection in visual space of a tetrahedron). However, in the case of Wallacian functionalism, there are no phenomenal qualia on either side of the equation. There are only sentences in our folk psychology assigning certain geometrical experiences to subjects, and so long as the interpretation of these sentences preserves their truth and their counterfactual inter-connections, we have met every constraint on a successful interpretation.

In the case of behavioristic functionalism, we have real connections between the subjects of psychological states on the one hand and the subjects of behavior on the other. We assign beliefs and desires to x in a way that corresponds rationally to the behavior of x. In the case of Wallacian functionalism, we have only the universal wave function on the side of the base theory. All real or fundamental behavior is ultimately behavior of that function, and so there are no localized constraints on the connections between belief and desire and behavior. It is the theory of the folkish world as a whole (with both human belief and behavior contained in a single package) that confronts the model of pure QM as a whole.

In addition, as I will argue in sections 3.4 and 4.1 below, Wallacian functionalists lack any of the resources used by metaphysical realists to meet the challenge of Putnam's argument: causal ties between emergent terms and their quantum-mechanical referents, specially eligible or natural properties (at a phenomenological level), so-called "reference magnets," or metaphysically primitive facts about semantics or reference.

There is one particular case of referential indeterminacy that is especially devastating to the Everettian interpretation: namely, it is indeterminate whether a particular emergent world is assigned to quantum states with a high or low amplitude. Thus, there is no objective fact of the matter about whether the quantum probability associated with a given "branch" is high or low. Thus, the problem is not just that of finding a reason for believing that we are in a high amplitude branch. The problem is that we cannot give any real meaning to the question itself. If a given emergent world is realizable in a low-probability segment of the wavefunction at a given time, there is another interpretation function (and thus, another realization of that world in that wavefunction) that assigns it to a high-probability segment, and vice versa. It is all a matter of performing the appropriate permutation of quantum objects.

If there is no objective matter of fact about the quantum probabilities corresponding to the various emergent realities, then the Deutsch-Saunders-Wallace strategy for defending the reliability of observed statistics is further undermined. Emergent realities in which the statistics radically disconfirm quantum mechanics will be, not only as real as our own, but possessing the same status in regard to the underlying quantum probabilities. They will have just as much right to claim to be realized in the high-amplitude sectors of the quantum wavefunction as do the QM-confirming branches.

3.5 Model-Theoretic Indeterminacy Guarantees the Truth of Our Emergent Theories

Let $T_{emergent}$ be one of our target theories of the world: folk psychology or a scientific theory of "emergent" phenomena. We can suppose that $T_{emergent}$ is internally ideal and that it has a realization in the model of quantum mechanics, $M_{true-QM}$. Let $I_{intended}$ be the "intended" interpretation of the theory $T_{emergent}$ in the model $M_{true-QM}$, with a domain consisting of the space-time regions and quantum subsystems of the quantum world and with the predicates of the language $L_{emergent}$ assigned appropriate intensions in the corresponding model $M_{true-QM}$.

Now consider a theory $T_{bizarro}$, whose intended model includes the same interpretation function $I_{intended}$ but includes a different, counterfactual model of the quantum world, $M_{counterfactual-QM}$. Both $M_{true-QM}$ and $M_{counterfactual-QM}$ have infinite models, and both T_{emerge} and $T_{bizarro}$ are semantically consistent with the hypothesis of a domain of infinite cardinality. By the Skolem-Löwenheim theorems, there is an interpretation $I_{bizarro}$ of $T_{bizarro}$ in the actual model of the quantum world, $M_{true-QM}$. Thus, the bizarro emergent world represented by $T_{bizarro}$ is realized in the actual quantum world in just the same way as $T_{emergent}$ is.

In fact, *all possible theories of emergent domains are actually true*: if they are logically consistent (in the logic of quantified counterfactual conditionals), and they contain no quantum-mechanical vocabulary and make no claims about the finite size of reality, then (by the Skolem-Löwenheim theorems), they have a model that extends $M_{true-QM}$.[4] In fact, just this point was made by H. A. Newman in 1928, as a criticism of Russell's structuralism.[Newman 1928]

In fact, the situation is even worse than this, since Wallace doesn't require perfect realization in $M_{true-QM}$—just a reasonable degree of approximation to such perfect realization. So, even *inconsistent theories* or theories that entail the existence of a finite domain or that entail falsehoods about the structure of spacetime will nonetheless have quantum realizations and so will be *actually true* theories of a world that emerges from the quantum world.

The upshot is this: we are free to believe and say *whatever we want* about the emergent world of macroscopic objects, and we are guaranteed to believe and speak the truth (so long as our stories are internally coherent and not massively inconsistent). As a result, every consistent story corresponds to a real, emergent world, on par with our own. This includes the world of Tolkien's mythology or that of H. P. Lovecraft, the world of Harry Potter or Greek mythology. They are all just as real as our own. And, even more importantly, our own theory of the emergent world is true by a kind of stipulation: true simply by virtue of satisfying our demands for its internal coherency.

But that is surely wrong. If our theory of the emergent world is true by a kind of stipulation, then it can't be interpreted realistically. To interpret it realistically is to take seriously the metaphysical possibility that it could be wrong. For example, all of the evidence we have for classical mechanics could be misleading ("could" metaphysically, not epistemically): it could have been produced by some other quite unknown mechanism. For example, the planetary orbits that led to Kepler's laws and ultimately to Newton's laws of motion could actually have resulted from the fact that the planets move on gigantic rails built by ancient aliens. For the theory of classical mechanics to be a substantive theory of the world, it must have the metaphysical possibility of being wrong. But Wallace's functionalism denies it that chance.

3.6 *Epistemological and Pragmatic Consequences*

If classical physics is understood as true by stipulation, this undermines any rational confidence we might have in quantum mechanics, since a large part of our evidence for QM consists in its agreement with classical mechanics when interference terms are small.

In fact, Wallace's functionalism leads quickly to an epistemological catastrophe: if we cannot interpret our theories of the emergent world realistically, then no belief in such a theory can count as objective knowledge. And yet, all of our knowledge of the truth of quantum mechanics depends on our having objective knowledge of experimental data that belong to an emergent domain. So, in the end, Wallacian functionalism is epistemologically self-defeating, destroying the only grounds we have for believing that quantum mechanics is true at all, to say nothing of believing that it exhausts the fundamental level of reality.

In fact, we couldn't even interpret our emergent scientific theories as *instrumentally* valuable in an objective way, since *any* theory of our future experiences would be equally true (just one more realizable emergent theory). In addition, what counts as *the same* qualitative properties of experience is itself up for grabs via the interpretation function. By choosing a suitable function, we can make any set of predictions about future experience come out true.

Thus, pragmatism itself is inconsistent with radical indeterminacy of meaning, as Plato recognized in the *Theaetetus*:

> SOCRATES But, Protagoras (we'll say), what about the things which are going to be, in the future? Does he [the individual human being] have in himself the authority for deciding about them, too? If someone thinks there's going to be a thing of some kind, does that thing actually come into being for the person who thought so? Take heat, for example. Suppose a layman thinks he's going to catch a fever and there's going to be that degree of heat, whereas someone else, a doctor, thinks not. Which one's judgment shall we say the future will turn out to accord with? Or should we say that it will be in accordance with the judgments of both: for the doctor he'll come to be neither hot nor feverish, whereas for himself he'll come to be both?
> THEODORUS. No, that would be absurd.
> [McDowell 2014, p. 56, 178c1–10]

Every claim about the future, practical consequences about believing and acting on an emergent theory of the world will itself be part of some emergent theory of the world. I have shown that every such theory, so long as it is not massively inconsistent and doesn't entail the finitude of the universe, will be realizable in $M_{true-QM}$ and so will be true. Thus, we cannot appeal to pragmatic considerations (like avoiding being eaten by a tiger) as grounds for preferring some theories over others.

3.7 The Argument's Upshot in a Nutshell

To sum up, there are four disastrous consequences for Wallacian functionalism:

- Radical indeterminacy of content, via Putnam's paradox: there an infinite number of alternative interpretation functions mapping our actual theory of the world into M_{QM}. In particular, there is no fact of the matter as to the quantum probability associated with any given emergent world.
- Every consistent and internally coherent story (more precisely, every story consistent with an infinite domain) represents an emergent reality in M_{QM}, on a par with our current best theories about the macroscopic world.
- So, we can't go wrong in proposing theories about the emergent world we inhabit, so long as our theories are consistent with an infinite domain, and so long as they are internally coherent from a semantic and epistemological point of view.
- These facts undermine any claim to know that quantum mechanics is true, on the basis of experiments and observations that depend in any way on the emergent. The impossibility of objectively false theories of the emergent domain makes objective knowledge of that domain impossible, including objective knowledge of the data and observations upon which we ground our claims to know the truth of quantum mechanics itself. Therefore, Wallacian functionalism is epistemologically self-defeating.

4 Putnam's Paradox: The Problem of the Missing External Constraints

The central problem for Wallace's proposal is that we are missing all the constraints that were available for phenomenalism or Analytic Functionalism. We have no counterpart to the laws of perspective that rigidly tied the physical world to the phenomenal data (in the case of Phenomenalism) or to the overt verbal behavior that rigidly tied (via a principle of charity) the acceptable psychophysical theories to the physical facts. All the constraints we have are constraints of consistency and coherency, but as Putnam's paradox shows, these are not sufficient to delimit the range of emergent worlds to any significant degree. These considerations demonstrate that no functionalist theory can bridge the gap between the quantum and emergent levels. What is needed is some additional, metaphysical constraint.

In this section, I will argue that the Oxford Everettian approach lacks the resources to build in such constraints. I will proceed by a process of elimination, showing that each of six plausible candidates cannot supply the necessary constraints on the interpretation of emergent theories in the Everettian setting. There are two possible constraints that we can eliminate quite quickly: an appeal to the concept of *emergence* itself, and a brute preference for the emergent theories that we actually endorse.

1. Can the concept of *emergence* fix the interpretation?
    Could we hope that the concept of *emergence* itself could somehow provide powerful constraints on what counts as an acceptable interpretation function for a candidate theory of an emergent world? It doesn't seem so: all that we can say is that the emergent world must be realized by some such interpretation function. There just aren't any a priori or analytic constraints on what that function must be like.
2. Can we use "our" actual language and folk theory?
    No, because first we would have to establish that such things exist, and that they are relatively determinate and well-defined. But that's just the problem. It seems that any coherent theory about what language we do in fact speak and what beliefs we do in fact hold will turn out to be equally true.

In the following subsections, I consider four additional candidates: (1) causal constraints plus the category of *natural* or especially *eligible* properties, (2) an appeal to facts about spacetime locations and trajectories, (3) the use of the *simplicity* of our emergent theories or of their quantum interpretations (or both), and (4) the use of facts about decoherence.

4.1 Can Causal Constraints or Natural or Eligible Properties Fix the Interpretation?

One standard realist approach to the Putnam paradox, defended by Michael Devitt [Devitt 1983, 1991] and Hartry Field [Field 1998], is to appeal to a causal theory of reference. Such a causal theory could constrain the range of acceptable interpretation function by requiring that the atomic truths including a given predicate be assigned to a property in the model of such a kind that an appropriate causal mechanism can be found between occurrences of that property in the world and uses of the predicate by speakers of the language. To apply this idea to Wallace's functionalism, we would have to require that there exist causal connections of the right kind between language use (or concept deployment) on the one hand and the emergent properties that we language users are supposed to be representing.

However, this strategy just won't work here, for three independent reasons, two having to do with the base theory and the other to do with the emergent theories. First, human language and concepts do not even exist in the quantum world to begin with, so the question of whether our language use or concept deployment is causally connected to anything at all cannot arise, independently of assuming the truth of a given emergent theory and the correctness of a given interpretation function. In contrast, classical phenomenalism and Russellian structuralism included concepts and concept-use within the base theory, and so the issue of what properties that concept use could be causally connected to could constrain the interpretation of emergent theories. And, although Analytic Functionalism did not include concept-use within the base theory, it did include language users and their overt linguistic behavior, which could be used to tie concepts to fundamental causal connections. In Wallace's functionalism, any causal connections that involve concepts must be inextricably part of the emergent story itself. Hence, they can be of no help whatsoever in linking the language of the emergent theory to the underlying quantum world. The emergent causal connections that exist occur entirely within the story or theory that defines the emergent world. And so, at best, we can simply add another coherence condition to our emergent-world stories: namely, that *in the story* there are the right sort of causal connections between (emergent) environmental conditions and (emergent) language- and concept-use.

Second, there are no causal connections between facts in Wallacian functionalism in the base model: all we have are counterfactual conditional dependencies. There is good reason for this, since the pure formalism of quantum mechanics contains no non-Humean information about *causation* over and above the facts about what counterfactually depends on what. But counterfactual dependence is not sufficient to fix reference determinately, as we have seen.

Third, emergent theories did not historically and still do not generally include any information about the quantum realm. Hence, we cannot even require ideal or coherent emergent theories to include causal connections between concepts and underlying quantum processes.

A standard approach to resolving Putnam's paradox, championed by David Lewis [Lewis 1983], is to appeal to *perfectly natural* or *eligible* properties that can serve as "reference magnets" for the interpretation of predicates. However, to qualify as *perfectly natural* a property must be one of the fundamental properties of the world. In the case of Wallacian functionalism, this would not enable us to move far enough away from the austere properties of pure quantum mechanics to define the properties and relations of macroscopic branches, to say nothing of biological and psychological properties.

4.2 Can We Use Spacetime to Restrict the Interpretation Function?

As I mentioned above, I am willing to embrace, for the sake of argument, the Wallace-Timpson theory of spacetime realism.[Wallace and Timpson 2010] So, there will be some common vocabulary between many emergent theories and the theory of quantum mechanics: both will have the vocabulary and the axioms needed to characterize the spacetime continuum.

However, beyond the vocabulary of space and time, there will presumably be no other overlap between our quantum and emergent vocabulary. It's clear that there is no such overlap in "our" emergent world (if it exists). This still leaves us with a very weak constraint on true emergent theories: they must not entail anything false about the structure of spacetime. This is very unlike the situation in the case of classical phenomenalism, where we could assume a common vocabulary about what regions of space are *occupied* by bodies at what points in time. We could use simple laws of perspective to link bodies in public space with sense-data in private spaces.

Couldn't we restrict the interpretation function by requiring that macroscopic properties assigned to regions of spacetime must be interpreted by quantum properties assigned to the same region? But the problem of cosmic entanglement ensures that there are no quantum properties that are localized to any finite region. If we take spacetime seriously, we must understand the quantum wavefunction to be what Peter Forrest calls a *polyfield*, in which fundamental magnitudes are assigned to N-tuples of widely separated spacetime points (for a very large N, representing roughly the number of fundamental particles in the universe). [Forrest 1988] There is, therefore, no simple function from quantum events in a space-time region and macroscopic object-events or phenomenal appearances.

Furthermore, appeals to spacetime won't help to blunt the Putnam-style argument for referential indeterminacy. We can still get complete indeterminacy for every term and predicate except for the geometrical ones. For example, Pruss's argument based on the 12-billlion-year time-shift will still work. We can insist that the events of the emergent world be located somehow in the real spacetime continuum of $M_{true-QM}$, but there is no way to ensure that they line up in any fixed or determinately intended way with the pattern of quantum events in that same continuum.

Here again, the contrast with classical phenomenalism is instructive. A classical phenomenalist could, in effect, locate phenomenal qualia in regions of public spacetime and then stipulate that the sensory qualities of any physical object located in that same region by the emergent theory correspond to those of the qualia. However, for the Wallacian functionalist qualia are themselves just parts of the emergent theory. Moreover, they are parts that presumably share no intrinsic features with the underlying quantum events.

4.3 Why Not Simplicity?

Wallace can legitimately complain that I have ignored the constraint that he mentions explicitly: the constraint of *simplicity*. Wallace could impose on the extended interpretation function a condition related to Lewis's condition of *naturalness* [Lewis 1983] by requiring that the function map emergent predicates onto relatively simple sets of n-tuples of relatively simple quantum entities.

But before we examine the utility of a simplicity requirement, we must ask a more fundamental question? Why should *simplicity* be of any relevance to the *metaphysical* and *ontological* question of defining emergent reality? Simplicity is plausible as an epistemological constraint: other things being equal, the fact that one theory with wide scope and a high degree of accuracy is much simpler than all of its competitors with comparable scope and accuracy seems a good (if defeasible) reason for thinking that the simpler theory is objectively true. Simplicity may be an indicator of *probability of truth*, but it does not seem to be a criterion of existence or reality. How could the real existence of an entity be a function of its simplicity?

The metaphysical deployment of simplicity is especially implausible once we remind ourselves that the very definition of *simplicity* is difficult and contentious. Simplicity, like beauty, seems to be in the eye of the beholder. Different theories of the emergent world will attribute different standards of simplicity to the scientific community. More fundamentally, we can ask: what is the truthmaker for the "correct" account of simplicity here? And what is the metaphysical ground for imposing any simplicity constraint at all?

There are two independent parameters of simplicity to consider: (1) the simplicity of the *theory* of the emergent world (is it, for example, finitely or recursively axiomatizable, or at least approximately so?), and (2) simplicity of the *interpretation function* that interprets the non-quantum vocabulary in the quantum model.

4.3.1 Maximizing Simplicity

It might seem that simplicity provides a solution to Putnam's paradox: take the correct interpretation function to be the *simplest* extension of the base interpretation that verifies the emergent theory $T_{emergent}$. However, this will only work if we are given the complete theory of some emergent reality. The problem is that there is at the quantum level no set of privileged theories of emergent domains. And any consistent theory, no matter how bizarre, will have *some* realization in the quantum model and so, in all likelihood, some simplest realization.

Can we pick out the privileged emergent theories by focusing on the *simplest* theories that are realizable in the quantum model? We can't maximize simplicity of both theory and interpretation simultaneously, since the simplest possible interpretation function is just the original interpretation function of the quantum model (with no addition), and the simplest possible theory is the just theory of the original quantum model (the totality of purely quantum truths, including the theorems of logic).

In other words, the simplest emergent theory is just the null theory: the theory consisting of nothing but the theorems of logic. That will obviously give us no help in fixing the interpretation function. The simplest extension of the interpretation function is just identity, and that will give us no help in fixing the emergent theory. Maximization is futile in either case.

Maximizing the simplicity of one parameter or the other might be useful if we could first fix the metaphysically correct value of the other parameter. For example, if we could independently fix the true theory (and language) of the emergent world, it might make sense to try to maximize the simplicity of the extended interpretation used to realize that theory in the model of quantum mechanics. Or, if we could independently restrict the set of interpretation functions to a very narrow class, we could then use the degree of simplicity of the resulting emergent theories as a way of selecting the "best" emergent world. However, we have been unable to find any independent constraint. How, for example, do we select the right emergent theory to use? For any consistent theory of the emergent world, no matter how bizarre, there will be some interpretation that maximizes the simplicity of its truth-makers in $M_{true-QM}$.

4.3.2 Setting a Minimal Degree of Simplicity

Perhaps instead of *maximizing* the simplicity of one or the other, the Wallacian functionalist could just require that the simplicity of one or the other (or both) has to meet a certain fixed standard, allowing any theory whose realization meets that standard to count as really emergent. Let's focus on the simplicity of the interpretation function. I see three problems here.

1. Any requirement of relative simplicity will have to be quite loose and permissive, since we know that the entities and properties of the emergent manifest image are, under the most optimistic assumptions, far from natural. See recent work on color and color-experience, for example ([Hardin 1993, Pautz 2014, Tye 2000]). Neurological and other biological properties will be highly disjunctive and gerrymandered from the viewpoint of fundamental quantum mechanics, and phenomenological, intentional, and semantic properties even more so. If the requirement of simplicity is too strong, it would give us *no* emergent worlds at all; if too weak, it would give us far *too many*.
2. What could be the truthmaker or metaphysical ground of the correct standard of minimum simplicity? How are we supposed to explain the connection between complexity and unreality?
3. There has to be some counterweight to simplicity, or we should embrace an eliminativist theory (a no-emergence theory), in which our theory of the "manifest" world just is fundamental quantum mechanics. We need some reason not to set the standard at the maximum level of simplicity. So, what is the counterweight? Usefulness? Apparent truth? But these criteria only make sense given a manifest theory. We need people, organisms, perceptions, beliefs, purposes, etc. in order to make these judgments.

Again, Plato's *Theaetetus* point applies again. Usefulness cannot be both part of the emergent picture and the criterion for which picture is really emergent. If there's nothing to balance against simplicity, then simplicity becomes either too strong a constraint (eliminating all emergent worlds) or a completely vacuous one. If we are going to be pragmatists, there has to be some kind of bridge between the emergent world we posit and our actual experiences and beliefs (how the world *appears to us*—in both a non-epistemic and epistemic sense of the phrase). But the phenomenal world, as we might call it, is also *part* of the emergent world, and so just another part of the theory it is supposed to be used in evaluating.

4.3.3 Degrees of Reality

Could we talk about degrees of reality, as measured by the simplicity of the isomorphism into the wavefunction? Instead of requiring maximum simplicity of our emergent reality, and instead of arbitrarily setting some minimum standard of simplicity, we could adopt a kind of sliding scale: the simpler a theory and the simpler its realization in the one true quantum model, the *more real* the corresponding emergent world would be.

But can we make any sense of one world being *more real* than another? Isn't reality (whether emergent or fundamental) always a simple matter of Yes or No? In addition, why should we accept any level of reality short of the highest one? What counter-pressure could make us prefer a theory that is to any degree unreal? As we have seen, neither conservatism nor pragmatism provide any such independent counter-pressure.

Finally, there would still be a huge number of alternate emergent realities that will count as equally real (by this standard) as our own world. In fact, each of the bizarro permutations of the "intended" interpretation function $I_{intended}$ will be just as simple, considered as mathematical functions, as the original interpretation function.

4.3.4 The Best Option for Everettian Functionalists

There might be some emergent theories that have a uniquely simplest realization in the quantum model: a realization that is *much simpler* than the second-best realization. We could stipulate that it is exactly such theories that represent an emergent reality.

However, as Wallace has pointed out, it is unlikely that any of our quasi-classical branch theories have a uniquely simplest realization. They are all afflicted with a significant degree of vagueness and indeterminacy. Still, we might hope that such theories have a *relatively compact* class of simplest realizations, each of which is *significantly* simpler than any realization outside of the class. In other words, there might be a fairly sharp peak of simplicity associated with certain realizations of the theory. We could stipulate that only such theories represent emergent realities and that only interpretations at or near the unique peak of simplicity count as correct interpretations of the theory. To be more exact, we should look at the class of simplest *near* realizations of the theory: extensions of the interpretation that verify most of the sentences of the emergent, or most of the most important ones.

There will be some considerable amount of work to be done to show that the quasi-classical theories of the Everettian branches actually meets this new, more stringent condition of emergence. First, the condition might be too broad. For all we know, there are many fantastic theories that have a simplest interpretation at the quantum level (nota bene: this simplest interpretation could be horrendously complex, so long as all the others are even worse). Imagine a panpsychist or even pan-voluntarist theory, according to which all physical entities have sensations and make free choices. This would count as a real emergent world, so long as there is one interpretation function for the fantasy that is significantly simpler than the others. Second, the condition might be too narrow. We know very little about the class of possible interpretations of our emergent theories. We know that in many cases the intended interpretation of the emergent theory (e.g., geology, thermodynamics, psychology) is extravagantly complex. Is that intended interpretation always uniquely simplest of all the possible interpretations? It is hard to tell. In fact, Alexander Pruss's argument (recounted in section 3.4 above) shows that the intended interpretation for our theory of astronomy is not significantly simpler than an alternative interpretation that shifts our descriptions of the current state of the cosmos backward 12 billion years.

But perhaps the technical viability of this solution can be verified. There remain three philosophical difficulties. First, the criterion of simplicity is highly subjective and variable, probably even conventional. Simplicity is in the eye of the beholder, but the beholder is part of the emergent world, and so simplicity cannot provide an independent, exogenous constraint on the identification of real emergence. There are possible theories of the world in which the conscious inhabitants have radically different standards of simplicity from our own, and so the *simplest* interpretations of the emergent theories by their lights will be quite different from our own.

Second, this solution raises severe epistemological worries. How could we know that our current theories of the emergent world satisfy the constraint of having a uniquely simplest interpretation in the quantum field, without having independent access to the level of quantum reality? We can presumably know that the truths, both categorical and subjunctive, of our favored emergent theories are verified somehow or other by the quantum level, but how could we know whether they are verified according to an interpretation that is substantially simpler than any alternative?

Finally, the criterion of simplicity is inherently vague and indeterminate. How could reality itself select one precise version of simplicity as the ground of determinate interpretation? It seems that we would have to look for that property of simplicity that is uniquely *natural* or *eligible*. But, the simplicity of emergent theories is not a feature of the underlying quantum reality, and so the Everettian have no grounds for attributing extreme degrees of naturalness or eligibility to such emergent realities. We could always pick out that precise form of simplicity that favors our current theories of emergent science, but such a tactic would be evidently ad hoc and metaphysically unmotivated.

4.4 What About Decoherence?

I haven't said much yet about decoherence. Surely that's a problem, given the prominent role that decoherence plays in Wallace's account of the emergence of the macroscopic world. What exactly does decoherence tell us? It tells us that under certain circumstances, quantum systems can mimic the dynamics of classical mechanics, because of the way in which environmental interactions suppress the interference terms in the systems' Hamiltonian operators.

How is that relevant to our theories of the emergent world? It's relevance seems to depend on two assumptions:

1. The dynamics of the emergent world must be (approximately) that of classical (Newton-Maxwell) mechanics.
2. The dynamics of the emergent world should closely mimic those of the underlying quantum reality.

Given these two assumptions, decoherence would indeed be crucial, since it would be needed to explain how and why the emergent world can exist, given that the underlying quantum reality does *not* generally obey classical mechanics. Decoherence gives us reason to believe that, under most normal circumstances, the dynamics of actual quantum systems will effectively *approximate* those of classical mechanics.

But what *metaphysical* grounds do we have for accepting either of these assumptions? It seems clear, in fact, that they are both false. It has never been obvious that the emergent world of everyday macroscopic objects obeys Newton-Maxwell dynamics. If it had been obvious, it would not have taken scientists millennia to transcend the limitations of Aristotelian, Archimedean, and Galilean mechanics. Even today, most physical phenomena do not apparently obey classical laws, as Nancy Cartwright pointed out in *How the Laws of Physics Lie*. [Cartwright 1983] In addition, there are many special sciences, including especially social sciences like politics, economics, and sociology, in which classical mechanics plays little or no role.

In addition, why should we assume that *any possible* emergent world must be classical in its dynamics? What basis is there for such a stipulation? It doesn't seem to be built into the concept of *emergence* in any way.

Turning to the second assumption, there seems again to be no reason to suppose that the dynamics of the emergent systems must mimic those of the underlying physical reality. Surely, it is sufficient if they are functionally realized by that reality.

In any case, even if we grant both assumptions and thereby limit the emergent world to conditions in which decoherence obtains, this is still going to result in constraints that are far too weak. Any theory whatsoever that is consistent with classical mechanics and which entails substantive content only under conditions under which decoherence would obtain will have an acceptable interpretation in $M_{true-QM}$ and so will represent a genuine emergent world, as real as our own world. This would include bizarre Dan-Brown or John-Bircher conspiracy theories, theories of Aryan racial supremacy, the alternative histories of Philip K. Dick, UFO realism—so long as these respect classical dynamics in ordinary conditions, there will be a real emergent world corresponding to each.

Of course, if we keep loading up conditions on a *proper* emergent world, we will eventually isolate the theory we want. The following five conditions might work:

1. Extend the model $M_{true-QM}$ of quantum mechanics to a model $M_{QM+branches}$ with a branch parameter, each branch being assigned a probability weight, a period of time, and a set of particles in each world.
2. Privilege the basis consisting of position and momentum by having the extended model $M_{QM+branches}$ assign definite but branch-relative position and momentum to each particle that belongs to the branch and at each time assigned to the branch.
3. Require that the sum of branch-probabilities corresponding to a set of particle positions (or momenta) be a good approximation to corresponding sum of probability amplitudes in the original model $M_{true-QM}$.
4. Require that the dynamics assigned to particles by branches approximate a dynamic theory that is both simple and relies on highly localized, separable quantities (i.e., very like classical mechanics) .
5. Require that the branch structure include as many particles and as much time as possible, given the other constraints.

We can then stipulate that the only real emergent worlds correspond to the set of truths verified by some such extended model $M_{QM+branches}$. In addition, we could count as an emergent theory a theory that is expressed in a reduced language of $L_{absolute}$, reduced by the replacement of each branch-relativized predicate with an absolute version of the same predicate (including location and momentum predicates). A theory $T_{absolute}$ in the reduced language could count as *realized* by $M_{QM+branches}$ just in case there is a consistent assignment of branch-parameters to the formulas of $T_{absolute}$ results in a theory that is verified by $M_{QM+branches}$.

However, such a move has four disadvantages:

1. The account is no longer tied to and no longer provides a *general theory* of emergence. Consequently, we would have to deny the emergent reality of other special sciences, like chemistry, thermodynamics, biology, psychology, and the social sciences.

2. We would be offering no account of why these conditions are of great *metaphysical* significance. Why must an emergent world satisfy just these conditions to count as real?

3. We would give up the claim that decoherence generates the privileged basis by itself. Instead, we would simply be stipulating what we shall count as the correct basis. If we try to get around this by deleting conditions (2) and (3), we will be unable to dissolve the Putnam paradox, since permutations of the intended model of the emergent world will meet the other three conditions. This would leave Pruss's time shift argument, with the superfluity of minds in obviously mindless regions, untouched.

4. We would be making the many-worlds or many-branches structure of emergent reality true by stipulation.

5 The Solution: Real Essences and Extra-Conceptual Grounding

5.1 Two Forms of Grounding

My real complaint is with Daniel Dennett's "Real Patterns" [Dennett 1991], which is the original inspiration for Wallace's functionalism. I believe, in fact, that my arguments in section 3 above would apply with almost equal force to *any* functionalist account of the emergent world, including Bohmian interpretations of quantum mechanics. And the deterministic version of Bohm's theory (in which everything occurs with probability 1 or 0) also has difficulty accounting for the quantum probabilities. (I leave this extension as an exercise for the reader.)

The problem for Dennett is this: what makes a pattern *real?* As we have seen, mere realizability in the true quantum model of the world does not suffice. Here is the crux of the problem: we have to construct both the theory and its semantics simultaneously, with an aim toward maximizing simplicity (in both dimensions). What, then, keeps us from simply collapsing into the identity isomorphism and the trivial theory? If we had a fixed theory and had to find the simplest semantics, or if we had a fixed semantics and had to find the simplest theory, the problem would be well defined and constrained.

We need some *top-down* constraint. It also has to be an *ontological* constraint: a set of natural kinds of emergent entities, each with a fixed real essence. These essences can then constrain both the story and the story's semantics. But to carry this out, we can't be eliminativists or physicalists— we can't limit the fundamental structure of reality to what can be described exclusively in quantum-mechanical terms. When things have real essences, they must be real.

In *The Atlas of Reality*, my co-author Tim Pickavance and I argue for a distinction between *conceptual* grounding and *extra-conceptual* grounding. [Koons and Pickavance, pp. 62–5] On our view, *grounding* is an explanatory relation between truths or facts. Along with Kit Fine [Fine 2012] and Gideon Rosen [Rosen 2010], we take all grounding facts to be underwritten by facts about some real essence or essences of entities involved. When we say that one truth is grounded in another, there are two importantly different cases to consider. First, it could be that the truth of *p* is grounded in the truth of *q* because of relations between the essences of some of the concepts or logical operators appearing in the two propositions. So, for example, the truth of a disjunction (*p*∨*q*) is grounded in the truth of the atomic proposition *p* (if both are true) by virtue of the essence of the logical operator of disjunction, ∨. Similarly, the truth of *John is a bachelor* may be grounded in the truth of the proposition *John is a never-married adult male human being* by virtue of the essence of the concept *bachelor*. In other cases, however, we must appeal to the essences of extra-conceptual entities, entities that are not essentially part of some abstract object of thought. For example, the existence of the singleton {*Socrates*} is wholly grounded in the existence of Socrates, but this grounding relation depends on certain facts about the essence of a singleton set like {*Socrates*}. Thus, the grounding of the existence of the singleton set in the existence of its member gives us no reason to eliminate sets from our ontology. We explain the distinction further in *The Atlas of Reality*:

> The distinction between conceptual and extra-conceptual grounding turns on a very subtle difference. Compare the following two claims, where '[*Fa*]' abbreviates the proposition *a* is *F* and '[*Ga*]' abbreviates *a* is *G*:
>
> • [*Fa*]'s truth is grounded in [*Ga*]'s truth.
> • [*Fa*]'s truth is grounded in *a*'s being *F*, and *a*'s being *F* is grounded in *a*'s being *G*.
>
> In both cases, the truth of [*Ga*] is in some sense prior to, more fundamental than the truth of [*Fa*]. In the first case, the dependency is propositional or conceptual, in the latter case, extra-conceptual. To distinguish between the two, we have to look carefully at what licenses or justifies the explanatory connection between *a* is *F* and *a* is *G*: is it licensed by the essence of the property designated by the predicate *F* or the concept the predicate expresses? Is the essence involved in something in the mind-independent world, or is it merely in the mind?
>
> [Koons and Pickavance 2017, pp. 63–4]

The two kinds of grounding have very different ontological import. Conceptual grounding gives us reason to think that the entities putatively designated by the concepts in the proposition whose truth is grounded in other propositions (lacking those same concepts) do not really exist. We can safely eliminate them from our ontology. However, as Pickavance and I explain, the eliminativist option breaks down in the case of extra-conceptual grounding: if the entities in the grounded fact didn't exist, they couldn't have an essence, and without an essence, the relation of extra-conceptual grounding between that fact and its ground could not exist.[Koons and Pickavance 2017, p. 64]

It would be incoherent to say both that all grounding is conceptual and that reality is purely quantum-mechanical, because concepts and propositions do not appear at the fundamental level of quantum mechanics. And yet, they must have non-trivial essences if the project of conceptual grounding is to work. This simple fact explains the incoherence of Dennett's attempt to reduce our existence to the reality of *patterns* in things as they appear to *us*. It's impossible for us concept-wielders to be both the grounds of emergent reality and merely another component of that emergent reality.

What's needed in a coherent account of a world that emerges from QM is the extra-conceptual grounding of emergent entities in the model of quantum mechanics. And the very existence of extra-conceptual grounding falsifies the claim that the quantum-mechanical exhausts the fundamental structure of reality. The real essences of emergent entities must be co-fundamental with the quantum-mechanical facts. These essences must themselves be ungrounded or perhaps zero-grounded, to use Fine's term. [Fine 2012]

Once we posit new entities with their own essences that can explain (in a top-down fashion) those entities' grounding in the quantum domain, we open up the possibility that these composite, macroscopic entities (and not just their essences) are also *ontologically fundamental*: metaphysically dependent on but not *wholly* grounded in (not fully explainable in terms of) the micro-quantum realm. In particular, there could be a diachronic and causal component to the answer to van Inwagen's Special Composition Question [van Inwagen 1990, pp. 21–2]: the existence of a composite macro-object on a certain branch of the cosmic wavefunction might be causally dependent on the prior existence of composite objects in that branch in the immediately preceding period, that is, partly dependent on composite objects in that branch with the required fundamental causal powers. This would, of course, fit well with Aristotle's vision of the world, in which the generation of new composite substances is always the result of the "corruption" of pre-existing substances, with the processes of corruption and generation being explainable in terms of the exercise of active and passive causal powers by participants in the processes. On a *hylomorphic* interpretation of Everettian quantum mechanics, the state of the quantum wavefunction (in particular, the presence of decoherent branches) is the *material cause* of the existence of certain composite, macroscopic entities. But the quantum function by itself is not a metaphysically sufficient ground or explanation: in addition, there must be a *formal cause*, reflecting the real essence of a natural kind of macroscopic, composite object. The presence of such a substantial form in a branch of the cosmic wavefunction at a particular place and time would have a diachronic causal explanation, which could make reference to earlier facts at the emergent scale in the branch in the spatial neighborhood of the persisting or newly generated composite substance.

To put the point more formally, we must enrich our base model, representing fundamental reality, by supplementing $M_{true-QM}$ with a set of natural-kind essences K and a fundamental composition relation $COMP$. The new model, M_{QM+HM} (HM for "hylomorphism") would be defined over a language that contains constants for each of the natural kinds in K along with a four-place *part-of* predicate P, where $P(k, p_1, p_2, t)$ represents the fact that both particles p_1 and p_2 are at time t proper parts of a substance of kind k. The truth-conditions for the P predicate will be given by the fixed $COMP$ relation in the model, with the stipulation that, for fixed time t, each particle can be part of at most one substance. That is, a particle cannot satisfy the P predicate at the same time for two different natural kinds, and the binary relation of being two parts of a substance of a given kind will be an equivalence relation on particles (reflexive, symmetric, and transitive).

The natural kinds will make a real difference by virtue of constraining acceptable models to connect substances of each kind with appropriate branches in the branching-extension of M_{QM} defined by the five conditions in section 3.7 above.

M_{QM+HM} is an **acceptable model of the emergent world** if and only if there is a branch extension of the base model $M_{QM+branches}$ meeting the five conditions in 3.7 such that, for any kind k in K, for any world w in W, for any particles p_1 and p_2 and time t in $D(w)$, if $M_{QM+HM} \models P(k, p_1, p_2, t)$, then there is a branch b in $M_{QM+branches}$ such that p_1 and p_2 belong to b in w at t, and the tuple $\langle p_1, p_2, t, b \rangle$ satisfies all of the metaphysical conditions associated with natural kind k.

The appeal to natural kinds of substantial forms enable us to overcome the four problems we identified at the end of section 3.7:

1. The account is a general theory of emergence. Every emergent domain depends on an appropriate set of natural kinds (macrophysical, thermodynamic, chemical, biological, etc.).
2. It is the fundamental existence of the emergent natural kinds that lends metaphysical significance to the constraints.
3. The essences of the natural kinds select the privileged basis of operators, by requiring a range of values for the corresponding parameters.
4. The essences of the natural kinds provide the ground for the truth of the multiple-branch structure within the wavefunction, since each essence requires the existence of a branch of an appropriate kind for its actualization.

5.2 Bonus: Restoring the Real World's Unity

Once we have real essences at the macroscopic level, we have the real possibility of diachronic, horizontal causation at that same level. In particular, we can consider positing a dynamic element to the solution of van Inwagen's *Special Composition Question*. We can call the result the *traveling forms* interpretation of quantum mechanics. This interpretation has some similarity to Jeffrey Barrett's *single-mind* interpretation [Barrett 1995], except where Barrett has a cohort of conscious minds traveling through the branching structure of the Many Worlds Interpretation, the traveling forms interpretation has instead a cohort of composite macroscopic objects. In Barrett's interpretation, all branches but one are occupied by zombies, by living human bodies that lack consciousness. In Barrett's picture, it is the presence of real consciousness that picks out the uniquely actual branch.

On my traveling forms interpretation, in contrast, all branches but one are occupied by pluralities of particles that fail to compose *anything at all*. We might call these pluralities of fundamental quantum particles "compositional zombies." Although they have, from the microphysical perspective, everything that is needed for the potential existence of macroscopic objects (stars, planets, organisms, macro-molecules), no actual composite entities correspond to these branches. They are occupied wholly be compositional zombies.

Thus, the traveling forms version is not committed to anything like substance dualism: it is consistent with the supervenience of the mental on the physical, so long as the physical includes facts about which particles compose larger physical wholes. It does, however, deny that the *compositional* facts about physical entities supervene on the microphysical or quantum facts alone. Whether a branch corresponds to a domain of *actual* composite physical objects will depend on two factors: (1) does the branch satisfy decoherence to a sufficient degree of exactitude (a degree determined by the real essences of composite physical kinds)? And, (2) has the branch been occupied by composite objects in the immediate past, composite objects that are disposed either to persist in time or to generate new composite objects? In addition, there is a third, indeterministic factor: whenever a branch splits into two potential macro-worlds, the substantial forms responsible for composition jointly actualize macroscopic composition in just *one* of the new branches, with a probability normally determined by Born's rule (i.e., a probability proportional to the square of the amplitude of each branch).

I talk of traveling *forms* because I want to relate this interpretation to the Aristotelian theory of *substantial forms*. A substantial form is something (a principle or process) that is responsible for making what would otherwise be a mere cloud or heap of smaller entities into a single composite substance. For Aristotelians, a *substance* is an entity that is primary in the order of existence, an entity that has unity to the highest degree (per se unity) and which is the bearer of fundamental causal powers. It is an axiom of Aristotelian metaphysics that no substance is composed of other substances. Substances stand at the top of the compositional hierarchy. (For more details, see the chapter in this book by Alexander Pruss and the appendix below.)

6 Conclusion

Clearly, what's needed to rescue Wallace's picture is some further constraint on the interpretation function. Wallace's intention is surely that this extra constraint should have something to do with decoherence, with a linkage of some kind between macroscopic and quantum dynamics. However, as we have seen, the functionalist model that Wallace adopts, following the lead of Dennett's "Real Patterns" [Dennett 1991] won't deliver what is needed.

Ultimately, the extra constraint must have a top-down salient to it: it must derive somehow from the real essences of macroscopic *substances* (in Aristotle's sense, primary beings). But once we add these new elements of constraint, it will be hard to resist the temptation to move still farther away from the Everettian picture. The essences of macroscopic substances can give rise to novel causal powers at that level, causal powers that can determine which branch of the quantum wavefunction is really occupied by composite substances of the appropriate kind, restoring a single, unified world to the picture and thereby avoiding the twin problems of possible trans-world values and of anti-Darwinian branches.

Notes

1 I would like to acknowledge the support during the 2014–15 academic year of the James Madison Program in American Ideals and Institutions at Princeton University (for a Visiting Fellowship) and the University of Texas at Austin (for a faculty research grant).
2 Strictly speaking, the function isn't unique, but any two acceptable functions will be such that each is a linear transformation of the other.
3 Let A be a set of formulas that is logically consistent (in Lewis's conditional logic). Extend A to a maximum consistent set C. Given our assumptions about the model (namely, that the system of spheres is dense, with a sufficient number of worlds in each sphere-membership equivalence class), we can find an interpretation function I that verifies all of C (and, therefore, also A) in M_{base}. We can use C to impose a partial ordering on the formulas of the language: $\phi \leq \psi$ iff '$((\phi \lor \psi)\Box\!\!\rightarrow \phi)$' $\in C$. We can then use the Axiom of Choice to find a 1-to-1 function from this ordering into the system of spheres S in such a way that the smallest sphere containing a ψ-world contains the smallest sphere containing a ϕ-world iff $\phi \leq \psi$. Now take every set G of atomic formulas such that for every χ consisting of a conjunction of members of G and of negations of atomic non-members of G, the formula '$\neg(\phi\Box\!\!\rightarrow \neg\chi)$' belongs to C. Select a world from the designated set of closest ϕ worlds and place it in the extension of exactly the members of G. By repeating this for every formula and every set of atomic formulas G, we will build the appropriate interpretation function I.
4 Remember: I assumed above that $M_{true\text{-}QM}$ has a sufficiently rich structure that we can embed in it the canonical model for the logic of counterfactuals. That means

that any consistent theory in that language has an interpretation in $M_{true\text{-}QM}$, aside from issues of the cardinality of the world.

References

Barnes, E. (2010) 'Ontic vagueness: A guide for the perplexed', *Noûs* 44:601–27.
Barrett, J. (1995) 'The single-mind and many-minds versions of quantum mechanics', *Erkenntnis* 42:89–105.
Cartwright, N. (1983) *How the Laws of Physics Lie* (Oxford: Oxford University Press).
Cartwright, N. (1989) *Nature's Capacities and their Measurement* (Oxford: Oxford University Press).
Davidson, D. (1973) 'Radical interpretation', *Dialectica* 27:314–28.
Dennett, D. (1991) 'Real patterns', *Journal of Philosophy* 88:27–51.
Deutsch, D. (1999) 'Quantum theory of probability and decisions', *Proceedings of the Royal Society of London* A455:3129–37.
Devitt, M. (1983) 'Realism and the renegade Putnam: A critical study of Meaning and the Moral Sciences', *Noûs* 17:291–301.
Devitt, M. (1991) *Realism and Truth* (Princeton: Princeton University Press).
DeWitt, B. and Graham, N. (eds.). (1973) *The Many-Worlds Interpretation of Quantum Mechanics* (Princeton: Princeton University Press).
Ellis, B. (2001) *Scientific Essentialism* (Cambridge: Cambridge University Press).
Everett, H. (1956) *Theory of the Universal Wavefunction* (Princeton University: PhD dissertation).
Field, H. (1998) 'Some thoughts on radical indeterminacy', *The Monist* 81:253–73.
Fine, F. (2012) 'Guide to ground', in Correia, F. and Schnieder, B. (eds.) *Metaphysical Grounding* (Cambridge: Cambridge University Press): 37–80.
Forrest, P. (1988) *Quantum Metaphysics* (Oxford: Basil Blackwell).
Gell-Mann, M. and Hartle, J. (1990) 'Quantum mechanics in the light of quantum cosmology', in Zurek, W. (ed.) *Complexity, Entropy, and the Physics of Information* (Redwood City, Calif.: Addison-Wesley):425-58.
Gell-Mann, M. and Hartle, J. (1993) 'Classical equations for quantum systems', *Physical Review*, D47:3345–82.
Griffiths, R. (1984) 'Consistent histories and the interpretation of quantum mechanics', *Journal of Statistical Physics* 36:219–72.
Hardin, C. (1983) *Color for Philosophers, 2nd edition* (Indianapolis: Hackett).
Koons, R. and Pickavance, T. (2017) *The Atlas of Reality: A Comprehensive Guide to Metaphysics* (Oxford: Wiley-Blackwell).
Lewis, D. (1966) 'An argument for identity theory', *Journal of Philosophy* 63:17– 25.
Lewis, D. (1972) 'Psychophysical and theoretical identifications', *Australasian Journal of Philosophy* 50:249–258.
Lewis, D. (1973) *Counterfactuals* (Cambridge, Mass.: Harvard University Press).
Lewis, D. (1980) 'Mad pain and Martian pain', in Block, N. (ed.) *Readings in the Philosophy of Psychology, Volume I* (Cambridge, Mass.: Harvard University Press): 216–22.
Lewis, D. (1983) 'New work for a theory of universals', *Australasian Journal of Philosophy* 61:343–77.
Lewis, D. (1984) 'Putnam's paradox', *Australasian Journal of Philosophy* 62:221–36.
McDowell, J. (2014) *Plato's Theaetetus* (Oxford: Oxford University Press).
Newman, H. (1928) 'Mr. Russell's "Causal theory of perception"', *Mind* 37:137–48.
Omnès, R. (1988) 'Logical reformulation of quantum mechanics', *Journal of Statistical Physics* 53:893-975.

Pautz, A. (2014) 'Philosophical aspects of colour', in Bayne, T., Cleeremans, A. and Wilken, P. (eds.) *Oxford Companion to Consciousness* (Oxford: Oxford University Press): 150–4.

Price, H. (2010) 'Decisions, decisions, decisions: Can Savage salvage Everettian probability?', in Saunders, S., Barrett, J., Kent, A., and Wallace, D. (eds.) *Many Worlds? Everett, Quantum Theory, and Reality* (Oxford: Oxford University Press):369–90.

Pruss, A. (2015) 'Everettian quantum mechanics and functionalism about mind', at http://alexanderpruss.blogspot.com/2015/05/everettian-quantum-mechanics-and.html [last accessed 28.7.17]

Putnam, H. (1978) *Meaning and the Moral Sciences* (London: Routledge & Kegan Paul).

Putnam, H. (1980) 'Models and reality', *Journal of Symbolic Logic* 45:464-82.

Putnam, H (1981) *Reason, Truth, and History* (Cambridge: Cambridge University Press).

Ramsey, F. (1929) 'Theories', in Braithwaite, R. (ed.) *The Foundations of Mathematics and other Logical Essays* (London: Littlefield and Adams):212–36.

Ramsey, F. (1988) 'Truth and probability', in Gärdenfors, P. and Sahlin, N. (eds.) *Decision, Probability, Utility: Selected Readings* (Cambridge: Cambridge University Press):19–47.

Rosen, G. (2010) 'Metaphysical dependence: Grounding and reduction', in Hale, B. and Hoffman, A. (eds.) *Modality: Metaphysics, Logic, and Epistemology* (Oxford: Oxford University Press):109–36.

Russell, B. (1927) *The Analysis of Matter* (London: K. Paul, Trench, Trubner & Co.).

Saunders, S. and Wallace, D. (2008) 'Branching and uncertainty', *British Journal for the Philosophy of Science* 59:293–305.

Savage, L. (1988) 'The sure-thing principle', in Gärdenfors, P. and Sahlin, N. (eds.) *Decision, Probability, Utility: Selected Readings* (Cambridge: Cambridge University Press):80–85.

Sellars, W. (1962) 'Philosophy and the scientific image of man', in Colodny, R. (ed.) *Frontiers of Science and Philosophy* (Pittsburgh: University of Pittsburgh Press):35–78.

Tye, M. (2000) *Consciousness, Color, and Content* (Cambridge, MA: MIT/Bradford).

van Inwagen, P. (1990) *Material Beings* (Ithaca, NY: Cornell University Press).

Wallace, D. (2007) 'Quantum probability from subjective likelihood: improving on Deutsch's proof of the probability rule', *Studies in the History and Philosophy of Modern Physics* 38:311–32.

Wallace, D. (2010) 'How to prove the Born rule', in Saunders, S., Barrett, J., Kent, A., and Wallace, D. (eds.) *Many Worlds? Everett, Quantum Theory, and Reality* (Oxford: Oxford University Press):227–263.

Wallace, D. (2011) *The Emergent Multiverse* (Oxford: Oxford University Press).

Wallace, D. and Timpson, C. (2010) 'Quantum mechanics on spacetime I: Spacetime state realism', *British Journal for the Philosophy of Science* 61:697–727.

Zeh, D. (1973) 'Toward a quantum theory of observation', *Foundations of Physics* 3:109–16.

Zurek, W. (1982) 'Environment-induced superselection roles', *Physical Review* D26:1862–80.

Appendix

A The Traveling Forms Interpretation

My version of the Traveling Forms interpretation draws heavily upon the decoherence program and the work of the Oxford Everettians. I take decoherence as defining a set of branches, each of which constitutes the *potential* existence of an emergent realm of composite objects. The potential existence of emergence objects is a product of two things: the decoherence of a branch of the wave function, and a fixed inventory of macroscopic essences. However, the combination of these two is not *sufficient* for existence of any composite physical entity. In addition to the material cause (the branch of the wave function) and a formal cause (the macroscopic essences) there must also be an efficient cause at the emergent level: some pre-existing composite substances whose causal powers are responsible for jointly *actualizing* the potential of one of the branches to be the material substrate of further emergent composite substances.

When the actualized branch of the world splits into two or more potential successors, the substances making up the actual branch determine, through the exercise of indeterministic active and passive causal powers, which one of the successor branches shall be actual. Each exercise of such a causal power is a thoroughly local affair, but by actualizing one of the potential successor-states in its environment, each substance contributes to choosing a single branch for the entire cosmos. Hence, all of the substantial forms *travel* together through the branching structure of the decohering quantum world.

A.1 Traveling Forms and Ontic Vagueness

There is an obvious objection to this wedding of the Oxford Everettian QM with Aristotle's hylomorphism. The processes of decoherence produce branches that are only *approximately* classical. The emergent entities occupying each branch are only approximately localized in a spacetime region, and, in fact, the very number of branches is indeterminate, depending on how fine-grained a set of macroscopic descriptions we deploy. How, then, can it be a metaphysically fundamental fact that only certain macroscopic substances exist and exist in a way that corresponds to just one branch? The emergent entities of the Oxford school's version of Everettian QM are irreducible vague, and vague things cannot be fundamental.

There are two possible responses. First, one could hold that the substantial forms added to the theory by the traveling forms interpretation are able, in and of themselves, to fill in the indeterminacies left by the quantum wavefunction. So long as something sufficiently branch-like exists in the quantum wavefunction, the wavefunction is enabled to play its role as the material cause of the existence of macroscopic objects, with the substantial forms supplying definite locations for composite wholes, locations that are more determinate than the sum of the locations of their parts.

But there is a second solution that I think is preferable: simply assert that vague objects can be fundamental. Many philosophers have defended a thesis of *ontic* vagueness, vagueness in the world that is not merely the by-product of ambiguity or linguistic looseness. We could, for example, model such ontic vagueness by postulating that there can be more than one actual world: see Barnes 2010 or Koons and Pickavance 2017, pp. 275–9. These multiple actual worlds must be "bunched" together pretty tightly (at least, at the macroscopic scale): no cases of cats that are both alive and dead are allowed. Indeed, as Pickavance and I argue, ontic vagueness seems unavoidable, since linguistic or conceptual vagueness entails ontic vagueness (since meanings and thoughts are things).

A.2 Three Bonuses for Traveling Forms

First, the traveling forms interpretation ensures that everything a rational agent could care about exists in only one branch. Hence, Savage's Sure Thing Principle applies with exception, given the supervenience of value on being. Consequently, we can explain why rational agents must assign probabilities to the various possible branches in a way that respects classical probability theory. We can then appeal, quite legitimately, to the sort of physical symmetries noted by Deutsch and Wallace, as grounds for identifying objective chance with the square of the quantum wave amplitude.

Second, the traveling forms interpretation solves the problem of anti-Darwinian branches. Once we return to a one-world interpretation of probability, we can again treat possibilities that have astronomically low probabilities as *close to* impossibility. We don't have to imagine that they are all equally denizens of reality. This feature of one-world probability is equally applicable to retrospective and prospective uses of probabilities. Hence, we can justify our a priori confidence that we do not inhabit a madly anti-Darwinian branch.

Third, the traveling forms interpretation enables us to ground scientific knowledge in the interaction with thoroughly local causal powers. Scientific knowledge is possible only because objects have active causal powers and we (and our instruments) have corresponding passive causal powers. Actually, experimentation depends on causal powers running in both directions, so we can suitably *prepare* the experimental situation by undertaking the appropriate interventions or manipulations. (See Nancy Cartwright, *Nature's Capacities and their Measurement* [Cartwright 1989], and Brian Ellis, *Scientific Essentialism* [Ellis 2001].)

The multi-world Everettian functionalist has no room for localized causal powers. The quantum wave function is essentially non-separable. It can perhaps *simulate* localized causal powers, but simulation is not realization. Simulated experimentation is not experimentation. It can yield only the simulacrum of knowledge, not real knowledge. If simulated experiments were as good as real experiments, we could save a lot of money by abandoning our laboratories with their expensive equipment and just run all our experiments in CGI!

4 A Traveling Forms Interpretation of Quantum Mechanics

Alexander R. Pruss

1 From Quantum Mechanics to Many-Worlds to Many-Minds

Interpretations of quantum mechanics divide into two classes: no-collapse interpretations where the wavefunction always evolves deterministically according to the Schrödinger equation (i.e., there is a unitary family of time-evolution operators) and ones where this evolution is interrupted by "collapse". No-collapse interpretations have the benefit that they have a simple and precisely defined dynamics for the evolution of the wavefunction, with a minimum of free parameters. Collapse interpretations, on the other hand, need to have an additional account of (a) the mathematics of the collapse phenomenon and (b) what triggers it. Copenhagen-style collapse interpretations say that collapse is triggered by measurement, but it is famously difficult to give a precise account of what counts as a measurement. Newer collapse theories like the Ghirardi-Rimini-Weber (Ghirardi et al. 1986) theory give a precise stochastic dynamics for what triggers collapse, but this both complicates the physics and introduces a new free parameter that controls the frequency of collapse.

Of course, our epistemic preference for simpler theories can be overridden by other considerations, such as predictive or explanatory power, but it is at least worth seriously exploring simpler theories to see where they lead. For this reason, in this paper I will focus on no-collapse theories. As regards the evolution of the wavefunction, no-collapse theories are simplest. However, some no-collapse theories supplement the simple account of the evolution of the wavefunction by some additional dynamics. Most famously, Bohm's no-collapse theory supplements the account of the wavefunction with a deterministic account of the movements of particles. The "Traveling Forms" theory that I will explore in this paper is in this vein: it accepts a simple Schrödinger dynamics for the wavefunction, but supplements it with a story about the dynamics of Aristotelian forms. I will leave it to the reader to judge how the additional complexity in this story measures up against the additional complexity in the collapse theories.

The "Traveling Forms" interpretation will be explored with a line of thought that starts with the very simplest of no-collapse theories. Where it ends isn't all that simple, with the notable exception of maintaining the simple Schrödinger dynamics of the wavefunction. But I hope that the step-by-step development of it will be plausible at least to those with Aristotelian sympathies.

Throughout this paper, I will assume that something like quantum mechanics, appropriately interpreted, provides a comprehensive global description of physical reality. One could instead take a view like Cartwright's (Cartwright 1983, 1999) that our laws of physics describe special and fairly localized situations. I have no new argument for this comprehensiveness: I am just more sympathetic than Cartwright to taking aesthetic considerations like elegance and simplicity as a guide to truth. Readers not sympathetic to this such considerations may take the project to be Quixotic.

I am putting off the question of what the metaphysics of the wavefunction or quantum state is until Section 6. Until then, as is common practice, I will often ignore the difference between the wavefunction considered as a mathematical object—a function (i.e., a set of ordered pairs) from an interval of real numbers to vectors in a Hilbert space—and the time-varying physical reality that is represented by this mathematical object.

The physically and mathematically simplest of the no-collapse theories is Everett's many-worlds interpretation (MWI). On this interpretation, the wavefunction evolves deterministically over time in accordance with the Schrödinger equation, and encodes the whole truth about physics. After an electron in a superposition of spin-up and spin-down states passes through a magnetic field in the Stern-Gerlach experiment, the wavefunction encodes it as being in a superposition of two different positions, say, an upper position corresponding to the spin-up state and a lower position corresponding to the spin-down state. When an observer then checks whether the electron is in the upper or lower position, the wavefunction encodes the observer as being in a superposition of a state of observing the upper position and a state of observing a lower position.

But in fact, we never perceive ourselves to be in a superposition of two different observational states. To explain that, the Everett interpretation notes that the final state can be described as split into two superimposed branches: one where the electron has spin-up, is in the upper position, and is observed as being in the upper position, and the other where it has spin-down, is in the lower position, and is observed to be in the lower position. These branches can be thought of as worlds inside a multiverse, and so there is an observer in one branch who unambiguously observes the upper position, and an observer in the other branch who unambiguously observes the lower position.

A standard objection to MWI is that it does not do justice to prospective probabilities. Suppose that the initial electron state is so prepared that the spin-up state has significantly higher weight than the spin-down state, which nonetheless has non-zero weight, and that I am the observer. Thus, the state is

$$a \left| \uparrow \right\rangle + b \left| \downarrow \right\rangle \tag{1}$$

where $|a| \gg |b| > 0$ and $|a|^2 + |b|^2 = 1$. Then, according to the Born rule, which is empirically central to quantum mechanics, I should assign probability $|a|^2 > 1/2$ that I will observe the electron as being in the upper position and that I will not observe the electron as being in the lower one. But it seems that there are two final branches: one with an observer observing the upper position, and one observing the lower position. The two observers are ontologically on par, and they each derive from me.

It seems that given the metaphysics of the situation, I should make one of four predictive judgments:

(1) Each observer is a future me, so I should assign probability one to *both* observations.
(2) Neither observer is a future me, so I should assign probability zero to *both* observations.
(3) By indifference, as the observers are metaphysically on par, I should assign probability 1/2 to each observation.
(4) This is a situation where probability assignments make no sense, so I should take each observation to be a probabilistically non-measurable[1] event.

But none of these are what the Born rule requires of us, which is an asymmetrical assignment of a high probability to the upper position observation and a low probability to the lower one. Granted, the wavefunction assigns a higher weight to the branch with the upper position observation. But this weight does not describe either an objective chance or an uncertainty of any proposition (whether *de dicto*, *de re*, or *de se*). Imagine that you were going to branch into two future persons, one of whom had *literally* a higher weight—i.e., was fatter—than the other. You wouldn't epistemically privilege the events that will happen to the fatter one. Why should you epistemically privilege the events that will befall the one with the higher wavefunction weight?[2]

Defenders of MWI have two main answers to this. The first is to say that in reality the branches are not neatly delineated, and we cannot say that there are only two relevant branches. That challenges option (3), but strengthens option (4), and leaves (1) and (2) unaffected. And (3) was already the weakest of the options, because of its reliance on the dubious principle of indifference.

The second response is to offer axioms of decision theory and prove, given the axioms, that probabilities should be attached in accordance with the Born rule (the pioneer is Deutsch 1999). This response suffers from two serious difficulties. The first difficulty is that it assumes that if we were to find ourselves in the metaphysical scenario described by MWI, there would be a rational decision theory available to us giving rational ways to assign values to uncertain future outcomes. But the plausibility of (1) undercuts that assumption.

The second difficulty is a reliance on axioms that are dubious in fission situations. For instance, consider the principle that adding an additional payoff v_2 to each outcome of a game E_1 with value v_1 results in a game with value $v_1 + v_2$.[3] But now suppose that playing E_1 results in a metaphysically symmetric fission of the player into n branches, where $n > 1$. Then, by symmetry, the player either survives in all branches or in none. If the player survives in all branches, then the value with the extra payoff will be $v_1 + nv_2$, as the player will get the extra payoff twice. If the player survives in no branch, then the value with the extra payoff will be just v_1, which will be the value of certain death. Only in the special case where $v_2 = 0$ does the composite game have value $v_1 + v_2$. And if we add that there is no fact of the matter as to what the number n of branches is, then things only get worse: there is no fact of the matter about the total value received.

In order to solve the probability problem, Albert and Loewer introduced the many minds interpretation (MMI) of MWI (Albert and Loewer 1988). The idea is that there are infinitely many minds associated with each branch and (conscious) brain pair in the branching multiverse. When branching occurs, infinitely many of the minds go into each outgoing branch. However, each individual mind has objective chances of going into a particular outgoing branch defined by the Born rule. Thus, in the above spin case, each of my infinitely many minds independently has chance $|a|^2$ of going into a branch where it observes the upward position and chance $|b|^2$ of going into a branch where it observes the downward position. If I am identified with one of these minds, my credence in the two observations should be $|a|^2$ and $|b|^2$, respectively, as the Born rule requires. Nonetheless, equal infinite numbers of these minds populate the outgoing branches.[4]

However, even though MMI solves the problem with *prospective* probabilities, it suffers from a different problem. The uncollapsed wavefunction of the universe includes many strange branches. The brains in some of these branches inhabit sceptical scenarios. For instance, there will be brains in vats and Boltzmann brains—brains that appear suddenly out of thermodynamic chaos, live for a short time, and go back to chaos. Some of these brains will have phenomenal states exactly like ours. And because of the infinities involved in MMI, infinitely many of the minds with phenomenal states exactly like yours right now inhabit a sceptical scenario and infinitely many of them do not. Moreover, the infinities are supposed to all be of the same cardinality, that of the continuum. So it seems that you cannot say that it's more likely than not that your mind is in the non-skeptical scenario set.

Moreover, one can find a pair of sets of continuum-many minds phenomenally indistinguishable from yours, such that (a) your mind is in one of these sets, (b) no two minds ever occupied the numerically same brain but instead the sets pick out minds from completely separate branches, and (c) all the minds in one set are associated with skeptical scenarios and none of the minds in the other set are. Because of (b), one cannot use the branching chances that MMI uses to solve the prospective probability problem to say that it is more likely that your mind is in the non-skeptical set than the skeptical set. It thus appears that MMI leads to skepticism.[5]

There are also ethical problems with MMI. Suppose Alice, Bob, and Carl are suffering from an equal pain, and only one full dose of a painkiller is available. However, Alice and Bob will gain complete relief from a half dose, while a full dose is needed to give Carl relief. If all three are innocent strangers, and I am choosing between giving a half dose to each of Alice and Bob or a full dose to Carl, I should choose to relieve two people's suffering rather the suffering of one. But on MMI, whether I give (a) a half dose to Alice and Bob each, or (b) a full dose to Carl, the same infinite number of minds have relief from pain, since $c + c = c$, where c is the cardinality of the continuum. So, it seems, I have no moral reason to give the half dose to Alice and Bob over the full dose to Carl, which is absurd.

2 From Many Minds to Traveling Minds

What if instead, we suppose that there is at most one mind per brain, so that when branching happens, that mind goes to one of the outgoing branches, with chances given by the Born rule? This is called the single-mind view, and it is rejected by Albert because it leads to the "mindless hulk" problem (Albert 1992: 130). When branching occurs, all but one of the branched brains will be mindless hulks. As the minds spread out through the branches, eventually only one brain in the vicinity of a given mind's brain will be minded, and all the other brains around will be mindless hulks. Thus, on this view, we are probably surrounded by zombies, which is absurd, and makes much of ethics useless.

There are at least two ways of fixing this problem. One way is to suppose that when branching happens and a mind goes along a branch, new minds come into existence to populate the brains in the other branches. Prospective probabilities will still be given by the Born rule, but there are no more mindless hulks around us.

But this leads to two problems. First, it means that probably we (or at least our minds) are much younger than we think we are. Given the exponential explosion in minds on this picture, most minds come into existence at some point well advanced in life. Second, it means that although the brains around me aren't the brains of zombies, their minds are probably not the same ones that I met with yesterday. And this may well create ethical problems by undercutting promises.

The second solution to mindless hulk problem is to say that while there are many mindless hulks, there are none around here. The idea is that the minds are fellow-travelers. When one goes down a branch, the others all come along. So our friends and family (and strangers and enemies) all go with us. The brains around us have minds, and indeed the same minds that we encountered in the past. This traveling minds (TM) view was first offered by Squires (Squires 1990) and then by Barrett (Barrett 1995). It avoids the mindless-hulk problem, without creating the diachronic identity problems that the constant creation of minds view faces. It solves the probabilistic problem facing the MWI. It is not subject to MM's skepticism problem, as there no longer guaranteed to be infinitely many minds, some in skeptical scenarios and some not, with the same phenomenal state as me. Nor is this subject to the ethical problems of MM, as Alice, Bob, and Carl each have exactly one mind.

There are two apparent costs of TM as a version of MWI. The first is that the law of nature requiring the minds to travel together may seem *ad hoc*. In Section 5 we will see that this is a merely apparent difficulty: there is a very natural way to develop TM out of modal views.

The second is that TM is a dualist theory, and hence more ontologically complex than plain MWI. Moreover, it is *more seriously* dualist than MM. For on MM, facts about the states of minds supervene on the wavefunction: each brain in each branch is occupied by the same cardinality of minds. But on TM, some branch-brain pairs correspond to one mind and some to none. However, as is well known, MM is still a pretty seriously dualist theory. For although global facts such as that a cardinality κ of the minds transitioned from state A to state B supervene on the wavefunction, there are primitive facts about the identities of minds such as that mind m_{20} will transition from state A to state B, which do not supervene on the wavefunction.

And TM is a surprisingly attractive theory, filling a niche in logical space that has largely been assumed to be unavailable. It can be elaborated to be a dualist theory where the physical world is causally closed, but yet there is robust and non-overdetermined mental causation, thereby solving the interaction problem. It can likewise be elaborated to yield robust libertarian free will together with a causally deterministic physical universe.

Here is how these tricks are done. The physical basis of the story is MWI: a causally deterministic physical multiverse. If one so wishes, one can further specify that this physical universe is causally closed: no physical state is even partly caused by any non-physical state. But in addition to the physics of the physical multiverse, there is a mental dynamics. The minds are connected to particular portions of the multiverse, and travel through it following the Born rule. We can then specify that what portion of the multiverse a given mind is connected to at a given time is at least partly determined by its mental state (and maybe some additional primitive relation). The current mental state of your mind together with the wavefunction of the multiverse then causally affects where in the multiverse your mind will be attached in the future, with a dynamics obeying the Born rule.

But because our minds are constrained to travel together, when your mind takes a branch, mine comes along. Thus, your mental states affect which portion of the multiverse *my* mind is connected to, and *vice versa*. Which portion of the multiverse my mind is connected to affects my experience. But it also affects my bodily state. It does this by affecting which three-dimensional body slice in the multiverse counts as *my* current body slice. Hence, your mental states can robustly causally affect what my mental and physical states are. This causation does not contradict the causal closure of the physical, and does not overdetermine any physical states. Rather, it affects which physical states are whose if anybody's.

Furthermore, the mental dynamics are indeterministic. We can suppose them to be under the agent's control but nonetheless in accord with Born's rule. This control could either proceed through agent causation (with chances of action corresponding to the probabilities in Born's rule) or using a variant of Kane's non-causal libertarianism (Kane 1996). Thus, we have robust libertarian free will together with a causally deterministic physical universe.

Of course, one does not have to accept causal closure of the physical along with TM. One might suppose, for instance, that there is a non-physical first cause of the universe, making an initial-state exception to causal closure. Or, one might allow the minds to affect the wavefunction of the universe, but of course, then one no longer has a solution to the interaction problem and the simplicity benefits of not having wavefunction collapse largely disappear.

3 From Traveling Minds to Traveling Forms

In Aristotelian hylomorphism, human minds are forms of bodies. But all material substances have forms, not just human bodies. So if we are to build an Aristotelian version of MWI, we will need to decide what happens to the forms of other substances. Bare MWI doesn't have forms in it, so it is not satisfactory from an Aristotelian point of view.

Each of the dualist views building on MWI has an obvious analogue where the claims about minds are extended to all forms. Doing this does not, however, resolve any of the difficulties we saw facing MM, single-mind or one-mind-per-brain. If anything, the difficulties multiply.

For instance, the Aristotelian analogue to MM will say that there are infinitely many forms associated with each appropriately shaped chunk of matter. But in addition to the epistemological and ethical problems, which are in no way helped by extending the theory beyond minds, we now may have the problem of multiple forms informing the same matter, something that Aristotelians tend to deny. Or, once the single-mind view is extended to include forms, we have the problem that not only most of the human-shaped chunks of matter around us are mindless hunks, but the chunks of matter that would seem to fit other forms are mere formless heaps. Quite likely, the earth has no elephants or oak trees, but only heaps shaped like them but literally formless. But then the Aristotelian apparatus is not very useful for studying the world around me—perhaps the only biological substance I ever met is myself.

We resolved the problems facing MWI and other dualist extensions of MWI by going with TMs. We can now craft an Aristotelian analogue to TM: the traveling forms (TF) interpretation. In addition to there being minds—i.e., forms of thinking animals like us—that travel together through the branching multiverse, there will be other forms, and all of these will travel together, sticking to the same branch.

In this picture, in all but one branch of the multiverse, the macroscopic chunks are formless chunks or heaps or waves rather than substances. These formless quantum systems may correspond to the same aspects of the global wavefunction that a donkey or an oak tree does, maybe even having an exactly similar effective system wavefunction, but because of the lack of form all there is is the wavefunction. Only here, in our branch, does the wavefunction come together with an asinine or quercine form and produce a donkey or an oak, with its distinctive biological function, and, in the case of the donkey, its mental life. In the other branches, eliminativism about the biological and the mental holds sway.

One might also consider a more ontologically austere version of the view, where reference to forms is replaced by composition (cf. Koons 2016). There are primitive facts about which pluralities of particles compose a whole. And these facts are so arranged that non-trivial wholes travel together, while compositional nihilism holds in all the other branches of the universe. This is akin to Markosian's view that facts about composition are brute (Markosian 1998).

I am cautious, however, about the sense in which the constituents of the "macroscopic chunks" can correctly be identified. What there fundamentally is at the level of the physics is the wavefunction. One can *talk* as if there were particles in a branch (though the exact identification of branches is itself problematic), but it is far from clear that the particles are there in the ontology to be composed into wholes. One might do better to talk of partial constitution. Some facts about the wavefunction may be partially constitutive of the existence of substances. But only partially: what substances there are does not supervene on the wavefunction. There are fundamental contingent facts about which "aspects" of the wavefunction partially constitute a substance, with relevantly similar aspects in one branch (namely, our branch) giving rise to a substance but not so in another.

Nonetheless, plausibly there are metaphysically explanatory benefits of the full Aristotelian apparatus of forms, and so I shall develop the view in terms of traveling forms rather than, say, traveling constitution. But the interested reader can try to adapt the ideas to the more austere view.

It is essential to this view that forms be taken seriously as fundamental constituents of substances rather than as grounded in the arrangement of the matter. The forms are responsible for the characteristic behaviors of the substances that they are the forms of, rather than being a mere summing up of those behaviors (for more of what I think about forms, see Pruss 2013). The Aristotelianism of TF is a serious dualism, even though it is not a substance dualism because form is not a substance, but a central constituent of substance. Given an Aristotelianism that takes forms seriously in this way, and given that we have good reason to think that substances, and hence their forms, can persist even if their matter completely changes, one needs a dynamics that explains what chunks of matter each form is attached to at various times. If we have—as I think we do—independent reasons to accept such a strong Aristotelianism, the addition of a dynamics for forms to the base MWI story is a *necessary* addition and hence not a serious cost to the theory.

4 What Has Form?

It is uncontroversial in Aristotelian metaphysics that all living organisms have form. But does anything else? Aristotle attributes form to artifacts in a derivative way. Instead of a house having an intrinsic form like a donkey does, a house's form is found in the mind of the architect. Since the TF picture has minds in it, it can take up this story about the forms of artifacts.

It is only our branch of the multiverse that has human (or alien) artifacts. In other branches, there will be quantum systems whose effective wavefunctions behave much[6] like the effective wavefunctions of people, and, corresponding to that behavior, there will be quantum systems whose effective wavefunctions behave much like those of houses. But there won't be any people in these other branches. Thus, there won't be any designers in these branches with forms of houses in their minds. And hence, there won't be any houses, unless they are made by *immaterial* substances, like God or angels, that are not localized to a branch.

What about individual fundamental particles, like electrons or quarks? Do they have form? This way of asking the question is potentially misleading, as it suggests that there *are* such things as particles, and then queries whether *they* have form. But as we saw when considering the "traveling composition" view, it is far from clear whether bare MWI should be read as implying that particles are in the ontology. The wavefunction could be the whole ontology, and talk of the existence of particles could be a convenient *façon de parler* akin to talk of average plumbers, holes, and shadows.

If there are no particles in the bare MWI ontology, then insofar as TF builds on that ontology, the question is whether TF will suppose particle forms, which combine with aspects of the wavefunction to constitute particles.

It is simplest to say that there are not, and hence develop a version of TF in which the ontology does not include particles.

If, on the other hand, we allow for particle forms in TF and suppose that bare MWI does not have particles, then it will be most elegant to suppose that just as the forms of larger things travel together, the forms of particles travel with them. On this version, the formless branches have no particles at all, just a wavefunction. There is something rather attractive about this picture, since it means that we do not have the strange spectacle of heaps of particles that behave just as donkeys and oaks but are mere "zombie" donkeys and oaks. On this view, instead, it is only in our branch that there are particles, and hence there are no "zombie" donkeys and oaks in any other branch—and presumably none in our branch either.

This picture has much in common with Bohmian interpretations of quantum mechanics. On Bohmian interpretations, there is an uncollapsed wavefunction, with many branches, but only one branch has real particles (often called "corpuscles" in the Bohmian literature) with fully defined positions, and the particles travel together through the multiverse. The motion of the particles then is governed by the wavefunction through a "guiding equation". This guiding equation is carefully chosen so that if the arrangement of the particles initially satisfies some statistical constraints, the evolution of the system will match the Born-rule predictions of orthodox quantum mechanics. Traditionally, the particular guiding equation that is chosen is also deterministic. However, the constraint that the evolution of the system should match the empirical predictions of orthodox quantum mechanics appears to be consistent with other guiding equations, and Bell has given an indeterministic one (Bell 1987: 173–80). An indeterministic Bohmian interpretation has a certain advantage over a deterministic one: quantum mechanical explanations are statistical in nature, and extracting statistical explanations from an underlying deterministic dynamic is always at least *prima facie* problematic. The indeterministic Bell-Bohm interpretation is very similar to a TF with particle forms, except that the Bell-Bohm interpretation privileges particles over other substances.

If, on the other hand, there are particles in the bare MWI ontology, then we can ask whether on TF (a) all (in all branches), (b) some but not all, or (c) none of these particles have forms. Supposing that all particles, throughout all the branches, have forms makes for an inelegant system, given that this is not true of other substances on TF. Moreover, on this view, we will have heaps of formed particles making up "zombie" donkeys and oaks in other branches, and perhaps it is a slightly lesser departure from common sense if the "zombie" donkeys and oaks are made from more "metaphysically shadowy" formless particles, ones wholly constituted by aspects of the wavefunction. Given TF, the view that some but not all of the particles have forms—presumably with the specification that the particle forms travel together with other forms—makes for greater elegance, and I will dismiss the "all" view. But if formless particles have physical behavior indistinguishable from that of formed particles, it may seem needlessly complex to saddle some particles with form, and so the "none" view appears simplest.

Whether or not there are formless particles in the bare MWI ontology, then, we have a choice to make between two views of formed particles: either there are none, or they travel along with formed macroscopic things.

While it is in an important sense simpler to suppose that there are no formed particles, formed particles help solve a problem that faces TF as well as TM. On TM, minds came into existence in correlation with the brains in one particular branch out of many. What process selected that branch and destined the others to be full of zombies?

An initially attractive hypothesis would be that the *first* branch to get a brain got a mind. But that doesn't seem right. For presumably, there were some extremely low-weight branches where, very early on in the universe, particles (at least in a manner of speaking, if they aren't in the MWI ontology) quantum-tunneled into a Boltzmann brain. It seems implausible to suppose that that freak accident was what ensured that billions of years later, non-Boltzmann brains, like human ones, would get minds in our branch. And it is not clear what would happen if there were no earliest brain in the multiverse (say, because before each Boltzmann brain, there was an earlier, or because there was a tie).

Now, TF can solve the branch selection problem for the initial minds that TM faces. Minded animals evolved from mindless animals. But mindless animals are still organisms and hence have forms. The law of nature that ensures that forms stick together in a branch could be taken to ensure that it is in a branch that already has forms—say, of plants and mindless animals—that the forms that are minds of minded animals would arise. We might even say that the forms of plants and mindless animals causally contributed to the existence of the forms of minded animals.

Of course, this only pushes the problem back. Why did forms of plants and mindless animals arise in our branch but not in others? Now if we have formed particles in TF, then we can answer that question: for the same reason that minds only came into existence where there are forms of plants and mindless animals, the forms of primitive organisms only came into existence where there were forms of particles.

If there were particles from the beginning of the universe, this pushes the problem to the beginning of the universe. Moreover, at the beginning of the universe it need not be a *selection* problem. For it may well be that at the beginning there is only one branch. Granted, there would still be a problem of explaining the origins of the universe and of a one-branch wavefunction at the beginning. But that's not an additional problem: the problem of *initial* or *boundary* condition faces every physical theory. But a mysterious branch selection *in media res* seems more problematic.

We don't know much about the very, very early universe. Maybe there were no particles there. But even if there were no particles, perhaps there were other primitive entities that had a form—maybe a field, say—and perhaps a similar solution can be invoked then. But if only the entities of higher-level sciences like biology have form, then we are stuck with form appearing late in the history of the universe, and it is puzzling where it appears.

It is natural to think that forms could causally contribute to other forms' existence, and so pushing the existence of forms to the (admittedly mysterious) beginning of the universe will reduce the mystery over the rest of time.

The best version of TF so far, thus, holds that particles need forms to exist and that they exist along with other forms once the other forms come into existence during cosmic evolution.

There is, however, a further question about particles. Particles with forms are substances. But Aristotelian metaphysics holds that no substance has substantial parts. Are there, then, particles that are parts of macroscopic substances like donkeys and oaks?

There are three interesting options. First, we could simply answer in the negative. On this view, defended by Scaltsas (Scaltsas 1994) and Marmodoro (Marmodoro 2013), particles cease to exist when accreted into a substance, and the excretion of a particle by substance is an instance of the generation of a new particle. Second, we could answer in the affirmative, denying the maxim that substances are never composed of substances.

Third, and perhaps most interestingly, we could answer in the affirmative while maintaining the maxim. This may initially seem impossible, but there are at least three ways of telling this story. On all three stories, the particle's categorial status changes from being a substance to being something else, say an accident. On the first way, the particle's particle form perishes, but the metaphysical work done by its form is taken over by the form of the larger substance. On the second way, the particle loses its particle form, but the metaphysical work done by its form is taken over by a new accidental form within the larger substance. On both of these two ways, the post-intake particle is constituted by aspects of the wavefunction relevantly just as in the case of the pre-intake particle, together with a different form from the previous. We can further subdivide these two ways as to whether the particle can survive such a change of form. The third way is that when the particle becomes a part of the larger substance, its form remains, but changes from being a substantial form to being some other kind of form, say an accidental form.

We leave for further investigation which option to take.

5 Values of Observables

Let me sketch a way of making the TF story more precise. Modal interpretations of quantum mechanics are no-collapse interpretations that single out a collection of privileged mutually commuting observables and posit that the privileged observables have definite values. Which observables are privileged can vary over time.[7] The evolution of the values of the observables over time is guided by the wavefunction.

Bohmian mechanics is a modal interpretation where particle positions are privileged, but where the evolution of the values of the privileged observables is deterministic. The determinism in Bohmian mechanics requires that the probabilistic nature of quantum predictions be grounded in our ignorance and track back to the statistical features of the initial conditions of the universe. Such a deterministic ground for probability is philosophically problematic and can be avoided by combining the privileging of positions with an indeterministic wavefunction-guided dynamics given by Bell (Bell 1987: 173–80). Bacciagaluppi and Dickson then extended this indeterministic dynamics to other sets of observables and adopting an indeterministic dynamics appears to be more typical among modal interpretations (Bacciagaluppi and Dickson 1999).

Now, the Aristotelian form of a substance defines the kind of thing the substance is, its metaphysical species. We can then think of a substance of a certain kind as having certain determinables in virtue of being the sort of thing it is. For instance, in virtue of being the sort of thing they are, conscious animals have the determinable *being phenomenally some way* (the determinate might be "null", when the human is unconscious), many organisms have the determinable *sex*, spiders have eight leg-state determinables (with a null determinate when the leg is detached), and so on. Call these *species-based determinables*.

We can then suppose that a given species-based determinable D of a substance x then corresponds to a physics observable $O(x, D)$, in such a way that x's having a particular well-defined maximally specific determinate C of D requires $O(x, D)$ to have a particular value $o(x, D, C)$. In the simplest case, $O(x, D)$'s having the value $o(x, D, C)$ is sufficient to nomically or causally ensure that D has the determinate C. But perhaps there are higher-level properties that do not nomically or causally supervene on values of physics observables. In that case, O_D's having the value $O(o, D, C)$ will only be a necessary condition for the substance to have C, and there will be no more specific physics observable that yields such a necessary condition.

Further, we shall suppose that there is a null value of the determinable D at times t at which the substance x doesn't exist. Hence, Alexander's war horse Bucephalus now has null sex, null leg-states, etc. This simplifies the story a little by not requiring the relevant observables to vary with time, though modal interpretations can handle such variation.

We finally suppose that the observables $O(x, D)$ will always commute, and choose a modal interpretation on which they always have well-defined values. And then we choose an indeterministic dynamics for the values of the observables, say that of Bacciagaluppi and Dickson (Bacciagaluppi and Dickson 1999).

I now claim that this modal theory is actually a precisification of TF. It has the uncollapsed wavefunction in it, which constitutes the multiverse part of TF. It has forms in it. The only thing more it needs is for the forms to travel between branches.

And they do. Consider the set O of all the observables $O(x, D)$ that correspond to the substances x and their determinables D. "Branches" of the multiverse now correspond to eigenvectors of all the observables in O. At any given time t, the actual values of the observables in O define a vector $|a_t\rangle$ in the Hilbert space corresponding to the wavefunction. This vector uniquely defined by these two properties: (a) it is an eigenvector of all the observables in O corresponding to the values that these observables actually have at t, and (b) its projection on the orthogonal complement $E_O^\perp$ of that eigenspace E_O equals the projection of the actual full-state vector on $E_O^\perp$. Then, $|a_t\rangle$ corresponds to a particular branch of the multiverse.

Then, what makes it be the case that a particular form inhabits a branch is that the actual maximally specific determinates of all the species-based determinables pick out a set of values of all the observables in O, and a set of values of all the observables together with the actual value of the wavefunction picks out a joint eigenvector vector $|a_t\rangle$ of the observables in O that corresponds to a "branch".

It now trivially follows that the forms travel together. No additional law of nature, besides the dynamics of the values, is needed to get the forms' togetherness. Rather, the togetherness is simply a consequence of the fact that the determinates of the species-based determinables of all the substances *jointly* pick out the branch (with the help of the wavefunction).

6 The Ontology Behind the Wavefunction

There is one final gap in the story: what is the metaphysics behind the wavefunction itself? We can think of the wavefunction as a mathematical object, namely a function ψ from an interval I of real numbers to a Hilbert space H. Thought of that way, the question of the metaphysics of the wavefunction is a question in the philosophy of mathematics rather than of physics. But assuming a realist philosophy of science, this mathematical object represents some aspect of reality. The interval of real numbers represents a sequence of times, while the value $\psi(t)$ of the function ψ at a given real number t represents some feature of the physical configuration of the universe at a time represented by the real number t. I will call the feature of the physical configuration of the universe represented by $\psi(t)$ the "physical wavefunction".

The physical wavefunction affects the dynamics. Plausibly, this requires the physical wavefunction either to have fundamental causal efficacy or to be grounded in something else that has fundamental causal efficacy (e.g., *winning a game* does not have fundamental causal efficacy, but is grounded in the activities and potentialities of agents that do have fundamental causal efficacy, and so it can still cause one to be happy). Moreover, even though the evolution of the physical wavefunction itself is deterministic, the physical wavefunction appears contingent: the initial conditions surely could be other than they were (e.g., the universe could have started in some pure state that always remained static).

In an Aristotelian picture, causal efficacy comes from substances and their accidents. We now have a choice. Either the ordinary formed substances—oak trees, people, and maybe particles—ground the physical wavefunction, or else the physical wavefunction is grounded in or identical with some other substance.

One could certainly have a theory on which, in addition to the determinables that correspond to observables, the ordinary substances had determinables which jointly determined the wavefunction (both physical and mathematical). Thus, facts about the physical wavefunction would be grounded in a miscellany of facts about determinables of ordinary substances, which facts would then be jointly represented by the mathematical wavefunction. For instance, one could suppose that I have determinables whose determinates encode the first ten components of the wavefunction (considered mathematically), while you have determinables whose determinates encode the rest of the components of the wavefunction, and that jointly you and I exercise, by virtue of the causal powers bound up with these determinates, all the causal power involved in the traveling form dynamics. This is not a plausible theory: first of all, you and I are not *that* special, and, second, we still have to answer which ordinary substances' determinates encoded the wavefunction before you and I appeared on the scene.

Presumably, thus, the theory would have to say that *each* ordinary substance encodes some aspect of the wavefunction—perhaps aspects that are particularly relevant in the vicinity of the substance. Sadly, this makes our theory become ungainly through many degrees of freedom as to how the information about the wavefunction is distributed among the ordinary substances. We could, for instance, suppose a Leibnizian story in which every single substance carries full global information, so that each substance has sufficient information to reconstruct the value of the wavefunction at any given time and, by two way determinism, throughout time. On this view, the value of the wavefunction is overdetermined by the information carried by the individual substances. Or, we could have the radical opposite, in which at any given time, one of the ordinary substances carries information about the global wavefunction, and then we have an arbitrary choice as to which substance that is. Or we could have something in between, whereby different particles carry different portions of information that allow the wavefunction to be reconstructed. There are many ways of setting this up. The degrees of freedom involved here make TF less elegant—but also give hope that there may be multiple solutions.

It seems to be overall more elegant simply to suppose a special global substance whose state grounds the wavefunction. Given that the state changes, this state will presumably be an accident of the wavefunction, though we might think that the special substance essentially has a time-varying determinable but the wavefunction state comes from the determinates of that determinable, and these determinates are accidents (compare how a horse essentially has mass, but the particular determinate of mass that it has is an accident). Then, the causal influence of the physical wavefunction is grounded in the causal powers bound up with these accidents.

Then, this special substance will causally affect the dynamics of all the ordinary substances. On the simplest version of the TF view, this interaction is unidirectional. One can also generate a more complex version on which some higher-level substances can affect the wavefunction in ways that violate the Schrödinger equation, but then we lose the main benefit of not having collapse.

7 What Is Matter?

There is a further and very difficult question for TF: what is matter? I have, after all, talked of "chunks of matter" in explaining the theory. It is tempting to think—as an anonymous reader suggests—of the quantum state or physical wavefunction as the matter corresponding to the form. But the TF story requires there to be a vast unoccupied region of the physical wavefunction—the region of what I called the "formless chunks". And yet the physical wavefunction has definite properties and causal powers, while unformed matter does not.

If the physical wavefunction is grounded in features of individual ordinary substances, without a special global substance to ground the wavefunction, then these features could perhaps be identified with the matter of these substances. But, first of all, this leaves the question of what grounds the unoccupied portions of the wavefunction. Maybe there is one special substance that does that, in addition to the ordinary substances grounding the occupied portions. The view on which the physical wavefunction was grounded in features of individual ordinary substances seemed more complex, and this is a yet further complexity. That doesn't rule out the view, but let's explore what we could say if take the view that there is only one special substance that grounds the whole wavefunction.

One move, then, is to depart from historical Aristotelianism and not take the matter particularly seriously. What there is are forms and the physical wavefunction. The forms have a certain relationship to features of the physical wavefunction, by picking out a branch of it as in Section 5, and the forms in their relationship to the wavefunction constitute substances. Standing in such a relationship to something is what plays the role in TF of the forms' "informing matter" (whether the "special substance" that I suggested might ground the physical wavefunction counts as having matter on this story is not clear), but there is no such *thing* as matter (this last is a statement that historical Aristotelianism will be friendly towards). "Chunks of matter" is, then, just a *façon de parler* for features of the physical wavefunction, and "formless chunks" for those features that are connected to the forms of ordinary substances.

Alternately, one might allow features of the physical wavefunction to serve as matter for the ordinary substances. This also departs from historical Aristotelianism by making a feature of one substance—the "special substance" grounding the physical wavefunction—be a constituent of other substances. What I called "formless chunks", then, are not literally formless—they correspond to features that, while not informed by the forms of the ordinary substances, are nonetheless structured by the form of the special substance.

There are no doubt other options. There is significant room for further research here.

8 Conclusions

The Traveling Forms account is a no-collapse interpretation of quantum mechanics that lets us take seriously non-microscopic levels of reality, as well as—we might add—higher-level laws, like biological ones, grounded in the forms of things. The account can be seen either as arising from many-minds interpretations via a generalization of the Many Minds interpretation which solves problems for the Everett interpretation, or as a natural modal interpretation of quantum mechanics that takes seriously the determinables that figure in higher-level laws.

The theory allows for an elegant story about how higher-level causes—including our will, but not limited to our will—can be genuinely and robustly efficacious, even if the microphysical level—the level of the wavefunction—is entirely closed. It is a story of robust higher-level causation that neither supervenes on lower-level causation nor requires downward-causation. At the same time, more fully accounting for Aristotelian matter will take further research—I've only sketches some possible avenues—and this part of the story is not at present elegant.[8]

Notes

1 In fact, saturated nonmeasurable, i.e., with neither a lower nor an upper probability.
2 Here's a potential answer: perhaps personal identity goes along the higher weight path. But on that view, you are *certain* to make the upper position observation and *certain* not to make the lower one, which does not match the Born rule when the spin-down state has non-zero weight.
3 In Deutsch's setting (Deutsch 1999: 3133), this principle is a direct consequence of his understanding of additivity, so an argument against this principle is an argument against additivity as he understands it.
4 It turns out that there is a technical problem here. In order to allow for uncountably many branchings, Albert and Loewer suppose that the infinity of minds associated with a branch-brain pair is the uncountable infinity of the continuum. However, there is no guarantee that if continuum many minds each independently has a non-zero chance $|a|^2$ of going into an up-branch and a non-zero chance $|b|^2$ of going into a down-branch, then continuum many will go into each branch. Not only is there no guarantee of this, but on the standard product probability measure model one cannot even say that the probability that each branch will get continuum many minds is non-zero—the event of each branch getting continuum many minds is non-measurable in the product measure. Perhaps the probability measure can be extended to solve this problem. (For more discussion, see Pruss 2016).
5 A possible solution is to say that there is no fact of the matter as to which brain—one of the skeptical or non-skeptical brains—your mind is attached to. But in that case, we have another problem: there is no fact of the matter whether you are or are not in a skeptical scenario. And that destroys realism.

6 Or maybe even exactly. This depends on whether minds might not have some special ability to affect the wavefunction or whether an appropriate causal closure doctrine holds.
7 And, in some versions, with context. But that is not an approach I will take.
8 I am grateful to Robert C. Koons for many discussions that have greatly helped to improve TF, as well as to an anonymous reader and William Simpson for comments that have enriched this paper.

References

Albert, D. Z. (1992) *Quantum Mechanics and Experience* (Cambridge, MA: Harvard University Press).
Albert, D. Z. and Loewer, B. (1988) 'Interpreting the Many-Worlds Interpretation', *Synthese* 77: 195–213.
Bacciagaluppi, G. and Dickson, M. (1999) 'Dynamics for Modal Interpretations', *Foundations of Physics* 29: 1165–201.
Barrett, J. A. (1995) 'The Single-Mind and Many-Minds Versions of Quantum Mechanics', *Erkenntnis* 42: 89–105.
Bell, J. S. (1987) *Speakable and Unspeakable in Quantum Mechanics* (Cambridge: Cambridge University Press).
Cartwright, N. (1983) *How the Laws of Physics Lie* (Oxford: Oxford University Press).
—— (1999) *The Dappled World: A Study in the Boundaries of Science* (Cambridge: Cambridge University Press).
Deutsch, D. (1999) 'Quantum Theory of Probability and Decisions', *The Royal Society, Proceedings: Mathematical, Physical and Engineering Sciences* 455: 3129–37.
Ghirardi, G. C., Rimini, A. and Weber, T. (1986) "Unified Dynamics for Microscopic and Macroscopic Systems." *Physical Review D* 34: 470.
Kane, R. (1996) *The Significance of Free Will* (New York: Oxford).
Koons, R. C. (2016) Manuscript.
Markosian, N. (1998) 'Brutal Composition', *Philosophical Studies* 92: 211–49.
Marmodoro, A. (2013) 'Aristotle's Hylomorphism, Without Reconditioning', *Philosophical Inquiry* 36: 5–22.
Pruss, A. R. (2013) 'Aristotelian Forms and Laws of Nature', *Analysis and Existence* 24: 115–32.
—— (2016) 'Uncountably Many Coin Tosses and a Technical Problem for the Many-Minds Interpretation', *Alexander Pruss' Blog*, at http://alexanderpruss. blogspot.com/2016/10/sequences-of-uncountably-many-coin.html [last accessed 11.10.17].
Scaltsas, T. (1994) *Substances and Universals in Aristotle's Metaphysics* (Ithaca, NY: Cornell University Press).
Squires, E. (1990) *Conscious Mind in the Physical World* (Bristol: Hilger).

5 Half-Baked Humeanism

William M. R. Simpson

1. The Resurrection of Causal Powers

Causal powers are considered to be certain features of reality that bring about change with some kind of natural necessity. A power is typically individuated by reference to a characteristic manifestation and depends upon certain stimuli for its activation: for example, hot water has the capacity to dissolve lumps of sugar.[1] However, a power may also exist independently of its manifestation: water retains its power to dissolve sugar, even when it is not being exercised by dropping sugar-lumps into a cup of tea.

The concept of causal power that arises within Aristotle's account of change is receiving increasing attention in contemporary metaphysics,[2] and has been adapted in different ways in other areas of philosophy besides, including the philosophy of science, of mind and of perception.[3] For many philosophers, this renewed preoccupation with powers proceeds from a perceived failure of Humean metaphysics to offer a satisfactory account of the natural world and of scientific practice. However, these concerns have not been confined to card-carrying Aristotelians. Toby Handfield has recently propounded a doctrine of *Humean dispositionalism* that claims to reconcile causal powers with broadly Humean convictions, allowing modern Humeans to put powers to work in good conscience and embrace necessary laws of nature.

In this paper, I offer some critical reflections concerning Handfield's account of causal powers,[4] partly because I wish to resist a form of dispositionalism I believe to be mistaken, and partly because I suspect there may be more general lessons to learn from the problems that arise in attempting to pour the *neo-Aristotelian* wine of causal powers into *corpuscularian* wine-skins that philosophers have been carrying since the seventeenth century.[5] According to corpuscularianism, the natural world is composed of a single set of basic constituents that are characterised by the fundamental laws of our best physics. This view is widely held by other dispositionalists besides Handfield, who typically reject Hume's ontological austerity.[6] For this reason, I think Humean dispositionalism may supply a convenient control in testing a variety of modern power ontologies.

I begin by presenting Humean dispositionalism as an attempt to construct a theory of causal powers without adopting the modal relations between powers and their manifestations that are commonly supposed to be required by contemporary dispositionalists (Section 2). Handfield suggests we should identify the *manifestations* of causal powers with types of *causal processes* discovered by the natural sciences, whilst identifying *powers* with properties that are parts of the *structures* of these processes. In so doing, he believes we can give an account of how causal powers may be connected to their manifestations without appealing to any objectively modal features that are incompatible with Humeanism.

I argue against the coherence of this compromise in the light of quantum physics (Section 3): Humean dispositionalism is unable to isolate causal processes in a world with quantum entanglement, and therefore unable to individuate causal powers. I think this problem persists across a wide variety of ontologies (Section 4). Insofar as physical reality is conceived in terms of a set of basic constituents that are completely characterised by quantum physics, it makes little odds whether Humeans adopt holism (Section 4.1), structuralism (Section 4.2), atomism, 'gunk', or an ontology of events (Section 4.3) in their dealings with quantum entanglement. In each case there seems to be small prospects for advancing an account of dispositions on the basis of causal structures.

I conclude that, whilst Handfield may be right to reject the claim that causal powers are connected to their manifestations by modal relations, Humean dispositionalism misconstrues the nature of causal powers, and that both Humeans and Aristotelians should reject Handfield's proposed compromise (Section 5).

2. Humean Dispositionalism

2.1 *In Search of Humean Dispositions*

Consider again the case of a sugar-lump that is dissolved when it is dropped into a cup of hot tea. This is a type of change that appears, under certain conditions, to be brought about by some kind of natural necessity. Dispositionalists claim that water has a *capacity* to dissolve sugar, whether or not any sugar-lump is dropped into a cup of hot tea. This capacity is part of the nature of water and renders this change intelligible. Handfield would like for Humeans to be able to say the same.

According to standard Humean accounts, however, it is a contingent fact that most instances of sugar being wet are followed by instances of sugar being dissolved. The world consists of many such regularities, and whilst general 'laws' may be formulated to account for particular 'dispositions', there are no natural necessities that need to be explained. Handfield is dissatisfied with this approach to laws and dispositions. He thinks such an account gets 'the order of explanation entirely the wrong way round'.[7]

Like David Hume, Handfield believes there are no necessary connections in nature. Unlike most Humeans, he does not believe that the absence of any objectively modal facts rules out an ontology of causal powers.

2.2 A Neo-Humean Theory of Composition

Hume's anti-modal convictions are compatible with a variety of views concerning the plurality and composition of what exists, including different kinds of atomism and holism. However, David Lewis has provided a way of translating Hume's philosophical intuitions into a theory of composition that is often identified with contemporary Humeanism. His strict neo-Humean account turns on two principles:[8]

Recombination: there are no objectively modal connections in nature   (Rec)
between distinct existences; any combination is possible.

Humean Supervenience: the whole truth about the world supervenes upon   (HS)
a spatiotemporal distribution of perfectly natural properties localised at
point-instances.

Together, Recombination and Humean Supervenience atomise reality into a mosaic of local and contingent matters of fact. There are distinct existences, but there is no modal glue to bind them together. There is structure, but it consists only of spatiotemporal relations. Fundamentally, the natural order is an arrangement of 'just one little thing and then another',[9] without any necessary connections between them.

It is a wide-spread belief among Anglophone philosophers in the analytic tradition that scientific explanations in general succeed only by reducing complex things to the sum of their atomic parts, down to the microphysical level (and beyond). This conviction has likely had much to do with the prevailing influence of neo-Humean principles. However, Handfield believes we have good reason for rejecting the second of these two principles (HS) on the basis of quantum entanglement.[10] According to Humean Supervenience, material reality is supposed to divide into arbitrarily small portions located at separate points in space, such that whatever may be true of our world must be true of any duplicate arranged with the same spatiotemporal relations. Two quantum-entangled particles, however, can be arbitrarily separated in space whilst remaining interdependent with respect to their measurable properties, such that a change in one will invariably be accompanied by a change in the other.[11] In such cases, the physical state of the 'entangled' system cannot be decomposed into a product of the constituent states associated with its spatially separated parts. The two particles appear to enjoy some kind of necessary connection.

For Handfield, the heart of classical Humeanism does not lie in the spatial separability of natural properties (HS), but rather with the rejection of irreducibly modal relations demanded by the principle of Recombination (Rec). By contrast, the principle of Humean Supervenience seems to him less motivated and less worth the effort of saving—a misnomer whose popularity may have more to do with the fact that many philosophers have not progressed beyond classical physics and secondary-school chemistry. Where spatial separability fails, Handfield seems to suggest *fusion* instead: whenever 'someone attempts to demonstrate a necessary connection between distinct things', he explains, we are free to posit instead that 'the putatively distinct things are not really distinct; the world is less loose and separate than you thought'.[12]

The principle of Recombination, however, appears to oppose any common sense account of causal powers. Whilst different theories of dispositions vary in the details, there is broad agreement that a power ϕ to R will require the truth of some non-trivial and objectively modal proposition concerning the power ϕ and its manifestation R—for example, a non-material conditional of the form:

> Whenever something instantiates the power ϕ to R when S, it is true that if it were the case that S, then it would be the case that R.

In our example, it seems the power of water to dissolve sugar and the manifestation of a sugar-lump dissolved in hot tea are connected by some form of natural necessity.[13] Yet by insisting on Recombination, Handfield sides with Humeans concerning the absence of any 'metaphysical glue' between distinct existences.[14] Nonetheless, he would like to permit causal powers a non-vacuous role in explaining necessary laws of nature that involve different things. In this sense, Humean dispositionalism is only *half-Humean*: Handfield wants to keep an ontology of powers without attributing any of the objectively modal features that characterise neo-Aristotelian approaches.

2.3 A Humean Theory of Structure

Handfield believes a distinctly Humean form of dispositionalism is feasible, just so long as we are able to distinguish certain causal processes in the world as natural kinds. In his opinion, it is the *structures* of these processes, uncovered by our best natural sciences, that can be used to vindicate the peculiar characteristics of causal powers.

In launching his theory of compositional structure, Handfield defends the rights of Humeans to implement Kripkean intuitions concerning *a posteriori* necessities, such as the claim that having six protons is a necessary property of Carbon, or the widely accepted belief that water is necessarily identical to H_2O.[15] Where such necessities exist in nature, they are amenable to explanation in terms of identity and rigid designation.[16] These kinds of necessities are not the sort that Humeans need deny: instead of saying there could be a world in which water is not H_2O, we should say that a world without H_2O is a world without water.[17] The particular coinstantiation of properties that comprise water can still be conceived as an accidental arrangement of determinate constituents: it is not *grounded* in anything else. However, both designators in the identity, water $\equiv$ H_2O, must pick out the same entity in every possible world, in order for it to be an *essential* property of water that it has the composition of H_2O.

This Kripkean kind of essentialism is integral to Handfield's 'richer conception' of structure, which includes more than external relations between spatiotemporal parts: for any complex entity with this kind of structure, 'there exist relations between the property of being that sort of structure and the properties of the constituent parts'.[18] Necessarily, anything that has the *structural property* of being H_2O, for example, has a part that has the property of being a hydrogen nucleus. Such relations are essential to being the type of thing that it is. However, Handfield insists that these '*internal relations*' need not add anything to the basic ontology, but may be seen to supervene upon the intrinsic nature of the relata taken separately. 'There is nothing else that is involved in grounding the relation'.[19]

Handfield recognises two kinds of internal relations. Just as a property like 'being H_2O' implies the presence of a part that is 'being a hydrogen nucleus', so a property like 'being NaCl' implies the absence of any such part. We can thus distinguish between '*relations of accommodation*' and '*relations of eviction*'. His central claim is that we should apply these relations to causal processes. Just as a composite entity like water is a natural kind with an essential structure that includes relations of accommodation and eviction, so a causal process may be said to constitute a natural kind with the same kind of structural elements. The process of sugar dissolving in water, for example, is a type of process that *accommodates* the property of being sugar. The property of being polythene, on the other hand, is *evicted* by all dissolution process-types.

2.4 A Humean Theory of Powers

Handfield seeks to persuade us that causal powers and their manifestations can be understood in terms of causal processes possessed of essential structure. His strategy is: (1) to identify certain types of *manifestation* with certain types of causal processes, and (2) to analyse the modal nature of a *power* in terms of an internal relation between the property identified as a power and the property of being the *process-type* in which this power is manifested. More schematically:

Given a manifestation-type *M* for a power ϕ, a necessary condition of an   (HP)
intrinsic property *P*'s conferring the power ϕ is that *P* be *accommodated* by *M*.[20]

For example, salt has the causal power to dissolve in water because it instantiates the intrinsic property of being an ionic compound that is accommodated by a distinct causal process-type. On the other hand, the property of being polythene confers no such causal power because no such process accommodates this property as one of its essential parts. These types of processes and manifestations are supposed to be identified *a posteriori*. In this way, a necessary connection between a power and its manifestation is secured, without appealing to any primitive modality, because a process-type that is a natural kind *necessarily* accommodates this property within its structure. The independence of a power from its manifestation is secured by insisting that the process-type is not always instantiated: the property of being salt is not invariably accompanied by the process of dissolving.

There is a certain ingenuity to Handfield's approach: by identifying manifestations with certain types of causal process, he circumvents the problem of deviant causal chains that has plagued other attempts to analyse powers. For example, an army of nanobots collecting molecules of sucrose involves a different process to the sort that takes place when water acts as a solvent for sugar by forming hydrogen bonds with the molecules in each crystal; however, it is only the latter process that Handfield identifies as the manifestation of water's power to dissolve sugar. Likewise, the problem of masking is similarly averted: two protons that fail to repel because of the strong nuclear force are caught up in a different process than simple Coulomb repulsion, but a process involving two contributions that explain the resultant manifestation. Moreover, the internal relation between a power and a process-type does not involve the kind of 'Meinongian' relation between a power and a possible manifestation that has often been criticised, in which powers that have not been exercised must somehow 'point' toward non-existent manifestations.[21]

There are doubtless a number of other features of this account concerning which we may be inclined to raise questions. I mean to occupy myself here, however, with certain types of causal processes that are identified with the manifestation of powers, and the robustness of their physical structures. As it stands, (HP) affords the potential for multiplying powers *ad infinitum*, since the set of processes involving a stimulus and a manifestation may be as large and gerrymandered as one might care to imagine. For example, adding sugar to tea might result in the release of burning hydrogen gas, if a proverbially mad scientist has configured a mechanism for electrolysis and oxidation that activates under this condition. Yet, it is surely false to say that a cup of tea has the power to ignite when exposed to lumps of sugar. To prevent the multiplication of spurious powers, Handfield is obliged to associate power-conferring with a sparse set of causal processes that are structurally 'pure'.[22] The full criteria need not concern us here, but a necessary condition of purity is the possibility of isolation:[23]

Any causal process M is *isolated* if and only if neither M itself nor any part of   (IS)
M is connected causally to any entity which is not itself a part of M.

This qualification, however, introduces a marked tension in Humean dispositionalism that I should like to examine in further detail.

3. The Problem of Individuating Powers

Handfield is apparently concerned that his account should apply to examples in physics: for instance, he discusses possible examples of causal processes in quantum field theory and gestures toward the notion of their conserving a quantity as a way of getting a grip on them.[24] Handfield motivated his decision to drop Humean Supervenience-by appealing to the phenomenon of entanglement (Section 2.1). If there are physical processes that are supposed to confer causal powers, on this account, they must be analysed as physical structures. It is revealing to apply Handfield's account to a simple case of quantum correlations. In what follows (Sections 3–4), I mean to demonstrate that Humean dispositionalism fails to offer a coherent way of individuating causal powers.[25]

3.1 A Simple Case of Quantum Entanglement

Consider the following Stern-Gerlach experiment involving two electrons,[26] each characterised in quantum mechanics by a spin state: spin-up $|\uparrow\rangle$, or spin-down $|\downarrow\rangle$.[27] Both particles are passed simultaneously through an inhomogeneous magnetic field, where their quantum states evolve according to the Schrödinger equation. In this case, the dynamics is characterised by a Hamiltonian with an interaction term that depends both upon the magnetic field and the direction of the particle's spin. It is known that a particle with spin-up will be observed as deflecting in the opposite direction to a particle with spin-down, but the electron beam is mixed between spin-up and spin-down states. Our apparatus has been calibrated so that particles in state $|\uparrow\rangle$ are deflected to the upper-half of a suitably positioned screen, where their impact is duly registered.

Suppose particle 1 is in state $|\uparrow\rangle_1$. According to Humean dispositionalism, we might say that particle 1 has a power to deflect upwards in the magnetic field. It has this power in virtue of instantiating a physical property *spin-up*, which is accommodated in a pure process $M\text{-}Up_1$ that *manifests* upwards-deflection, yielding the response Up_1 on the detector. We can tell a similar story about particle 2. Suppose we assign to one person the task of reporting the deflection of particle 1, using one such apparatus, whilst another reports on particle 2, using an exact duplicate of this setup. In this way, each person can distinguish two outcomes for their particle tied to two different processes (Table 5.1), totalling four combined outcomes. It is clear that the property and response fields are derivable from the set of manifesting processes; it is *these* facts, if you will, that are carrying the story.

Table 5.1 Measurements for particles 1 and 2.

	P_1	Response	Process		P_2	Response	Process
1	$|{\uparrow}\rangle_1$	Up$_1$	M-up$_1$	1	$|{\uparrow}\rangle_2$	Up$_2$	M-up$_2$
2	$|{\downarrow}\rangle_1$	Down$_1$	M-down$_1$	2	$|{\downarrow}\rangle_2$	Down$_2$	M-down$_2$

The situation can be complicated by the introduction of superposition states, in which distinct states are combined to form new states. For example, a particle can be in a superposition of spin-up *and* spin-down states:

$$|\varphi\rangle = 1/\sqrt{2}\,(|{\uparrow}\rangle - \exp(i\varphi)|{\downarrow}\rangle) \tag{Eq. 1}$$

A particle in state $|\varphi\rangle$ possesses an equal probability of being measured as having deflected up or down: it is in an indeterminate state. Nonetheless, such a state is distinguishable from spin-up or spin-down *simpliciter* through the observable phenomenon of interference.[28] A Humean dispositionalist might conceivably delimit the particle's powers by positing additional processes. However, suppose the two electrons, emerging from their common source, are anti-correlated with respect to spin in an *entangled superposition*:

$$|\psi\rangle = 1/\sqrt{2}\,(|{\uparrow}\rangle_1 \otimes |{\downarrow}\rangle_2 - |{\downarrow}\rangle_1 \otimes |{\uparrow}\rangle_2). \tag{Eq. 2}$$

For a system in state $|\psi\rangle$, if particle 1 is measured to be spin-up, particle 2 *must* be measured to be spin-down, and vice versa.[29] However, persons 1 and 2 were not involved in producing the entangled state. Unless they communicate their results, they have no way of knowing that their particles are entangled and will record the same results. Nonetheless, the total number of outcomes for a system in state $|\psi\rangle$ is no longer four but *two*: Up$_1$ and Down$_2$ *or* Down$_1$ and Up$_2$. What shall we say of the processes and powers in this case?

The uninitiated might reasonably question why the correlation described in (Eq. 2) should be taken to imply the kind of nonlocal connection physicists have typically associated with the state of being 'entangled'. For example, one occasionally encounters students in Cambridge who persistently wear odd socks: either red on their left feet and blue on their right, or vice versa; the choice is random. Suppose that, during a laboratory demonstration in the Chemistry department involving such a group, an unfortunate explosion takes place in which the feet of each participant becomes space-like separated, before their socks can be removed and their colours noted. Whilst the anti-correlations that emerge might well be put down to some sort of social signalling, we have no cause to attribute the redness of one sock and the blueness of the other to any kind of 'spooky' connection spanning the gap between them. Why should we view the case of our two particles as involving something essentially different?

There is an important distinction that must be made between these two cases. Once the devices being used to detect the spin of the two particles in our example are rotated in relation to their axis of polarisation, the probabilities of measuring *Up* or *Down* will fall somewhere between one or zero. Significantly, the 'classical' assumption that the spin of each particle is locally determined prior to measurement—like the colour of each sock—results in one set of statistical predictions concerning their correlations, but quantum mechanics produces another, and it is the quantum mechanical predictions that are born out in experiments. John Bell famously demonstrated that the quantum theory diverges from the classical theory in predicting the dependence of the correlations upon the *relative* angle between the two polarisers—a fact which neither of the particles, considered separately, should be in a position to 'know'. In its simplest form, Bell's theorem states that any theory in which the particles are locally assigned separate spins cannot reproduce the predictions of quantum mechanics.[30] This theorem and its implications continue to be debated, but quantum nonlocality is now a widely accepted feature of modern physics.

3.2 The Fusion of Causal Processes

What can we say about causal processes and their powers in this case? Humean dispositionalism leaves little room for manoeuvre: the principle of Recombination requires that connected properties *cannot* be treated as distinct existences. Suppose we take Handfield at face value, and insist that these properties must be fused. Let us assign the states of particles 1 and 2 to properties P_1 and P_2, and their power-conferring manifestation-processes to M_1 and M_2. We may then argue as follows:

i. If P_1 and P_2 are entangled, then P_1 and P_2 are instances of the same property.
ii. If P_1 and P_2 are instances of the same property, they are part of the same process.

Therefore:

iii. If P_1 and P_2 are entangled, they are part of the same process.

(By i. & ii.)

Alternatively, we might adopt the more conservative approach of assuming these properties to be caught up within the activity of the same quantum process, rather than numerically identical, and posit premise (*iii*) directly. A *pure* causal process, however, is supposed to be *isolated* (IS): its properties

Table 5.2 Total outcomes and processes.

	Quantum state	Manifesting process	Response	
1	$1/\sqrt{2} \, (\vert\uparrow\rangle_1 \otimes \vert\downarrow\rangle_2 - \vert\downarrow\rangle_1 \otimes \vert\uparrow\rangle_2)$	M-up$_1$–down$_2$	Up$_1$	Down$_2$
2		M-down$_1$–up$_2$	Down$_1$	Up$_2$

cannot be caught up in another process, if that process is to confer causal powers (HP). We might capture this uniqueness as follows:

iv. If P is part of a pure process M_i, then P belongs uniquely to M_i.
v. P_1 is a part of a pure causal process M_1.
vi. P_2 is a part of a pure causal process M_2.
vii. P_1 and P_2 are entangled.

Therefore:

viii. P_1 belongs uniquely to M_1.   (*By iv. & v.*)
ix. P_2 belongs uniquely to M_2.   (*By iv. & vi.*)
x. P_1 and P_2 are part of the same process M_i. (*By iii. & vii.*)
xi. $M_i = M_1 = M_2$.   (*By viii., ix., & x.*)

The conclusion of this line of argument is that the structures of processes involving entangled properties must be *fused*: properties P_1 and P_2 are accommodated by the same causal process (Table 5.2).

The fusion of causal processes has obvious implications for this account of causal powers. If process M-up$_1$–down$_2$ accommodates a property that confers a power, it is arbitrary to attribute that power as an intrinsic feature of either particle 1 or 2; it should belong rather to the delocalised quantum state, however far the two particles may be separated in space, since the process must be identified with the manifestation of the total system. A policy of fusion and isolation, then, together with entanglement, can only lead to a *reduction* in the number of Humean powers and their *delocalisation* in space, as the various structures upon which these powers depend leak into each other. This appears to be in conflict with Handfield's claim that 'there may exist causal powers which are both local and intrinsic properties'.[31]

3.3 The Limit of Global Entanglement

The Humean dispositionalist may have hoped to idealise the world described by physics simply as 'one little process and then another', like a crow's nest woven together from separate twigs that can be pulled apart. The quantum physicist, by contrast, sees such structures fusing to form something less separable. From her perspective, the world looks more like an anthill in

Figure 5.1
Copyright *danshafarms*. Used with permission

which one passage leads into another (Figure 5.1). If the universe as a whole began in a quantum state, as many theorists have surmised, there is nothing within the dynamics of the Schrödinger equation that can disentangle it into separable states.[32] I shall refer to this theoretical possibility as *the limit of global entanglement*. To escape it, it seems we must either modify the quantum dynamics, or deny that the world is made of one set of fundamental constituents that are characterised solely by quantum physics.[33]

Of course, my example concerns only a simple case of correlations in nonrelativistic quantum mechanics (Section 3), which is now commonly regarded as a low-energy approximation of a relativistic field theory. One might hope to solve the problem of individuating powers by uncovering separate structures in quantum field theory. Handfield has argued, for instance, that (sets of) Feynman diagrams might be associated with distinct causal processes and their essential structures.[34] However, I think this suggestion is unlikely to receive sympathy from philosophers of physics: Handfield is aware that Feynman diagrams are used as heuristic calculating devices in perturbation theory, but fails to note that they are *limited* in their application to the case of weakly interacting fields. It is also unclear how such 'processes' could be deemed to confer the status of a power on any property that is supposed to be part of their structures.[35]

I mean to maintain my discussion, for the most part, at the level of nonrelativistic quantum mechanics, making occasional comments regarding any relevant features of quantum field theory. The phenomenon of entanglement has been shown to be even more deeply entrenched in axiomatic formulations of quantum field theories, but our difficulties may be discussed more simply using basic quantum mechanics.[36]

4. Quo Vadis, Dispositionalism?

Humean dispositionalism seeks a basis of pure processes for conferring causal powers (HP), which are identified by their manifestations. Yet its practice of fusing processes together to accommodate quantum entanglement is in tension with its theory for telling causal powers apart. I shall refer to this as the *problem of individuating powers*. Insofar as Handfield conceives physical reality in terms of some arrangement of fundamental constituents that is characterised by quantum physics, I think the phenomenon of entanglement poses a difficulty for Humean dispositionalism that needs to be redressed: its basic building-blocks do not seem to be the sorts of things that could be structured as processes that can be identified as the manifestations of powers. In what follows, I argue that this problem is invariant with respect to a number of physical ontologies that are compatible with quantum mechanics. For instance:

i.   We might accept a single causal process and embrace a type of cosmic holism.
ii.   We could add some kind of structure to account for quantum entanglement.
iii. We might restore locality by endorsing a modified form of Humean Supervenience, and choose between atomism, 'gunk', or an ontology of events.

I discuss how to conceive of physical constituents in each of these ways, in the light of quantum entanglement. In each case, I express doubts concerning the prospects of Humean dispositionalism.

4.1 The Holistic ascent

Some dispositionalists will be willing to bite the bullet: if the fate of causal powers is tied to processes that merge across space-time, then we must embrace global dispositions. We should note, before proceeding, that such a move involves reconsidering at least one of the motivations for dispositions suggested by some philosophers of science. Nancy Cartwright, for example, has argued that a capacity plays a well-defined role in marking the permanence of a contribution to the phenomenon under scrutiny.[37] The better the shielding we can afford a capacity from its environment, the better it manifests its nature. On this account, scientists proceed on the assumption that there is some signal amidst the noise, pulling things apart in an attempt to secure the set of special circumstances in which the feature in question can be studied. These circumstances are salient because they are the conditions in which other hindrances are stripped away, and we are able to see what something can do *in virtue of* having that capacity. For nonlocal dispositionalists, however, this cannot be a *constitutive* feature of causal powers in general: shielding cannot literally separate global powers from the noise of the environment, because they are constitutionally spread across space and time.

Whilst some dispositionalists may be willing to embrace global powers, I have two problems with Humean dispositionalists adopting the notion of a cosmic process. My first difficulty is this: I do not see how a single cosmic process could possess an essential structure that is robust against recombination, in the way in which natural kinds supposedly achieve robustness by rigidly designating the same compositional structure across different possible worlds. If the structure of the cosmic process is *essential*, then any recombining of the properties that comprise it in a possible world could not be identified with the *same* causal process. If such a process is truly *cosmic*, then recombinations could not be contrived without changing the cosmic manifestation with which this process has been identified. Yet if an entity cannot survive any form of recombination, it is confined to only one possible world. This being the case, it is hard to see how it could have an essential structure or satisfy the criteria for being a natural kind.

In weighing this objection, we should recall that Humean dispositionalism attempts to *derive* the modal force of a causal power from an identity conceived between a type of process and a manifestation (Section 2.2). The two relata that comprise this identity must rigidly designate their referents across *every* possible world generated by recombination. For the global dispositionalist who embraces a cosmic process, however, there can only be *one* such identity. Any powers in such a world must be accommodated as essential parts of a single process.

It might be argued that the structure of a cosmic process could survive some recombinations by being assumed as part of a possible world that is 'large' enough to contain it. Yet I fail to see any coherent way of advancing this position. Simply embedding the same distribution of whatever natural properties comprised a cosmic process in one world as a subset within some 'larger' world would not guarantee the transworld identity of this type of process and its manifestation. If global dispositionalism is true in this larger world, then this distribution of properties would be subsumed as part of a *different* cosmic process. Alternatively, if global dispositionalism is false in this larger world, then this process would be but one among many processes that comprise the sum of its cosmic behaviour. In either case, it is unclear to me how the identity between a *single* process and a *cosmic* manifestation is supposed to be held fixed across more than one possible world.

If cosmic powers are to derive any modal force from their being parts of a global process, there appear to be only two possibilities for maintaining dispositionalism: either the distinct elements that comprise the structure of a cosmic process *cannot* be freely recombined, or the structure of such a process contains no distinct elements at all. The second option is not without motivation: if the ontological fusion of entangled properties were to go so far as to condense reality to the point at which there were *no* distinct existences, there would be nothing distinct within the structure of a cosmic process to 'recombine' into a set of possible worlds, and the principle of Recombination would be vacuous.[38] On the other hand, to own the first option would be to accept something about the nature of the cosmic process that prevents the elements that comprise it from being freely recombined. It is hard to see how this could be regarded as anything other than an objectively modal fact. Either way, global dispositionalism seems to threaten to mutate into some form of Spinozistic necessitarianism, in which the entire cosmos is *necessarily* compelled to be the way that it is—hardly a Humean hypothesis. In short, I do not see how the modal features of causal powers could be explained in terms of internal relations between a power and a cosmic process type.

My second difficulty with Humean dispositionalists adopting the notion of a single cosmic process can be stated more succinctly: if powerfulness consists solely in having the potential to act, this would not be a world in which powerful properties could ever be exercised. Consider the example of a stick of dynamite sitting idly in a quarry: it has the power to explode, even though that power has not been exercised. A cosmos that is identical to the manifestation of a power, however, would be like a stick of dynamite that is always exploding: for the cosmos to exist, such a power must *necessarily* be manifested. Yet for a process-type to count as the manifestation of a power, according to Handfield, it must be *possible* for it to be uninstantiated.[39] Indeed, Humean dispositionalism is no different from many kinds of dispositionalism in this respect: it associates powerfulness with the *potential* to act in some way, and identifies a power ϕ to R with one entity (in this case, a natural property) and its manifestation with another (for Handfield, the structural property of being a particular type of causal process). It does so in order to account for the distinction between the potential *to R* and the actual manifestation R. It follows that, if the world is identical to such a process, it is not a world in which causal powers can be exercised to produce numerically distinct manifestations.

4.2 An Appeal to Structure

Humean dispositionalism is threatened by quantum entanglement: the more Humeans like Handfield exercise fusion to save Recombination (Section 3.2), the less likely they are to isolate a basis of pure processes that can be used to account for causal powers. Yet perhaps there are stones as yet unturned in tackling the problem of individuating powers. The move toward holism involved the claim that the total history of the world is interconnected. If causal powers are to avoid going fully global, dispositionalists must qualify this move with the claim that there are causally separable islands in space-time. Instead of fusing entangled systems, they might separate some of these islands by introducing some kind of structure within the supervenience base instead. We might then reconceive entanglement in terms of the instantiation of extrinsic connections between distinct existences. This approach is advanced in Darby (2012).

It has been argued that this solution is only achieved by adopting a *weakened* version of Recombination (Rec*) in which we must make a special exception for the case of quantum entanglement.[40] Given the extent of entanglement suggested by quantum physics, however, such a move would involve *severely* restricting free combinatorialism. If that is so, it is hard to see how such constraints could avoid committing Humeans to objectively modal facts, in the form of objectively modal relations, and difficult to defend this strategy from the charges of being *ad hoc* and superfluous. Why should physicalists in a reductive mood seek to preserve an ontology of local entities with intrinsic powers, if quantum processes do not need them? Why should dispositionalists seek an elaborate analysis of powers in non-modal terms, if primitive modality is no longer deemed objectionable? In what sense could any kind of dispositionalism constructed along these lines still be regarded as authentically Humean?

It may be retorted that the relations that comprise the structure of entanglement could be conceived *nomologically* instead, without injecting any causal glue into our ontology or implicating Humean dispositionalism with objectively modal commitments. I think it makes little difference to the doctrine of dispositionalism either way, since I do not believe adding structure of any sort will produce a satisfactory solution to the problem of individuating causal powers. We should recall that powers are supposed to be properties that are identified by their manifestations. For Handfield, these manifestations are supposed to be isolated by our best sciences as causal processes. If it were impossible for these manifestations to be isolated in principle, however, then such powers would be inseparable from the rigid nexus of relations in which they are embedded. In that case, we should not be able to distinguish them as *powers*, since we could not assign any definitional content to *what* they are supposed to *manifest*. Yet if causal structure is to be deployed at the base-level to accommodate an unlimited degree of entanglement, and powers are individuated by causal structures, it seems such relations will be rigid in precisely this sense. It follows that the addition of relational structure cannot preserve an ontology of causal powers, but only an ontology of hidden quiddities. It safeguards its properties only by burying them beneath a scaffolding of global structure. We may consider ourselves justified in dropping these properties altogether and following James Ladyman in adopting some ontic form of structural realism instead.[41] In any case, I think we should stop calling them 'powers'.

4.3 Back to Local Constituents via Humean Supervenience

We still have one card left to play in our search for Humean dispositions: we might dispute the assumption that quantum entanglement must be a *fundamental* feature of the natural world, as Handfield seems to suppose.

Perhaps Humean dispositionalism is guilty of giving up on neo-Humeanism too quickly. The principles of Recombination and Humean Supervenience have turned out to be more intimately connected than Handfield imagined: without the power of Humean Supervenience to atomise reality into distinct existences, Recombination has been left in danger of having nothing to recombine. In fact, quantum entanglement does not pose the insuperable barrier to full-blown neo-Humeanism that Handfield supposes.

One way to reinstate a modified form of Humean Supervenience is by reifying the mathematical configuration space in which the wavefunction of the quantum system is defined. Such a strategy—adopted, for example, in Loewer (1996)—involves treating the entangled state $|\psi\rangle$ as a field that assigns complex numbers at each point in a high dimensional space, and regarding basic properties as intrinsic qualities of points in this space, thus restoring locality. Humean Supervenience is then redefined to apply to configuration space instead of space-time.

However, this strategy is objectionable on several counts. As a low-energy approximation of quantum field theory, the configuration spaces of different wave functions may be viewed as derivative of structures in quantum field theory that are defined on ordinary spacetime. Viewed from this vantage, the value assigned to a point in configuration spaces is not a local fact about that point, but rather depends on the global state of the quantum field.[42] This move, by itself, does not succeed in restoring locality. Perhaps more troublingly, such a slide into mathematical Platonism is in danger of dissolving a real distinction between an abstract model and the reality it is supposed to be modelling, and in so doing shifting the entire theatre of actuality away from space and time to a mathematical world of abstract properties. As Michael Esfeld observes, this strategy seems to involve giving up 'a central tenet not only of common sense realism, but also of all working science'—a move most would only make with cringing reluctance.[43] Moreover, such a strategy may prove to be only a Pyrrhic victory for dispositionalism. If the high dimensional configuration space in which the wave function is defined is to be regarded as fundamental, superseding the world of space and time, then none of the objects probed by the empirical sciences could be supposed to have causal powers, being but shadows cast by the 'Platonic forms' (if you will) of a reified configuration space.

As Esfeld has argued, there are more promising options available for reaffirming Humean Supervenience via so-called 'primitive ontology' approaches to quantum mechanics. Such theories are characterised by an ontology consisting of one distribution of fundamental constituents in physical space, and a physical law for its temporal development.[44] Its basic elements are 'local beables', which are fully actual and precisely localised in ordinary space-time: they do not admit superpositions, entangled or otherwise.[45]

The de Broglie-Bohm theory, more recently conceived as Bohmian mechanics, affords one example of an alternative theory of quantum phenomena that is amenable to this kind of treatment.[46] The primitive ontology, in this case, consists of particles localised in a classical three-dimensional space following continuous trajectories. The theory includes two basic laws: one that determines the temporal development of the position of a particle with a certain velocity, and the Schrödinger equation, which fixes the velocity of each particle at any time t, given the position of *all* the particles at t. The wave function, however, can be nomologically conceived as being part of this law,[47] thus avoiding any reification of configuration space. An obvious Humean adaptation of this theory is to insist that this law does not correspond to anything extra in the ontology, in addition to a fundamental arrangement of physical constituents, but supervenes upon the particles that compose it. To make this work, the law must be made to *supervene* upon the distribution of matter throughout the whole of space-time.[48]

It is also possible to construct primitive ontologies based on the quantum dynamics proposed by Ghirardi, Rimini, and Weber, which includes the textbook postulate of the collapse of the wavefunction, and posits a mechanism for spontaneously localising quantum systems, thus avoiding the spectre of a globally entangled cosmos.[49] Again, the mechanism involved need not be taken to correspond to anything extra in the ontology, over and above a distribution of fundamental constituents. One option is to consider the primitive stuff as a *continuous* distribution of matter (also known as 'gunk'). Using this ontology, the temporal development of the wave function can be encoded as variations in density of this primitive stuff at each point in space and time, and the spontaneous collapse of the wave function can be represented as contractions in the density around certain regions. The chief difference between this conception of reality and a Bohmian particle ontology is that all the points in space-time are occupied but admit different densities, instead of tracing out worldlines in a void. Alternatively, it has been suggested that we should conceive the localisations of the wave function using an ontology of events (called 'flashes') occurring at certain points in physical space. In this case, the points at which the primitive stuff is instantiated are separated by gaps in space-time.[50]

In all these accounts, the principle of Humean Supervenience can be reclaimed, but must be restated with greater stringency: the mosaic of local matters of fact, upon which the whole truth about the world supervenes, may no longer be conceived in terms of 'perfectly natural intrinsic properties which need nothing bigger than a point at which to be instantiated',[51] but must be pared down to space-time points that are either occupied by primitive stuff or empty. Esfeld describes this position as 'physicalism without properties'.[52] Since one cannot attribute any natural properties independently of some context of measurement, such predications cannot refer to anything intrinsic to any fundamental constituents. Instead, their truthmakers must supervene upon the *total distribution* of featureless stuff, like the law that describes its evolution. Natural properties are simply predicable ways in which quantum objects behave in different measurement contexts.[53]

Ironically, conceiving physical properties in this way gives rise once again to an effective form of holism that is likely to produce some curious results outside of the philosophy of physics. It may come as an unwelcome surprise to some philosophers of mind, for example, for whom mental properties have generally been supposed to supervene upon physical properties, that every mind must consequently supervene upon nothing less than the entire cosmos.[54] Nonetheless, Esfeld regards these qualifications as entirely in keeping with the spirit of Humean parsimony, producing a kind of super-Humeanism, freed from 'stock objections . . . from quidditism and humility'. However, there appears to be no prospect for any compromise with dispositionalism along this route. A metaphysic without properties can hardly be expected to support an ontology of causal powers, nor can local beables plausibly be construed as parts of an essential structure that confers causal powers (cf. Section 4.1).

4.4 Reflections

None of the strategies I have considered are clearly viable options for Humean dispositionalists who wish to espouse a doctrine of causal powers, whilst conceiving of physical reality in terms of some fundamental arrangement of constituents that are characterised by quantum physics. Holism threatens Handfield's account of causal processes with modal necessities whilst depriving the cosmos of causal powers (Section 4.1). The super-Humean bid to restore locality, on the other hand, comes with a similar price-tag: Humeans following Esfeld must adopt a stringent formulation of Humean Supervenience (HS*) that expels any natural properties that might serve as powers from its supervenience base (Section 4.3). To many dispositionalists, Humean or otherwise, this may seem too high a price to pay: we may be tempted to appeal to additional structure instead. Yet this move must be made at the cost of reducing causal powers to hidden quiddities by engulfing their manifestations (Section 4.2).

In short, whilst Humeanism is compatible with a variety of views concerning the plurality and composition of what exists, it seems the compromise of Humean dispositionalism cannot be coherently sustained in the light of quantum mechanics. Whichever way we turn, we are left with the problem of individuating causal powers. Moreover, this difficulty is not obviously redressed by abandoning Recombination (Rec) as a metaphysical dogma and injecting modal glue.

5. Concluding Remarks

In this paper, I have considered a recent attempt to render the doctrine of causal powers compatible with broadly Humean convictions (Section 2), using the concept of quantum entanglement to probe the ontological commitments of Humeans and dispositionalists (Section 3–4). Metaphysical compromises are often problematic, and so it has proven in this case: Humean dispositionalism assumes a corpuscularian conception of nature in which reality consists of a single set of basic constituents characterised by quantum physics that must somehow be related to one another in order to be powerful. In attempting to preserve the explanatory virtue of powers, Handfield seeks a basis of causal structures, supposedly isolated by our best natural sciences, for the purpose of connecting powers to their manifestations (Section 2). However, the compulsion to fuse causal processes together, arising from quantum entanglement, is opposed to the pressure to keep their physical structures apart, for the sake of individuating powers (Section 3). This tension remains unresolved, whatever ontology of constituents is adopted (Section 4). I suspect that this is the stress that arises from trying to have one's cake and eat it.[55] Half-Humeanism may turn out to be half-baked.

Acknowledgements

The author would like to thank Hasok Chang, Anna Marmodoro, Anna Alexandrova, Timothy Crane, Jeremy Butterfield, and Toby Handfield for discussions, and to acknowledge the helpful criticisms of a number of philosophers of physics and metaphysicians in Oxford (too numerous to name) to whom he presented a draft of this paper in October 2016. He would also like to acknowledge his college, Peterhouse, for financial support.

Notes

1 I shall use the terms power, disposition, and capacity interchangeably, since I am interested in what these notions may have in common metaphysically, rather than any technical distinctions between them.

2 See Marmodoro (2010), Tahko (2013), Greco and Groff (2013), Novotný and Novák (2014).

3 Eg. Cartwright (1999) (science), Heil (2013) (mind), and Marmodoro (2014) (perception).

4 I shall be focussing on two papers: Handfield (2008, 2010).

5 Concerning the displacement of Aristotelian metaphysics by corpuscularianism, see Pasnau (2011).

6 This view is rejected as 'fundamentalist' by neo-Aristotelians like Cartwright. See Cartwright (1999).

7 Handfield (2010: 108).

8 Ibid.: 107.

9 Lewis (1986: ix).

10 Handfield states: 'Humean supervenience is almost certainly false'. See Handfield (2010: 107).

11 For the original thought experiment, see Einstein et al. (1935). See also the discussion in Bell (1964).

12 Handfield (2010: 107).

13 The modal strength of this type of necessity has been conceived in different ways, eg. as 'conditional necessity' in Marmodoro (2015), or 'modal tending' in Mumford and Anjum (2011: Sec. 8.1).

14 'Metaphysical glue' is a term of art, taken here to refer to some kind of relation.

15 For critical reflections concerning commonly asserted identity claims in the sciences, see Chang (2012).
16 Kripke's theories are discussed in detail in Kripke (1980).
17 Handfield (2008: 114).
18 Ibid.: 116–7.
19 Handfield (2008: 117).
20 This condition is defined in Handfield (2008: 118).
21 For example, see Armstrong (1997).
22 Whilst Handfield does not require that *only* pure processes should count as manifestations, he does demand that any process in nature that confers a power should 'resemble' a pure manifestation-type. His account therefore depends on establishing a 'delimited class of pure processes' in Handfield (2008: 123).
23 Handfield refers to this as 'closure'. I have adopted the term 'isolation' instead, to avoid confusion.
24 See Handfield (2010). Cartwright has argued that Dowe's notion of conservation has little application outside of fundamental physics. See Dowe (1992) and Cartwright (2004).
25 It will be sufficient for my purposes to focus primarily on nonrelativistic quantum mechanics, thus avoiding the technical complexity of more fundamental theories in which the same problems are present.
26 For the original Stern-Gerlach experiment, see Gerlach and Stern (1922a, b).
27 These states are mathematically represented as vectors in separate Hilbert spaces: $|\varphi\rangle_1 \in H_1, |\varphi\rangle_2 \in H_2$.
28 Interference occurs in this state in virtue of the phase relation $\exp(i\phi)$ between its two component-states.
29 The Hilbert space of the composite system H_{12} is the tensor product of the separate Hilbert spaces, H_1 and H_2, and the vector $|\psi\rangle \in H_{12}$ that jointly describes the two particles in the composite system cannot be factored into the product of two constituent states $|\psi\rangle_1 \otimes |\psi\rangle_2$.
30 Bell (1964, 1987).
31 Handfield (2008: 114).
32 According to James Hartle, 'the central question of quantum cosmology' can be formulated as follows: 'The universe has a quantum state. What is it?' See Hartle (2003: 615).
33 We may, for instance, live in a nomologically 'dappled world'. See Cartwright (1999).
34 See the discussion in Handfield (2010: 123–7).
35 We should not take the language of 'particles' in contemporary physics literally (unless we are adopting a Bohmian or Bell-type approach to quantum field theory): the particles posited by the standard model are excitations (or 'quanta') of spatially extended quantum fields.
36 For further discussion of entanglement in the context of algebraic field theory, see Clifton and Halvorson (2001) and Ruetsche (2011).
37 See Cartwright (1992).
38 Schaffer argues for priority monism, as an alternative to existence monism. On this view, there are many concrete objects, but they only exist derivatively. See Schaffer (2009, 2010).
39 See Handfield (2008: 119).
40 Esfeld (2014).
41 For pioneering work on ontic structural realism, see Ladyman and Ross (2009).
42 Myrvold (2015).
43 Esfeld (2014). See also Monton (2006).
44 Allori et al. (2008), Allori (2015).

45 I take a local beable to be an element of a physical ontology that exists in a bounded spacetime region.
46 de Broglie (1928), Bohm (1951).
47 See Dürr et al. (2013: Sec. 11.5 and 12).
48 Esfeld et al. (2013), Miller (2013).
49 Ghirardi et al. (1986).
50 See Esfeld (2014) and Allori et al. (2014).
51 Lewis (1986: ix).
52 Esfeld (2014).
53 See the theorems of Gleason (1957) and Kochen and Specker (1967).
54 For a recent work that rejects microphysical assumptions in the philosophy of mind, see Goff (2017).
55 A suspicion, since I do not suppose myself to have exhausted every possible rejoinder. For example, one line of argument Humean dispositionalists may wish to develop is that causal processes, whilst 'impure' because of entanglement, might still *resemble* (in some sense) possible causal processes that are 'pure' (correspondence with Handfield).

References

Allori, V. (2015) 'Primitive Ontology in a Nutshell', *International Journal of Quantum Foundations* 1(3): 107–22.
Allori, V., Goldstein, S., Tumulka, R., and Zanghi, N. (2008) 'On the Common Structure of Bohmian Mechanics and the Ghirardi-Rimini-Weber Theory', *The British Journal for the Philosophy of Science* 59(3): 353–89.
——— (2014) 'Predictions and Primitive Ontology in Quantum Foundations: A Study of Examples', *British Journal for the Philosophy of Science* 65: 323–52.
Armstrong, D. M. (1997) *A World of States of Affairs* (Cambridge: Cambridge University Press).
Bell, J. S. (1964) 'On the Einstein-Podolsky-Rosen paradox', *Physics* 1(3): 195–200.
——— (1987) *Speakable and Unspeakable in Quantum Mechanics* (Cambridge: Cambridge University Press).
Bohm, D. (1951) *Quantum Theory* (Englewood Cliffs: Prentice-Hall).
Cartwright, N. (1992) 'Aristotelian Natures and the Modern Experimental Method', in *Inference, Explanation and Other Philosophical Frustrations*: 44–71.
——— (1999) *The Dappled World: A Study of the Boundaries of Science* (Cambridge: Cambridge University Press).
——— (2004) 'Causation: One Word, Many Things', *Philosophy of Science* 71: 805–19.
Chang, H. (2012) *Is Water H2O?: Evidence, Realism and Pluralism* (Dordrecht: Springer).
Clifton, R. and Halvorson, H. (2001) 'Entanglement and Open Systems in Algebraic Quantum Field Theory', *Studies in History and Philosophy of Science, Part B: Studies in History and Philosophy of Modern Physics* 32: 1–31.
Darby, G. (2012) 'Relational Holism and Humean Supervenience', *British Journal for the Philosophy of Science* 63: 773–88.
de Broglie, Louis (1928) 'La nouvelle dynamique des quanta', *Electrons et photons. Rapports et discussions du cinquième Conseil de physique tenu à Bruxelles du 24 au 29 octobre 1927 sous les auspices de l'Institut international de physique Solvay* (Paris: Gauthier-Villars): 105–132. English translation in G. Bacciagaluppi and A. Valentini (2009) *Quantum Theory at the Crossroads. Reconsidering the 1927 Solvay Conference* (Cambridge: Cambridge University Press): 341–71.

———— (2007) *Physical Causation* (Cambridge: Cambridge University Press).

Dürr, D., Goldstein, S. and Zanghì, N. (1995) 'Quantum Physics Without Quantum Philosophy', *Studies in History and Philosophy of Science Part B: Studies in History and Philosophy of Modern Physics* 26(2): 137–49.

Einstein, A., Podolsky, B. and Rosen, N. (1935) 'Can Quantum-mechanical Description of Physical Reality Be Considered Complete?', *Physical Review* 47(10): 777.

Esfeld, M. (2014) 'Quantum Humeanism, Or: Physicalism Without Properties', *The Philosophical Quarterly* 64(256): 453–70.

Esfeld, M., Hubert, M., Lazarovici, D. and Durr, D. (2013) 'The Ontology of Bohmian Mechanics', *The British Journal for the Philosophy of Science* 65(4): 773–96.

Gerlach, W. and Stern, O. (1922a) 'Der experimentelle Nachweis der Richtungsquantelung im Magnetfeld', *Zeitschrift für Physik* 9: 349–52.

———— (1922b) 'Das magnetische Moment des Silberatoms', *Zeitschrift für Physik* 9: 353–5.

Ghirardi, G. C., Rimini, A. and Weber, T. (1986) 'Unified Dynamics for Microscopic and Macroscopic Systems', *Physical Review D* 34(2): 470–91.

Gleason, A. M. (1957) 'Measures on the Closed Subspaces of a Hilbert Space', *Journal of Mathematics and Mechanics* 6: 885–93.

Greco, J. and Groff, R. (eds.). (2013) *Powers and Capacities in Philosophy: The New Aristotelianism* (New York: Routledge).

Goff, P. (2017) *Consciousness and Fundamental Reality* (Oxford: Oxford University Press).

Handfield, T. (2008) 'Humean Dispositionalism', *Australasian Journal of Philosophy* 86(1): 113–26.

———— (2010) 'Dispositions, Manifestations and Causal Structure', in A. Marmodoro (ed.) *The Metaphysics of Powers: Their Grounding and Their Manifestations* (New York: Routledge).

Hartle, J. (2003) 'The State of the Universe', in G. W. Gibbons, E. P. S. Shellard and S. J. Rankin (eds.) *The Future of Theoretical Physics and Cosmology* (Cambridge: Cambridge University Press): 615–20.

Heil, J. (2013) 'Mental Causation', in S. C. Gibb, E. J. Lowe and R. D. Ingthorsson (eds.) *Mental Causation and Ontology* (Oxford: Oxford University Press).

Kochen, S. and Specker, E. P. (1967) 'The Problem of Hidden Variables in Quantum Mechanics', *Journal of Mathematics and Mechanics* 17: 59–87.

Kripke, S. (1980) *Naming and Necessity* (Cambridge, MA: Harvard University Press).

Ladyman, J. and Ross, D. (2009) *Every Thing Must Go: Metaphysics Naturalized* (Oxford: Oxford University Press).

Lewis, D. (1986) *Philosophical Papers*, vol. 2 (Oxford: Oxford University Press).

Loewer, B. (1996) 'Humean Supervenience', *Philosophical Topics* 24(1): 101–27.

Marmodoro, A. (ed.). (2010) *The Metaphysics of Powers: Their Grounding and Their Manifestations* (New York: Routledge).

———— (2014) *Aristotle on Perceiving Objects* (Oxford: Oxford University Press).

———— (2015) 'Dispositional Modality Vis-à-Vis Conditional Necessity', *Philosophical Investigations* 39: 205–14.

Miller, E. (2013) 'Quantum Entanglement, Bohmian Mechanics, and Humean Supervenience', *Australasian Journal of Philosophy* 92(3): 567–83.

Monton, B. (2006) 'Quantum Mechanics and 3N-Dimensional Space', *Philosophy of Science* 73: 778–89.

Mumford, S. and Anjum, R. L. (2011) *Getting Causes From Powers* (Oxford: Oxford University Press).

Myrvold, W. C. (2015) 'What Is a Wavefunction?', *Synthese* 192: 3247–74.

Novotný, D. D. and Novák, L. (eds.). (2014) *Neo-Aristotelian Perspectives in Metaphysics* (New York: Routledge).

Pasnau, R. (2011) *Metaphysical Themes 1274–1671* (Oxford: Oxford University Press).

Ruetsche, L. (2011) *Interpreting Quantum Theories* (Oxford: Oxford University Press).

Schaffer, J. (2009) 'Spacetime the One Substance', *Philosophical Studies* 145: 131–48.

—— (2010) 'Monism: The Priority of the Whole', *Philosophical Review* 119: 31–76.

Tahko, T. (2013) *Contemporary Aristotelian Metaphysics* (Cambridge: Cambridge University Press).

6 Disentangling Nature's Joints

Tuomas E. Tahko

1. Introduction

In the neo-Aristotelian tradition, the category of substance is typically regarded as the most fundamental of the ontological categories. Substance is the starting point of neo-Aristotelian ontology, the root of being *qua* being.[1] The neo-Aristotelian can claim the support of common sense for this commitment to substance: the idea that there are such individual substances like people, cats, and trees is certainly part of everyday ontology. On the face of it, science might seem to support the view too, at least while we remain at a fairly high level of generality. As Hoffman and Rosenkrantz put it, a plausible understanding of 'thing' is just that it means 'individual substance'.[2] But it is well known that this type of common sense approach to ontology faces serious challenges from contemporary science and especially from those philosophers of science who defend *ontic structural realism*. A central tenet of this approach is exactly that science has shown the 'folk ontology' to be misleading, as nothing corresponding to the neo-Aristotelian substance can be found; every 'thing' must go.[3] Perhaps that's not exactly right. After all, even if the individual substances that we typically postulate have to go, there would still be one potential candidate: the world, the universe, the cosmos, the spacetime as a whole.[4] This would lead us towards some type of *monism*, an approach which is enjoying something of a revival, especially due to Jonathan Schaffer's influential work.[5]

How should the neo-Aristotelian friend of substance react to these developments? At the very least, the neo-Aristotelian should examine the relevant arguments. Some of the most influential arguments build on quantum theory and the phenomenon of *quantum entanglement* in particular. Examining these arguments and their consequences for the idea of substance is precisely what I will aim to do in what follows. The focus is on one of Schaffer's central arguments, which will be outlined in the second section. We will then proceed to look at the underlying science in some more detail in the third section and highlight some open questions, which put Schaffer's argument in new a light. It will become apparent that the upshot of Schaffer's argument, if successful, would be even more radical than it first seems, and this poses a challenge for the dialectic of the argument. In the fourth section, a reconciliation is sought and a novel way of understanding substance is proposed, where primitive incompatibility is introduced as a necessary condition for substancehood. Finally, in the fifth section, we will examine how primitive incompatibility might be traced to quantum ontology, with special attention to wave function realism.

2. Schaffer's Monism and the Argument from Entanglement

The type of monism that Schaffer defends is *priority monism*. This is a less radical view than monism understood as the view that exactly one thing exists, sometimes called *existence monism*, which Schaffer takes to be an uncharitable understanding of monism—he in fact refers to Hoffman and Rosenkrantz and associates this type of uncharitable approach with them.[6] Hoffman and Rosenkrantz write as follows:

> Monism [. . .] is inconsistent with something that appears to be an evident datum of experience, namely, that there is a plurality of things. We shall assume that a plurality of material things exists, and hence that monism is false.[7]

But recall that by 'thing', Hoffman and Rosenkrantz mean 'individual substance', whereas Schaffer's view is that there is exactly one *substance*, be it spacetime, the cosmos, or whatever we might want to call it.[8] For Schaffer, 'substance' is a fundamental entity[9], but Schaffer's monism does not deny that there could be other 'things' in the world, it's just that those 'things' are not (fundamental) substances, but rather mere arbitrary parts of the cosmos.[10] So, the view is that the cosmos, the integrated whole, is ontologically prior to these arbitrary parts.[11] Nevertheless, Schaffer thinks that there is exactly *one* substance and indeed exactly one 'thing', if 'thing' is understood to refer to substances: material objects are not a second, distinct kind of substance.[12] To be clear: Schaffer is not denying that there is a plurality of *dependent* 'things', but for him these 'things' cannot be substances, because substances are fundamental entities. Yet, the type of pluralism that, e.g., Hoffman and Rosenkrantz favour, requires that individual substances are not dependent on the cosmos in Schaffer's sense: they are to be regarded as ontologically independent.[13] In fact, Hoffman and Rosenkrantz note that Aristotle's view was precisely that individual substances are fundamental and everything else depends on them, not vice versa.[14] So on this view, pluralism about 'things' is a pluralism about the *fundamental* 'things' and hence in direct opposition to Schaffer's view.

Leaving aside these complications, we can proceed to what is clearly supposed to be one of Schaffer's central arguments for priority monism: the argument from *quantum holism*. We can approach the idea via Quantum Field Theory (or Theories) (QFT), as Schaffer himself does.[15] QFT is sometimes presented as if it is a natural ally for the monist, since it paints a picture of the cosmos as one consisting of *fields* rather than particles. This may seem like an immediate challenge for the pluralist, because if the material things that we typically associate with individual substances are just something like excitations of an underlying field,[16] then they would seem to be only an idealization, not something with sharp *boundaries*.[17] It should be noted that this may be a bit of a simplification, since there could be more than one fundamental field. Moreover, one might question the assumption that QFT is a fundamental theory.[18] It is often thought that some quantum theory is true and universal, and hence, all fields would be quantum fields.[19] In any case, there is little doubt that this type of view—that modern science and QFTs in particular should drive us to abandon pluralism about individual substances—is present in physics, and perhaps even more widespread among philosophers of physics. Schaffer himself mentions several examples, such as H. Dieter Zeh, who claims that the formalism of QFT has always suggested that we ought to abandon the 'primordial particle concept' and replace it with fields.[20] So Schaffer's argument has at least some support among the practicing scientists—we will return to this in the next section. What is the actual argument, though? It proceeds from the phenomenon of quantum entanglement to quantum holism and then straightforwardly to priority monism.[21] Here is a reconstruction:[22]

1. The quantum state of an entangled system contains information over and above the information carried by the quantum states of its components.
2. The cosmos forms one vast, entangled system.
3. Entangled systems are fundamental wholes.
4. The cosmos is a fundamental whole.

Premise 2 requires further support. Schaffer suggests that it can be supported by physics or mathematics.[23] Firstly, if the world begins with the Big Bang, where everything interacts, this initial state of entanglement is preserved if we assume that the world evolves in a linear fashion, e.g., in terms of Schrödinger's equation. Secondly, if we assume that there is a universal wave function, then 'it is virtually certain that it will be entangled since measure 1 of all wave-functions are entangled'.[24] Now, it is worth noting, as Schaffer does, that these reasons in favour of premise 2 are only plausible if wave function collapse is ruled out as a form of the universe's evolution. This clearly rules out collapse approaches to quantum theory, but since Schaffer is quite aware of these limitations, I will set them aside as well (although we will refer to collapse approaches later on).[25] So let us assume that premise 2 is true.

What about premise 3? Schaffer argues that 'Democritean pluralism'— the idea that there are particles that have intrinsic physical properties and stand in external spatiotemporal relations—cannot account for properties of the entangled system. In fact, this idea is already implicit in premise 1 (which Schaffer himself does not list as a separate premise). The idea is that the entangled system is 'emergent' and cannot be reduced to its proper parts: 'The physical properties of the whole are not fixed by the total intrinsic properties of any subsystems'.[26] This idea is elaborated on by Ismael and Schaffer, who propose that entanglement is analysed in terms of *nonseparability*.[27] This leads towards the type of quantum holism that we've seen Schaffer to favour, with some form of emergence as the upshot: the composite system seems to be more than the sum of its parts. As a potential pluralist rejoinder, Schaffer considers the possibility of introducing fundamental, external entanglement relations, which might enable the pluralist to retain particles (and hence individual substances). However, he rightly points out that it's not at all clear that particles could be retained in such a theory, for as we already saw there is no clear place for them in the ontology of QFT (this is assuming that QFT *is* a fundamental theory, of course). Assuming that external entanglement relations will not help in saving pluralism, the conclusion follows from premises 2 and 3.

3. A Tangled Argument

Let us now take a moment to consider the dialectic of Schaffer's argument for priority monism in more detail. In particular, I'd like to highlight the commitment to emergence, which is evident from premise 1 of the reconstructed argument. The background of the idea is in fact something that we can see in the physics literature as well, and since Schaffer draws on this literature, we ought to acknowledge this. One way to introduce the idea is via *decoherence*, which concerns the appearance of classicality when quantum coherence is removed.[28] As our starting point, we can take the double-slit experiment. To get the correct result for the probability of an electron passing through a particular slit, we have to take into account *interference*, which depends on both components of the wave that splits when it encounters the slits. This produces the familiar interference pattern. Now, decoherence becomes evident when this trademark feature of quantum systems, the interference, is not observed, and instead, we have a system that appears to conform to a classical interpretation. But the reason for this appearance of classicality is that a system will also interact with its environment and indeed it will become entangled with its environment. This phenomenon can be produced simply by performing the double-slit experiment and observing the slits. So decoherence gives us the appearance of wave function collapse without requiring that such a collapse really occurs. But decoherence can also emerge spontaneously, because the system unavoidably interacts with stray air particles, etc.

This produces a general problem: the world appears to be classical and to consist of individual and distinct macro-objects, but if there is classicality in *appearance only*, then can we have even 'emergent' macro-objects? If the answer is supposed to be 'yes', then at the very least, we need a story about how this works, ultimately, about how the facts about the 'emergent' macro-objects are grounded in the wave function of a system.[29] We will return to this in section 5, where we discuss the notorious *macro-object problem* in more detail.

Let me tie this idea back to Schaffer's argument. In his most recent work, with Jenann Ismael, one possible approach being considered is *wave function realism*, i.e., the view that 'the wave function is a fundamental object and a real, physical field on configuration space'.[30] This is also a topic that will be discussed in more detail in section 5, but the context of the discussion is the following idea:

> [Q]uantum mechanics seems to allow two entities—call them Alice and Bob—to be in separate places, while being in states that cannot be fully specified without reference to each other. Alice herself thus seems incomplete (and likewise Bob), not an independent building block of reality, but perhaps at best a fragment of the more complete composite Alice-Bob system (and ultimately a fragment of the whole interconnected universe).[31]

The quantum holist faces precisely the type of problem that we saw above with respect to the phenomenon of decoherence, namely, in what sense, besides in *appearance only*, are Alice and Bob individual and distinct at all, if they are fundamentally nonseparable? Ismael and Schaffer entertain various ways in which we might consider Alice and Bob to emerge as 'modally connected non-identical events' from a common portion of reality.[32] For the sake of brevity, I will here discuss just one of these, based on wave function realism. Ismael and Schaffer suggest that: 'For the wave function realist, assuming that there is even such a thing as familiar three-dimensional space, it is to be treated as a derivative (or emergent) structure, and not a fundamental aspect of reality'.[33] We should note here that even though the quantum holist may prefer to talk about 'derivativeness' rather than 'emergence', the basic problem—how do we explain the appearance of classicality and the apparent distinctness of three-dimensional macro-objects such as Alice and Bob—is nevertheless the same.

Returning to Schaffer's original argument, recall that a crucial—although implicit—premise of the argument is that the whole, the cosmos, is somehow emergent and contains information over and above the information carried by the quantum states of its components. It is important for Schaffer's project that even though the emergent whole is prior to its parts, the whole does indeed have parts, and hence, we can get the classical objects, Alice and Bob, out of priority monism: 'the monist can guarantee a complete inventory of basic objects'.[34] Schaffer's various allusions to decoherence suggest that it may be via decoherence that Schaffer hopes to guarantee—or at least motivate—the possibility of maintaining the inventory of basic objects.[35]

The work on decoherence has made it clear that 'nonseparable quantum systems nevertheless typically *approximate* separable closed classical systems very closely'.[36] Even if nonseparable quantum systems are ultimately holistic, they nevertheless seem to have features very similar to those of classical systems. But one might argue that this is not taking nonseparability and the project of quantum holism to its logical conclusion. This line of thought requires tackling the underlying quantum ontology in more detail.

To give an example of an attempted explanation of quantum holism with the required level of detail, one might endorse something like the Bohm-Hiley project for an 'undivided universe', which 'requires not only a listing of all its constituent particles and their positions, but also of a field associated with the wave-function that guides their trajectories'.[37] This is the idea of a 'pilot-wave' postulated in the de Broglie-Bohm theory, which connects again with the phenomenon of decoherence—although as Bacciagaluppi notes, in the de Broglie-Bohm theory, we effectively have *two* mechanisms connected to apparent collapse and hence the emergence of classicality.[38] So, the Bohmian project as well will have to answer the question of where the classicality emerges *from*, but the reason why it might appear attractive is that the de Broglie-Bohm theory can refer to decoherence when it comes to the emergence of classical structures and *also* provide an interpretation of quantum mechanics that 'explains why these structures are indeed observationally relevant'.[39]

The upshot for Schaffer's project is that none of the intriguing work inspired by decoherence is going to give the priority monist the objectivity that the account needs—not without a completed quantum ontology. One source of the problem specifically for Schaffer is that he wishes to avoid committing to existence monism as it has been traditionally understood, i.e., that there is just one *thing*. As we saw, Schaffer favours priority monism instead, and he takes it that there is a plurality of 'things', albeit only as *parts* of the whole—the cosmos. Chris Fields has recently challenged this.[40] Here is a representative passage from Schaffer:

> [Q]uantum entanglement is a case of emergence, in the specific sense of a property of an object that has proper parts, which property is not fixed by the intrinsic properties of its proper parts and the fundamental relations between its proper parts.[41]

The worry is that Schaffer's account of quantum entanglement as a case of emergence will not get off the ground without an objective sense of 'proper part'—essentially, to view parts as localized 'systems'.[42] Fields suggests that what would be needed here is some reason to think that 'the notion of universal entanglement is consistent with any coherent mereotopology that yields elementary particles or any other proposed propertied fundamental objects as persistently identifiable, localizable parts'.[43] The challenge, due to Fields, is that the phenomenon of decoherence appears to be of no help in achieving anything but the illusion of such objective sense of 'proper part'. Fields insists, quite correctly, that an explanatory gap remains, since the classical world is a 'world of discrete, identifiable, time-persistent *objects*'; decoherence does not give us any principled reason to regard some collection of elementary particles as this type of a discrete object over any other.[44] But some such reason—an account of when and which proper parts compose an object—would be needed if we are to, as it were, reconstruct the classical world. There have of course been attempts to do so in the literature. Fields considers Wojciech Zurek's idea of 'quantum Darwinism', where classicality emerges due to the environment encoding information about a system during the process of decoherence.[45] Zurek suggests that the process is based on a 'Darwinian' mechanism that is ultimately responsible for the fact that only some of the potential collections of elementary particles are recognized (by us, the observers) as objects. But we need not dwell on such speculative ideas, for Schaffer himself does not rely on anything of the sort. This should, at any rate, be enough to highlight that a further story is needed here.

The more general upshot is that given Schaffer's premises, there would appear to be reasons to favour a much more extreme version of monism than Schaffer hoped. This is due to the tension identified by Fields, i.e., taking the cosmos to be an emergent, entangled whole does not appear to be consistent with objective proper parthood. So, if Schaffer holds on to the idea of 'universal entanglement', then he may end up back with the 'uncharitable' interpretation of monism, i.e., that only one thing exists[46]:

> [I]f quantum effects such as entanglement do not disappear at large scales, then the "classical world" is not an approximation but an illusion, and our ordinary notions of objecthood, locality, independence and causation are not approximately right but rather straightforwardly wrong. Such a conclusion flies in the face of all of our intuitions, and is difficult even to consider as a theoretical option.[47]

It is worth highlighting that we seem to have ended up with a position equivalent with the version of monism that we saw Hoffman and Rosenkrantz dismiss as evidently inconsistent with experience. This is the view according to which there are 'no particles, pebbles, planets, or any other parts to the world'.[48] Even Schaffer admits this to be potentially as implausible as Fields suggests in the passage quoted above: 'Perhaps monism would deserve to be dismissed as obviously false, given this interpretation'.[49] The challenge is that Schaffer's argument, at least on the present interpretation of decoherence, may lead us towards this very view. If this is right, then the choice we face is between existence (rather than priority) monism and some version of genuine pluralism. Yet, we have seen that it's difficult to understand how genuine pluralism could be reconciled with quantum mechanics, at least if we assume a non-collapse approach. Let us see if we can make some progress.

4. Substances Disentangled: Primitive Incompatibility

Having observed the tangles in the argument for priority monism from quantum entanglement, one might think that the options are clear: either we should bite the bullet and endorse existence monism rather than Schaffer's priority monism, or we should follow Hoffman and Rosenkrantz and dismiss monism as altogether unpalatable. But perhaps this is uncharitable. After all, Schaffer has plenty of other arguments in favour of priority monism, and even the case presented above is far from conclusive. This is not the place to attempt an assessment of all the arguments in favour of priority monism. Instead, let us approach the issue from the opposite end, by trying to reconstruct a pluralist, neo-Aristotelian substance ontology, given the challenges raised by quantum theory. The goal here will be primarily just to motivate the idea—apparently challenged by some approaches to quantum theory—that there could be such things as individual substances: things, objects, entities with objective boundaries. I shall not attempt to be faithful to Aristotle or even the neo-Aristotelian tradition in what follows. I am merely interested in the basic idea of *joints* in nature, joints that distinguish one kind of thing from another kind of thing. The existence of such joints would not be a sufficient condition for the existence of individual substances, but their existence does at least seem to be a necessary condition.

In fact, the strategy that I propose is to focus on *properties* rather than substances, for if we have some reason to think that there are properties that cannot be had by one thing at the same time—simultaneously instantiated incompatible properties—then we would already have at least a *prima facie* reason to postulate more than one thing and hence some boundary or distinguishing feature between these two (or more) things. As a matter of fact, Aristotle's work is a natural starting point when it comes to this theme, even though he discusses the theme of incompatibility also quite separately from that of substance.

In *Categories*, Aristotle discusses a distinctive feature of substances: that they are able to receive contraries.[50] By this he means that a substance, like an individual man, can be hot at one time and cold at another time. The existence of contraries—properties like hotness and coldness—does not itself in any way imply that there is more than one substance. But it may help to get us started to think about the difference between contrariety and incompatibility, as we see it in Aristotle's work. The important idea is that if contrary properties are instantiated *at the same time*, then it would seem that they cannot be instantiated by one and the same substance. It is precisely this idea of incompatibility that I wish to draw on. I will propose that by taking incompatibility to be *primitive* and *worldly*, we can motivate the acceptance of a plurality of individual substances.

The idea of primitive incompatibility is certainly traceable to Aristotle, but I should note that the idea has crept up repeatedly in more recent literature, especially in well-known attempts to define negation and to come up with truthmakers for negative truths.[51] I will not discuss these attempts here, interesting though they are. All we need for the argument at hand is the idea of primitive incompatibility. In Aristotle, the idea of incompatibility is at its clearest in some of his formulations of the law of non-contradiction (LNC). The following is my personal favourite:

> The same attribute cannot at the same time belong and not belong to the same subject in the same respect.[52]

This formulation of LNC focuses on attributes, which make for a particularly intuitive example of primitive incompatibility: it seems plausible that, say, a particle cannot at the same time both be charged and not be charged. Once we have grasped this idea, we can expand on it to the case of mutually exclusive properties. A particle cannot only both have and not have a charge at the same time, it also cannot both have a negative and a positive charge at the same time—these properties are incompatible. In *Categories*, Aristotle writes about 'contrariety' rather than 'incompatibility', but as I noted, it is in fact the latter which will be our focus here. It may nevertheless be helpful to compare the two notions. There is an aspect of contrariety that differs significantly from incompatibility: the former would seem to come in degrees, whereas the latter does not. We might ask: how many contraries can a proposition have? Following the Aristotelian formulation of LNC, it would seem that any incompatible attributes are also contraries. But even if incompatibles are also contraries, contrariety clearly comes in degrees: saying of something black that it is not white is different from saying that it is not grey, even though whiteness and grey-ness are just as incompatible with blackness—we might think that being grey is closer to being black (and closer to being white) than blackness and whiteness are to each other. Or, if colours are confusing here, think of properties like being tall and being short. The idea is clear from Aristotle himself:

> Since things which differ from one another may do so to a greater or a less degree, there exists also a greatest difference, and this I call "contrariety".[53]
>
> It seems that in defining contraries of every kind men have recourse to a spatial metaphor, for they say that those things are contraries which within the same class, are separated by the greatest possible distance.[54]

Here it seems that Aristotle in fact suggests to use 'contrariety' to describe absolute dissimilarity—at least sometimes. But simply the suggestion that there could be a spatial metaphor of contrariety distances the notion from the understanding of primitive incompatibility that I have in mind.

We might illustrate contrariety as follows:[55]

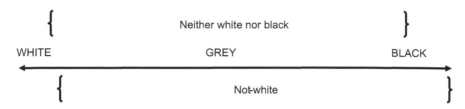

Figure 6.1

Based on *Figure 6.1*, we can define polar contraries (white and black) and immediate contraries (white and not-white), where immediate contraries are incompatible precisely in the sense that we are interested in.[56] If we wish to insist that primitive incompatibility itself does not come in degrees, which is indeed central to our argument, I now suggest to set contrariety aside and focus on incompatibility proper. This happens fairly naturally if we try to adapt the previous illustration to a case like charge:

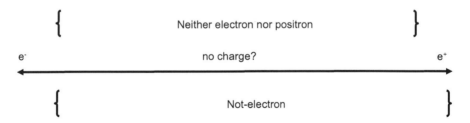

Figure 6.2

In *Figure 6.2*, there are no degrees of contrariety. Electrons have a unit negative charge (the elementary charge), which has an interesting role: all freely existing charged particles have a charge similar in magnitude either to the unit negative charge or an integer multiple of it. So it appears that the route from electron to positron could only involve neutral charge as an intermediate contrast; there are no degrees. Moreover, a neutral particle is a very different kind of thing than a charged particle (it is subject to different laws, for one thing), so it doesn't really perform the same job as 'GREY' does in *Figure 6.1*. This simple case of electron vs. positron gives us an easy example of incompatible properties. Since we are dealing (supposedly) with fundamental particles, this gives us at least a preliminary reason to think that incompatibility could be a fundamental feature of reality.[57] The case is really not that different from charge: macrophysical objects don't tend to manifest charge in the same way that subatomic particles do, even though their existence requires bonding behaviour that is enabled exactly by the charges of their constituent particles. When we study charge, it's exactly the unit negative charge of electrons—also known as the *fundamental unit of charge*—that turns out to be the most interesting. I will not attempt to construct a further argument here in favour of taking incompatibility as a primitive, worldly feature, but let me note that the idea has been (re-)gaining popularity. Francesco Berto's recent account is a good example; he thinks that there's evidence that incompatibility 'carves nature at its joints'.[58] Berto himself focuses on the semantic and logical aspects of the account and is interested in incompatibility as a basis for negation, but for our purposes, it is precisely the idea that incompatibility is a genuine feature of reality that is important. Does this idea survive when we move on to quantum ontology?

5. Substances Disentangled: Incompatibility and the Wave Function

So far, I've been speaking quite loosely about the idea of primitive incompatibility. One might, for instance, question the previous example concerning electrons, since it was assumed that electrons are *particles*—individual substances. But all this was just to get an initial understanding of incompatibility. In order to use the idea of incompatibility in an argument in favour of pluralism, we of course better not assume at the outset that there is a plurality of things. So, let us get back to quantum theory, which is what motivated Schaffer's argument in favour of monism. As we have seen, the key issue here is *how* classicality emerges at the macroscopic level, for the macroscopic world at the very least looks and feels as if it is classical and contains a plurality of objects with objective boundaries. Not that I propose to solve this problem here; we are effectively dealing with one of the central interpretive problems concerning quantum mechanics, what Alyssa Ney calls the *macro-object problem*: 'the problem is to explain how the wave function of a system could ground facts about macroscopic objects'.[59] One thing that is important to note here is that the solution to this problem—or more specifically, the possibility of pluralist substance ontology—does not necessarily depend on which approach to quantum mechanics is adopted. This is because there are versions of all the major approaches that may enable the type of ontology that the neo-Aristotelian seeks. In fact, it seems that we might be able to classify approaches to quantum mechanics that could do the trick on the basis of a single feature: whether or not the approach favours *wave function monism*.[60] This is the view that, fundamentally, a wave function is all there is; this view also assumes wave function realism, a view which we already mentioned in passing above: 'the wave function is a fundamental object and a real, physical field on configuration space'.[61]

However, as I will go on to speculate, even wave function monism may leave room for primitive incompatibility as it was introduced in the previous section.

According to wave function monism, there are no particles, except perhaps as derivative, emergent entities, in the sense that decoherence *may* suggest. But we have seen that there are reasons to think that this is, at best, an illusory sense, and if this is right, then wave function monism would really seem to amount to a type of existence monism. Note that, on the face it, the Bohmian approach, which we mentioned in passing above, will not be compatible with wave function monism. This is because the Bohmian approach does hold that there are, in addition to the wave function, particles with determinate locations in three-dimensional space.[62] But Ney observes that there are also versions of the Everettian and Ghirardi-Rimini-Weber (GRW) approach to quantum mechanics (although we will not discuss them here; GRW is a collapse approach) that are compatible with the denial of wave function monism.[63] However, there are similarly different versions of the Bohmian approach and it may be that only some of them would be compatible with a pluralist view—after all, we saw earlier that there is a clear element of universal 'connectedness' in the Bohm-Hiley version in particular. Regarding the GRW approach (and other collapse theories), David Albert notes that the world will consist of just one physical object, namely, the universal wave function.[64] While I cannot hope to analyse the ontological status of the wave function in detail here—partly because I simply lack the competence to do so and partly because the story will differ depending on the interpretation in question—I do hope to make it clear that the neo-Aristotelian pluralist has not yet been driven to a corner, whereby only some speculative approach to quantum mechanics would save the day.[65] All of the options mentioned here are live.

Let us take a closer look at what the wave function might tell us about incompatibility. Specifically, what does the wave function tell us about the case of seemingly distinct, objective 'particles' that at least look as if they could motivate pluralism about individual substances? Consider a two-particle wave function for a system of identical, indistinguishable particles. Quantum theory tells us that the probability density of the two particle wave function must be identical to the wave function in a situation where the particles have been interchanged. There are two ways that this can happen: the symmetric and the anti-symmetric case. In symbols, the symmetric case is $\psi(r_1, r_2) = \psi(r_2, r_1)$, and the anti-symmetric case is $\psi(r_1, r_2) = -\psi(r_2, r_1)$. Details of wave-function symmetry aside, it turns out that particles with symmetric wave functions in this scenario have integer or zero intrinsic spin—they are known as bosons—whereas particles with anti-symmetric wave functions have half-integer intrinsic spin—they are known as fermions. So, here we have what seems like a relatively simple but quite general case regarding the incompatibility of two types of fundamental particle, fermions and bosons. If this is correct, then it looks like the fermion vs. boson distinction is one of the best candidates for a distinction that 'carves nature at its joints'. If such fundamental incompatibilities do exist, then they may be seen as tracking worldly incompatibility in Berto's sense.[66] In other words, if incompatibility itself is fundamental, then fundamental incompatibilities like the fermion vs. boson distinction could be seen as grounded in this worldly incompatibility.

In the fermion vs. boson case, the incompatibility concerns the representations of the wave function, so one might still doubt whether we have arrived at genuinely incompatible properties that must be possessed by two distinct 'objects' (if we consider the wave function in the lines of wave function monism). So the problem is that those who favour wave function monism in particular might regard this story to be inherently misleading: we do not really have closed systems like the two-particle system described in this example. All we have is the universal wave function and fermions and bosons are just aspects of the universal entangled web, loosely speaking. Still, the incompatibility featured in this story about fermions and bosons must be rooted in something, and since it can be modelled with the wave function, even the wave function monist should be able to appreciate this. Forget about 'particles', what I wish to suggest is that they can be understood as a placeholder for features of the wave function that are ultimately responsible for the possibility of macro-objects. But insofar as one thing cannot possess two incompatible properties at the same time, it seems as if the wave function monist must either insist that this incompatibility itself is only an illusion, admit that there are at least two distinct things, or try to reduce the distinctness into something else. If the friend of substances is right, then we would also have the beginnings of a potential solution to the *macro-object problem*. On this approach, what grounds facts about macro-objects (and we can here think of any appearance of a 'particle' to be a macro-object) is the incompatibility that can be traced to the wave function itself.[67]

It is worth noting that the speculative approach being developed here assumes a form of wave function realism. I don't wish to make any commitments in this regard myself, but since the universal role of the wave function understood in a realist fashion is typically adopted in arguments that are causing trouble for neo-Aristotelian substance ontology, it seems that for the sake of argument it would be best to adopt this approach and see if we can deal with it. This type of realism about the wave function could be considered to suggest that the wave function is at least 'quasi-material', i.e., something close to what we regard as ordinary material reality, even if not quite the same thing.[68] As Goldstein and Zanghì note, it turns out to be quite tricky to determine the ontological status of the wave function—we would first need to know what quantum theory says. I cannot hope to make much progress with this issue here. Instead, I've focused on tracing the idea of incompatibility all the way down to the wave function, since I posited incompatibility as a necessary condition for pluralism about substances. But even if we regard the wave function as 'real'—presumably, as a field[69]—part of the problem that we now face is how to express incompatibility in terms of the 'parts' of one universal field. It might be suggested that this simply means that if *this* part of the wave function has a certain peak, then *that* part of the wave function cannot have a certain peak. Simplistic as this description is, it may be the only way that we can 'solve' the macro-object problem. Note that this would seem to apply even in the case of wave function monism. Ney herself suggests something quite similar, although without explicitly referring to incompatibility:

> Somehow the wave function grounds the existence of, for example, my desk. This is not because my desk is spread out all over the configuration space with parts corresponding here to a leg and there to a top. Rather, the persistence of peaks of the wave function grounds the existence of my desk. Point-sized regions of these peaks correspond to slightly different (classical) ways of there being a desk there, slightly different configurations of particles that could make up a desk, among other things. If instead, the peaks of the wave function were centered in a disjoint region of configuration space, then my desk would not have existed, or there would have been a very different kind of desk.[70]

This is certainly a controversial story about how we might reduce ordinary three-dimensional objects to a wave function in configuration space, but it seems clear that in order to address the macro-object problem, something on these lines is a promising way to go. By 'something on these lines', I simply mean the requirement that there are persistent peaks in one region of the wave function that will rule out certain other peaks, just like Ney describes. Consider another passage, from Jill North:

> For example, there being a table in three-space consists in nothing but the wave function's having a certain shape in its high-dimensional space. It's *true* that there is a table in three-space; it's just that this holds *in virtue of* some other, more fundamental facts. The truth about three-space (the grounded) is not a further fact beyond the truth about the wave function's space (the grounds)—that is, it isn't a fundamental fact— even though it is distinct from the grounds and is itself a real fact.[71]

This passage may remind someone of the earlier reconstrual of Schaffer's strategy to get a plurality of entities via the nonfundamental parts, while the cosmos as a whole is fundamental. We have seen that this strategy is problematic. Ney and North both appear to be in the business of explaining how we might get our beloved 'things' (why is it always tables?) back as nonfundamental entities that are grounded in or reduce to the universal wave function in $3N$-dimensional space. On any such view, one will have to postulate some *structure* that corresponds with the classical ways of being. North suggests that this structure could be specified, e.g., by the dynamical laws.[72] In Ney's story, the relevant structure seems to come from the persistence of the peaks of the wave function. If some story of this type turns out to be correct, then my thesis about primitive incompatibility is corroborated: you don't get even emergent, nonfundamental 'things' without fundamental incompatibility, because something like this must always be built-in to the relevant structure. Otherwise there could be no correspondence to the classical ways of being that Ney alludes to. I wish I could say more about what that fundamental, primitive incompatibility amounts to, but the problem here is that, being primitive, we might not be able to specify it very much beyond the type of stories that I've quoted above. Instead, we might attempt a *reductio ad absurdum*: consider what the world would be like if there were no incompatibility. I think that the best illustration of this might be something like Dummett's 'amorphous lump', a view according to which the world is structureless 'dough'.[73] We do not need to dwell on the anti-realist aspects of this idea—the point is simply that any kind of structuring will require some limiting principles, be it the dynamical laws or something else, and this immediately introduces an element of incompatibility. If this is right, then on any view where there is an underlying structure, there is also incompatibility. Presumably, Schaffer himself would not deny this, but the question is: who can give the most plausible explanation of this incompatibility?

Do we get the typical neo-Aristotelian conception of substance out of all this? Probably not. But nor do we get any other very clear ontological picture: the jury is still out on quantum ontology. Of course, we do know some things. For instance, it seems undeniable that on all the major approaches, some kind of holism will be present in quantum ontology. However, this does not necessarily entail monism, it only entails that there may be more dependence in the world than we once thought. To be sure, pluralism will need to be qualified at the end of the day, but if we can trace the source of incompatibility to quantum ontology, then there is at least some hope to settle the *macro-object* problem in a manner that preserves the key aspects of neo-Aristotelian pluralist substance ontology.[74]

Notes

1 For a general overview on substance, see Hoffman and Rosenkrantz (1994, 1997). I will not attempt to make any claims about Aristotle's own views regarding substances here, but for discussion linking the historical notion of substance and the neo-Aristotelian one, see Hoffman (2012); see also other articles in Tahko (2012a) for various views on categories. I should also add a caveat: although this article defends the possibility of maintaining the category of substance, I am myself thoroughly fallibilist about this issue. Should science conclusively show that substance is not a feasible category, I would be prepared to give it up. Hence, my own approach to the neo-Aristotelian tradition is very much methodological rather than substantial (see, e.g., Tahko (2013) for further discussion).
2 Hoffman and Rosenkrantz (1997: 1).
3 Ladyman and Ross et al. (2007).
4 Schaffer (2009).

5 E.g., Schaffer (2010a, 2010b).
6 Schaffer (2010a: 32), Hoffman and Rosenkrantz (1997: 77).
7 Hoffman and Rosenkrantz (1997: 78).
8 See Schaffer (2009).
9 Schaffer (2009: 131).
10 Schaffer (2010a: 49).
11 Schaffer puts forward an argument from common sense, which is supposed to establish that the many proper parts of the cosmos are arbitrary portions of the cosmos and hence that the cosmos is prior to its many proper parts. For a discussion and an attempted refutation of this argument, see O'Conaill and Tahko (2012); see also other articles in Goff (ed.) (2012) for further discussion on various versions of monism and their history.
12 Schaffer (2009: 133).
13 At least in one sense or another; see Tahko and Lowe (2015) for discussion.
14 Hoffman and Rosenkrantz (1994: 35).
15 Schaffer (2010a: 54).
16 Schaffer (2009: 143).
17 Schaffer (2010a: 48); compare with Tahko (2012b).
18 Thanks to the editors of this volume and especially Nic Teh for these observations.
19 Thanks to Alastair Wilson for pointing this out.
20 Zeh (2003).
21 When two systems become *entangled* via some physical interaction and then separate after some time, they cannot be described in the same way as before (Schrödinger 1935). Or as Schaffer (2010a: 51) puts it: 'An entangled system is one whose state vector is not factorizable into tensor products of the state vectors of its n components'. Importantly, entanglement is the cause of the seemingly unavoidable nonlocality, 'spooky action at a distance', which Einstein found unpalatable and which Bell's famous theorem demonstrated. See Bub (2015) for further details. Note that the argument at hand does not concern QFT as such—any approach to quantum theory must engage with the phenomenon of entanglement one way or another.
22 Based on Schaffer (2010a: 50–7); see premises 6–8.
23 Schaffer (2010a: 52).
24 Ibid.
25 Note also that we should distinguish between the Copenhagen interpretation -style collapse theories and objective collapse theories such as the GRW approach.
26 Schaffer (2010a: 53). Schaffer (2010a: 55) goes on to offer an argument from the possibility of emergence in favour of monism, but as we can see, that would not really be a distinct argument, as some sort of emergence is implicitly assumed already in the argument at hand.
27 Ismael and Schaffer (Forthcoming: §2.1.3).
28 For an in-depth introduction to decoherence, see Bacciagaluppi (2016). For an accessible account of the philosophical implications of decoherence, see Crull (2013). See also Wallace (2012) for an extensive discussion of decoherence and the emergence of 'macro-objects'. As Crull (2013: 879) notes sometimes 'classicality' is just understood as 'lack of interference'.
29 Thanks to Christina Conroy for pressing on this issue.
30 Ney (2013a: 37).
31 Ismael and Schaffer (Forthcoming: 1).
32 Ismael and Schaffer (Forthcoming: 12ff).
33 Ismael and Schaffer (Forthcoming: 17).
34 Schaffer (2010a: 57).
35 E.g., Schaffer (2010a: 53n31), Ismael and Schaffer (Forthcoming: 21).

36 Ismael and Schaffer (Forthcoming: 21).
37 Healey (2016: §9). See, also Bohm and Hiley (1993).
38 Bacciagaluppi (2016: §3.2.1).
39 Bacciagaluppi (2016: §3.2.1). For further discussion, see Pylkkänen et al. (2015).
40 Fields (2014a).
41 Schaffer (2010a: 55). Also quoted in Fields (2014a: 143).
42 Fields (2014a). See also Dorato (2016), where Schaffer's priority monism is put to test with reference to Rovelli's relational approach to quantum mechanics (RQM)—Dorato's conclusion is that RQM does not support monism.
43 Fields (2014a: 144); see also Fields (2014b).
44 Fields (2014a): 137.
45 Zurek (2003, 2009).
46 Schaffer (2010a: 32, 52).
47 Fields (2014a: 144).
48 Schaffer (2010a: 32).
49 Ibid.
50 Aristotle, *Categories* 4a10–21.
51 See, e.g., Demos (1917), Price (1990), Molnar (2000), and Berto (2015).
52 Aristotle, *Metaphysics* 1005b19–20.
53 Aristotle, *Metaphysics* 1055a4–6.
54 Aristotle, *Categories* 6a15–19.
55 Adapted from Horn (2001: 38).
56 On 'polarity', see also Beall (2000).
57 One might argue that this is just a nomic rather than a metaphysical feature of reality, in which case, incompatibility might not be fundamental. But even if this is the case, those who consider (some) laws of nature to be metaphysically necessary could arrive at the desired result. I will not discuss the status of the laws of nature here, but see Tahko (2015) for discussion. Thanks to Alastair Wilson for pointing out this issue.
58 Berto (2015: 11); see also Berto (2008).
59 Ney (2013a: 25–6).
60 See Maudlin (2010).
61 Ney (2013a: 37).
62 Ney (2013a: 38).
63 Ibid.: 42.
64 Albert (2013): 54.
65 The debate concerning the ontological status of the wave function is heated and interesting; see the essays in Ney and Albert (eds.) (2013) for a representative selection.
66 Berto (2015: 11).
67 As Nic Teh pointed out to me, there is an argument to be constructed here also on the basis of Pauli Exclusion Principle (PEP), which states that no two fermions in a closed system can have all the same quantum numbers at the same time: the PEP can be seen as a necessary condition for the stability of matter. From this we can infer that the fact that matter is extended is itself at least partially based on primitive incompatibility. However, I will not develop this argument in more detail here, as I have discussed very similar arguments based on PEP elsewhere, e.g., in Tahko (2009) and (2012b).
68 See Goldstein and Zanghì (2013) for a helpful survey.
69 See Ney (2013b: 169); 'The wave function is a field in the sense that it is spread out completely over the space it inhabits, possessing values, amplitudes in particular, at each point in this space'. Another problem here concerns three-dimensional space: we tend to think of material objects as existing in three-dimensional (or four-dimensional) space, but wave function realism would seem

to suggest that three-dimensional space is just an illusion, as we need to move to 3*N*-dimensional configuration space (Ney 2013b: 177ff.). For further discussion, see North (2013); North argues that although three-dimensional space is not fundamental, it can be seen to exist nonfundamentally rather than being merely an illusion (ibid.: 197ff.). So, there are some subtleties relating to high-dimensional quantum space, which is typically described as a configuration space. Strictly speaking, even this is not quite right, as configuration space is typically used to describe the coordinates of particles. I will leave these complications aside, as the idea of incompatibility does not depend on the number of dimensions. Similarly, I will leave aside complications relating to relativistic quantum mechanics, as do the authors of many of the articles that I have been discussing.

70 Ney (2013b: 180).
71 North (2013: 198).
72 North (2013: 200).
73 See Dummett (1981: 577).
74 I'd like to thank the editors of this volume, Christina Conroy, and Alastair Wilson for comments on previous drafts of this article. The research for this article was supported by two Academy of Finland grants (No. 266256 and No. 274715).

References

Albert, D. Z. (2013) 'Wave Function Realism', in A. Ney and D. Z. Albert (eds.) *The Wave Function: Essays on the Metaphysics of Quantum Mechanics* (Oxford: Oxford University Press): 52–7.

Aristotle. (1963) *Categories and De Interpretatione*, trans. with notes by J. L. Ackrill (Oxford: Clarendon Press).

——— (1984) *Metaphysics*, trans. by W. D. Ross, revised by J. Barnes (Princeton, NJ: Princeton University Press).

Bacciagaluppi, G. (2016) 'The Role of Decoherence in Quantum Mechanics', in E. N. Zalta (ed.) *The Stanford Encyclopedia of Philosophy*, Fall 2016 edition, at http://plato.stanford.edu/archives/fall2016/entries/qm-decoherence/.

Beall, J. C. (2000) 'On Truthmakers for Negative Truths', *Australasian Journal of Philosophy* 78: 264–8.

Berto, F. (2008) 'Adynaton and Material Exclusion', *Australasian Journal of Philosophy* 86: 175–90.

——— (2015) 'A Modality Called "Negation"', *Mind* 124(495): 761–93.

Bohm, D. and Hiley, B. J. (1993) *The Undivided Universe* (New York: Routledge).

Bub, J. (2015) 'Quantum Entanglement and Information', in E. N. Zalta (ed.) *The Stanford Encyclopedia of Philosophy*, Summer 2015 edition, at http://plato.stanford.edu/archives/sum2015/entries/qt-entangle/.

Crull, E. (2013) 'Exploring Philosophical Implications of Quantum Decoherence', *Philosophy Compass* 8/9: 875–85.

Demos, R. (1917) 'A Discussion of a Certain Type of Negative Proposition', *Mind* 26(102): 188–96.

Dorato, M. (2016) 'Rovelli's Relational Quantum Mechanics, Anti-Monism, and Quantum Becoming', in A. Marmodoro and D. Yates (eds.) *The Metaphysics of Relations* (Oxford: Oxford University Press): 235–61.

Dummett, M. (1981) *Frege. Philosophy of Language*, 2nd edition (Cambridge, MA: Harvard University Press).

Fields, C. (2014a) 'A Physics-Based Metaphysics Is a Metaphysics-Based Metaphysics', *Acta Analytica* 29: 131–48.

———— (2014b) 'Consistent Quantum Mechanics Admits No Mereotopology', *Axiomathes* 24: 9–18.

Goff, P. (ed.). (2012) *Spinoza on Monism* (New York: Palgrave Macmillan).

Goldstein, S. and Zanghì, N. (2013) 'Reality and the Role of the Wave Function in Quantum Theory', in A. Ney and D. Z. Albert (eds.) *The Wave Function: Essays on the Metaphysics of Quantum Mechanics* (Oxford: Oxford University Press): 91–109.

Healey, R. (2016) 'Holism and Nonseparability in Physics', in E. N. Zalta (ed.) *The Stanford Encyclopedia of Philosophy*, Spring 2016 edition, at http://plato.stanford.edu/archives/spr2016/entries/physics-holism/.

Hoffman, J. (2012) 'Neo-Aristotelianism and Substance', in T. E. Tahko (ed.) *Contemporary Aristotelian Metaphysics* (Cambridge: Cambridge University Press): 140–55.

Hoffman, J. and Rosenkrantz, G. (1994) *Substance Among Other Categories* (Cambridge: Cambridge University Press).

———— (1997) *Substance: Its Nature and Existence* (London: Routledge).

Horn, L. (2001) *A Natural History of Negation* (Stanford, CA: CSLI Publications).

Ismael, J. and J. Schaffer, "Quantum Holism: nonseparability as common ground," *Synthese* (Sept. 9, 2016), Doi 10.1007/s11229-016-1201-2 <https://doi.org/10.1007/s11229-016-1201-2>

Ladyman, J., Ross, D., Spurrett, D. and Collier, J. (2007) *Every Thing Must Go: Metaphysics Naturalized* (Oxford: Oxford University Press).

Maudlin, T. (2010) 'Can the World Be Only Wave Function?', in S. Saunders, J. Barrett, A. Kent and D. Wallace (eds.) *Many Worlds?: Everett, Quantum Theory, and Reality* (Oxford: Oxford University Press): 121–43.

Molnar, G. (2000) 'Truthmakers for Negative Truths', *Australasian Journal of Philosophy* 78: 72–86.

Ney, A. (2013a) 'Introduction', in A. Ney and D. Z. Albert (eds.) *The Wave Function: Essays on the Metaphysics of Quantum Mechanics* (Oxford: Oxford University Press): 1–51.

———— (2013b) 'Ontological Reduction and the Wave Function Ontology', in A. Ney and D. Z. Albert (eds.) *The Wave Function: Essays on the Metaphysics of Quantum Mechanics* (Oxford: Oxford University Press): 168–83.

Ney, A. and Albert, D. Z. (eds.). (2013) *The Wave Function: Essays on the Metaphysics of Quantum Mechanics* (Oxford: Oxford University Press).

North, J. (2013) 'The Structure of a Quantum World', in A. Ney and D. Z. Albert (eds.) *The Wave Function: Essays on the Metaphysics of Quantum Mechanics* (Oxford: Oxford University Press): 184–202.

O'Conaill, D. and Tahko, T. E. (2012) 'On the Common Sense Argument for Monism', in P. Goff (ed.) *Spinoza on Monism* (New York: Palgrave Macmillan): 149–66.

Price, H. (1990) 'Why "Not"?', *Mind* 99(394): 221–38.

Pylkkänen, P., Hiley, B. J. and Pättiniemi, I. (2015) 'Bohm's Approach and Individuality', in A. Guay and T. Pradeu (eds.) *Individuals Across the Sciences* (Oxford: Oxford University Press): 226–49.

Schaffer, J. (2009) 'Spacetime the One Substance', *Philosophical Studies* 145: 131–48.

———— (2010a) 'Monism: The Priority of the Whole', *Philosophical Review* 119: 31–76.

———— (2010b) 'The Internal Relatedness of All Things', *Mind* 119: 341–76.

Schrödinger, E. (1935) 'Discussion of Probability Relations Between Separated Systems', *Proceedings of the Cambridge Philosophical Society* 31: 555–63.

Tahko, T. E. (2009) 'The Law of Non-Contradiction as a Metaphysical Principle', *The Australasian Journal of Logic* 7: 32–47.

—— (ed.). (2012a) *Contemporary Aristotelian Metaphysics* (Cambridge: Cambridge University Press).

—— (2012b) 'Boundaries in Reality', *Ratio* 25: 405–24.

—— (2013) 'Metaphysics as the First Philosophy', in E. Feser (ed.) *Aristotle on Method and Metaphysics* (New York: Palgrave Macmillan): 49–67.

—— (2015) 'The Modal Status of Laws: In Defence of a Hybrid View', *The Philosophical Quarterly* 65(260): 509–28.

Tahko, T. E. and Lowe, E. J. (2015) 'Ontological Dependence', in E. N. Zalta (ed.) *The Stanford Encyclopedia of Philosophy*, Spring 2015 edition, at http://plato. stanford.edu/archives/spr2015/entries/dependence-ontological/.

Wallace, D. (2012) *The Emergent Multiverse: Quantum Theory According to the Everett Interpretation* (Oxford: Oxford University Press).

Zeh, H. D. (2003) 'There Is No First Quantization', *Physics Letters A* 309: 329–34.

Zurek, W. H. (2003) 'Decoherence, Einselection, and the Quantum Origins of the Classical', *Reviews of Modern Physics* 75: 715–75.

—— (2009) 'Quantum Darwinism', *Nature Physics* 5: 181–8.

Part 2

The Philosophy of the Life Sciences

7 Structural Powers and the Homeodynamic Unity of Organisms

Christopher J. Austin and
Anna Marmodoro

1. Introduction

Each biological denizen that populates our humble neighbourhood of the cosmos is a veritable world unto itself whose complex construction autonomously navigates the development and maintenance of its own intricate machinery. And although composed of an uncountable number of constituents, each of these multi-layered microcosms is a fundamentally unified being—each is in some way *one*, rather than *many*. But in virtue of what, metaphysically, are organisms more than merely bundles of biological bits whose diachronically disparate collections are continually washed away in a Hericlitean flux? In other words, what secures, metaphysically, an organism's continued persistence *as one* over time?

In this chapter, we present an account of organismal unity centred on a neo-Aristotelian conception of causal powers, introducing a novel type of power—a 'structural' power. In examining the empirical data of contemporary developmental biology, we make a transcendental case to show that structural powers are no mere philosophers' fancy, but rather are an integral ontological feature of unified, living beings—hypothesising their existence fills what would otherwise remain a glaring explanatory gap. The unique teleological nature of these powers, we argue, enables them to function as the proper metaphysical ground of the unity of organisms. According to our account, that unity is displayed not in the capacity of an organism to sustain the diachronic *stasis* of its morphological features (and their constituents), but in the persistence of its specified capacity for the *dynamically* adaptive re-organisation of those features.

2. To Be One: A Dynamic Disposition

If the mereological complexities involved in determining whether and under what conditions artefacts possess sufficient unity to be admitted into our ontology are overwhelming, the question of in virtue of what biological organisms enjoy that metaphysical status is even more so. The composition of something as simple as a common fruit fly, for instance, consists in an intricate hierarchy of causal-*cum*-functional dependence that holds among a multitude of anatomical and eidonomical levels of organisation. And what is perhaps more puzzling is the fact that whatever complex relation or set of relations is ultimately responsible for grounding the synchronic unity of these compositional elements—that is, at any particular point in time during an organism's existence—is plausibly incapable of performing that role diachronically, and for an important reason: those constituents, and, thus, whatever their relations' putative contribution in establishing the unity of an organism may be, are largely temporally transient. For living entities, characteristically, are loci of constant material exchange, their bodies never remaining strictly mereologically identical even over the smallest time-scales.[1]

Thus, the problem of accounting for the *diachronic* unity of organisms is one importantly distinct from that of accounting for their *synchronic* unity. The unity which organisms possess *at any time* might be grounded in the obtaining of a specialised spatial-*cum*-causal relation (or set of relations) which holds among the members of its constitutive mereological makeup or, perhaps in a more sophisticated account, by those members' unified achievement of a particular functional organisation. But the problem of accounting for the unity of an organism *over time* concerns the significantly more complex issue of determining what the unity *of those synchronically unified temporal stages* consists in. To our minds, the hurdle at which existing accounts of the unity of organisms fall is that they fail to recognise that there is an important distinction between the unity which materially constituted mereological sums—elaborately and complexly structured as they may be—possess and that which belongs to organisms: the diachronic unity of the former, but not the latter, consists in nothing more than the temporally successive invariance of whatever it is that grounds its synchronic unity— its mereological members retaining the same spatial-*cum*-causal structure, or functionally oriented organisation, etc. Organisms, on the other hand, diachronically persist *in spite of* the alteration over time of the structural organisation, or functional orientation of their mereological make up: whatever it is that grounds their being *one at a time* therefore cannot also ground their being *one over time*.

We know, however, that certain sets of temporally successive synchronically unified mereological collections *do* make up *single* organisms—but in virtue of what? In order to provide an answer here, a natural preliminary question is: on what are our judgements of the diachronic unity of organisms based? *Prima facie*, we make those judgements because although the compositional elements of organisms may be transient, their morphological profiles—that is, their general anatomical and eidonomical structures—are surprisingly diachronically stable. Plausibly, then, the ontologically privileged status which we recognise that organisms enjoy is one conferred upon them primarily in virtue of their being invariantly disposed to continually maintain the causal production of their particularised morphology. It is in this respect, then, that organisms remain *one* even in the face of *multiplicity*: they exhibit a singular, dynamically coordinative impetus which shapes the compositional character of a multitude of distinct, synchronically unified mereological collections over time. Thus, in the case of organisms, we propose that the diachronic unity which a set of those collections possesses consists in their members being subsumed within the operation of a complex teleological structure which we call a 'structural power'.[2] In the account we will offer, the unity of an organism consists in the persistence of the activity of its structural power which, in spite of the continual flux of its mereological makeup and so, the continual alteration of its synchronically unified composition, consistently causally conforms the shape and structure of its morphology.

More specifically, the function this power performs consists in continually organising the constituents of an organism (both anatomically and eidonomically) in a specific structural arrangement to produce and maintain the complete morphological profile particular to the natural kind to which it belongs.[3] Structural powers, to put it another way, are causally responsible for the developmental specifications of an organism's parts (which ones get produced, and by what means), and the structural complexities of their arrangement (the three-dimensional architecture of their modular components, and their spatio-functional relational hierarchy). Importantly however, although the particularised organisational complexity of an organism's morphology *at any time* certainly *results* from the causal operation of a structural power, it would be a mistake to identify any specific structural relation (or complex set thereof) among an organisms constituents with the manifestation of that power.[4] Instead, we propose that the manifestation of a structural power consists in a kind of *self-directed activity*: the goal toward which they are teleologically directed *just is* the performance of their unification role in the structural organisation of an organism's elemental constituents over time, the natural *result* of which is those constituents (synchronically) composing a particular morphological profile. Structural powers are therefore intrinsically dynamic—like any dispositional property, they are defined by what they do, but unlike most causal powers which are individuated by their role in the production of some further state, what they do *is doing*.[5]

But what exactly do structural powers *do* in order to function as the ground of the diachronic unity of organisms, and in what sense is that function *self-directed*?[6] We propose that the self-directed unifying role of structural powers consists in a kind of specialised *cyclopoietic* (literally "cyclically-productive") activity:[7] in traversing a kind of causal loop among the constituents of an organism, each is tied together in a continual diachronic cycle of co-production and maintenance.[8] According to our account, the manifestation of a structural power is the causal integration among an organism's constituents wherein each contributes to the generation and proper functioning of one another in the service of their cooperative construction of a particular organismal morphology. The self-directedness of structural powers therefore consists in the *recursive* nature of this unifying activity—they are "self-oriented" precisely because the goal of that activity is to establish the cyclical perpetuation of its own operation.[9] This operationally dynamic conception of organismal unity, now enshrined in the tenets of 'systems biology',[10] has long been recognised as a mark of the unique ontological status which organisms *qua* living beings possess.[11] In giving a *metaphysical* analysis of organismal unity, we propose then that the ontological boundaries of the causally iterative processes of generation, regeneration, and auto-regulation which sculpt the joints of the denizens of the natural world are themselves carved out by the dynamic activity of structural powers.

Nowhere is the teleological dynamism of structural powers more prevalent than in the homeostatic phenomenon of generative robustness, where the constituents of an organism are diachronically redirected toward the reproduction of a particular morphological structure in response to perturbation. As a homeostatic phenomenon, the processes which characterise generative robustness are both *persistent*—able to maintain the causal production of an end-state by means of compensatory changes within the system—and *pleonastic*—able to bring about an end-state *via* a number of alternative pathways.[12] 'Generative robustness', a phenomenon acknowledged to be both nearly ubiquitous in the biological realm[13] and a *sine quo non* of the evolutionary process,[14] encompasses two closely related types of homeostatic processes: *redundancy* and *degeneracy*.[15] In the event that the causal architecture of a biological system malfunctions due to the (ontological or functional) uncoupling of one of its constitutive members from the others, its proper functioning in producing a particular end-state can be restored in virtue of its possessing isomorphic 'redundant' elements which take-up the slack of the missing/disabled ones.[16] More interesting perhaps are cases of degeneracy,[17] where homeostasis is achieved without the aid of duplicate elements, but by biological systems "re-wiring" their causal-*cum*-regulatory architecture in such a way that its non-isomorphic elements become isofunctional (with respect to the "missing" element), thereby causally mirroring the required role within the perturbed network.[18]

As we see it, properly accounting for the phenomenon of generative robustness—one wherein while the constituents of an organism (and thus the character of its synchronic unity) may undergo significant variation over various timescales, the causal orientation of its organisation toward the production and maintenance of a particular morphology does not—is a job which requires the resources of our metaphysical toolbox: the continual binding together of the compositional elements of an organism toward a particular anatomical and eidonomical organisation over time is an expression of the causal architecture established by the cyclopoietic activity of structural powers. It is our claim that it is the persistence of this goal-directed activity which metaphysically grounds the diachronic identity of an organism: it is in virtue of the stability of its specialised dynamic operation that an organism remains *one* throughout the continual flux of its mereological makeup.

Importantly, the phenomenon of generative robustness suggests both what the metaphysics of the diachronic unity of organisms *cannot* and *must* consist in. With respect to the former, given that organisms *persist*, and persist *as one* throughout the exercise of such phenomena, its prevalence suggests that any metaphysical account of the diachronic unity of organisms founded on an extension of the grounds of their synchronic unity—that is, on either the persistence of a specific set of their compositional elements or some particular structural arrangement thereof—must ultimately be untenable. With respect to the latter, and given the former, the phenomenon suggests that the ontological ground of the homeostatic capacities of organisms must be *extra-compositional*—that is, grounded in neither the compositional elements of an organism, nor in any of their functionally unified synchronic configurations.[19] In light of this, and because the nature of structural powers is principally *programmatic*—in that their activity provides a directive structure which specifies the temporal situation-sensitive succession of an organism's synchronic states—we propose that we ought to conceive of them as being *encoded* within organisms: present, though irreducible to any physical element, or the functional coordination of a set of such elements. We picture the structural power of an organism as its *symphonic software*: its conductorial activity ensures that each collection of notes (compositional elements) which are harmonised in each measure (synchronically unified, both causally and functionally) flow together over time to compose a coherent set of thematic movements (morphological features).[20] While we acknowledge the novelty of this conception of the ontology of causal powers, we maintain that providing the proper ground of the homeostatic phenomena exhibited by organisms and subsequently accounting for the diachronic unity of organisms *via* an appeal to the teleological directedness of their morphology requires it.[21]

3. Morphological Variability and the Nature of Structural Powers

According to our account, the diachronic unity of an organism is secured by the persistence of the active functioning of a causal power—a 'structural power'—which is dynamically directed toward the continual productive organisation of its mereological constituents according to a particular morphological profile. We have suggested that the causal consequences of this power's activity within an organism are exhibited most clearly in the phenomenon of generative robustness, where the intransience of the tendency of an organism's mereological makeup to be reconfigured on the occasion of perturbation toward the restoration of that profile reflects the persistence of that power, and thus, of that organism as a unified being.

In our account, then, an organism's remaining *one* over time is intimately linked with the diachronic *stability* of its morphological profile: the activity of one and the same structural power over time is evidenced by the invariance of a teleological directedness toward the *same* morphology. In other words, in our account, the diachronic unity of an organism is displayed in (though not strictly identified with) the unchanging specificities of its homeostatic activity. However, here we must confront an important complication for, according to contemporary research in developmental biology, there is a significant sense in which the particularities of an organism's capacity for robust re-organisation are *themselves* capable of changing over time. This is perhaps most strikingly illustrated in the well-studied phenomenon of 'phenotypic plasticity', exhibited in the ability of organisms to adopt a wide range of morphological variability in response to intra-/ inter-/extra-cellular "environmental" signalling, or causal influences.[22] Such morphological variability ranges from being relatively minor, as when, for instance, butterflies adopt distinct wing patterning in response to seasonal weather signals,[23] to rather extreme, as displayed in water fleas' (*Daphnia*) development of large, helmet-like spikes in response to receiving chemical signals from nearby predators,[24] or in the caste system of ants, where hormonal signals distributed among a population produce radical changes in the morphology of its members, creating everything from winged and wispy drones to the thickly carpaced, reproductively charged queens.[25]

Given that the phenomenon of phenotypic plasticity is no conceptual outlier, but is instead acknowledged to be not only ubiquitous in the biological realm, but central to our understanding of both developmental and evolutionary processes, we must treat its theoretical consequences with ontological sincerity.[26] With respect to this discussion, one in particular stands out: organisms don't have a *single* and *unchanging* homeostatically maintained morphological profile but are instead equipped with an entire 'morphospace' consisting of *multiple* such profiles which may be adopted and robustly maintained at various points in their lives in response to extrinsic stimuli.[27] For an account of the diachronic unity of organisms like ours, in which *oneness* is intimately correlated with the persistence of a particularised impetus toward morphological *stability*, the reality of morphospaces poses a potential problem—for if the homeostatic phenomena displayed in the directedness of an organism's constituents toward the productive maintenance of a particular morphology is transient, then it seems so too must be the existence of the structural power whose cyclopoietic activity is causally responsible for that phenomena. And if that were the case, structural powers would be no more temporally *stable* than the constituents of an organism, and hence, no more able to ground the diachronic unity of an organism than they are.

Of course, this is only a problem for our account if a shift in an organism's exhibited morphology, and thus a corresponding alteration of its homeostatic maintenance toward a particular morphology, amounts to the corruption or loss of its structural power. Is this a plausible inference? We think not. While we certainly don't wish to deny that there are cases where this sort of alteration *is* a consequence of the loss of a diachronic structural power, we recommend caution in making that judgement—for not *every* instance of this sort of alteration is such a case. What, then, are the criteria to be used in properly discerning these cases? True to our Aristotelian roots, we suggest that temporal variation in an organism's morphological profile and the homeostatic maintenance thereof amounts to the change in, or loss of its structural power just in case that variation is not "of the nature" of that power.[28] Discerning what's "of the nature" of a power is no esoteric enterprise—it merely requires an appeal to the same standard we utilise in judgements of this sort in the empirical sciences: namely, non-random regularity. The thought here is simple: if, for any particular case, with respect to a particular structural power, the morphological divergence in question can be shown to be a non-random, repeatable occurrence, we have *pro tanto* justification for the judgement that such divergence *isn't* a genuine deviation from the "nature" of that power. And in phenomena of phenotypic plasticity, we can show precisely that.

In studying that phenomenon, contemporary research in developmental biology is actively engaged in discovering not only the scope and breadth of the plasticity of organismal morphology, but also the detailed mapping of the causal conditions under which that plasticity occurs within the morphologies of particular organisms. These mappings are known as 'norms of reaction', (or 'reaction norms'): drawing on a wealth of empirical data, these rather precise graphs detail the connections repeatedly and regularly observed between specific qualitative variations in an organism's morphological profile and various specific intra-/inter-/extra-cellular environmental conditions which that organism may be subject to.[29] That one can experimentally investigate and quantifiably catalogue in a rigorous fashion the 'plastic potential' of an organism's morphology, we suggest, lends credence to the notion that a sizable set of the "atypical deviations" from the characteristic morphological profiles associated with organisms are, strictly speaking, neither 'atypical', nor 'deviations'.[30] Rather, in the parlance of our account, they ought to be understood as the non-random and regularly repeatable exhibitions "of the nature" of their structural powers.

4. Stability Redux: Homeostasis vs. Homeodynamism

One might reasonably ask why, as we have just suggested, one should hold that it is "of the nature" of structural powers to diachronically organise the constituents of an organism toward the generation and continual maintenance of not only a single morphological profile, but of a wider, more fine-grained set of qualitatively diverse profiles. Given the diversity inherent in plastic phenomena, would it not be just as coherent to suppose, for instance, that the lesson we ought to learn from the repeatability and regularity of these morphological variations is that a single organism reliably possesses *multiple* distinct structural powers at distinct times, and in distinct causal contexts? Of course, adopting such a position would call into serious question our account of the diachronic unity of organisms, wherein the temporal persistence of a single structural power functions as its ontological ground. However, we think that would be the wrong lesson to draw—and showing why that is so will allow us to further elucidate a central aspect of our account.

The reason we think one ought to hold that it is "of the nature" of a *single* structural power to developmentally direct the synchronic states of an organism toward *multiple* teleological goals is that it is the view strongly suggested by two important features which characterise this multiplicity. The first is that the various morphologies that an organism is capable of exhibiting (and robustly maintaining) as adaptive displays of their phenotypic plasticity, qualitatively dissimilar as they may be, are causally "reachable" from one another *via* some series of transformations—e.g., changes in chemical thresholds, genetic expression levels, etc. Indeed, in experimental morphospace modelling,[31] the exhaustive collection of these particularised morphological possibilities is conceptualised as composing a single 'landscape' whose sub-regions' distance relations reflect their relative degree of transformational accessibility from one another.[32] The second feature stems from the fact that the regions of this landscape are not all created equal: some regions of that space are more heavily *weighted* than others. Some regions of that space, for instance, are "surrounded" by a kind of pleonastic pathing in virtue of which they have a higher probability of becoming occupied than others: the morphologies they represent are ponds into which many possible developmental tributaries flow.[33] Other regions "carve out" deep valleys which, once occupied, have a higher probability of remaining occupied: deviation from the morphology they represent *via* traversal to other regions on that landscape can only done with some degree of difficulty (if it all, in some cases). As we hope should by now be clear, the relative *weights* assigned to various regions in morphospace are measures of the *robustness* of the morphologies those regions represent.[34] Importantly however, the degree to which each of these morphologies is generatively robust—and so the extent to which they are homeostatically maintained in the event of perturbations, and thus represent significantly distinct available morphological states of an organism—can only be established as the inverse of the measure of the relative causal "strength" which the organism's composition must exhibit in order to achieve the aforementioned transformation to some other morphological state.

Taking both of these features of an organismal morphospace seriously inclines us to think that the plastic potential it enshrines characterises a single structural power. For the morphological forms which comprise it are only *superficially* separated from one another, as no particular form within that space is an inaccessible island, but rather, in forming a continuous, dynamically connected landscape, the regions which they occupy are, in a certain sense, only so many permutations of a single morphological type. We suggest then that the manifestation which defines a structural power ought to be understood as being metaphysically 'multifaceted'—that is, as consisting of a wide variety of quantitatively and/or qualitatively distinct permutations of a single manifestation-type within a wide range of causal contexts[35]. Of course, these permutations, and indeed, the entirety of that multifaceted landscape is only a static reflection of the dynamics which characterise the cyclopoietic activity of a structural power, mapping out, as it were, the full spectrum of its contextually sensitive teleological tendency to produce and maintain the production of an entire class of morphologies. Thus, metaphysically, the manifestation-type of a structural power is defined by an entire landscape of distinct, though dynamically united morphologies, each region of which (statically) represents a contextually particularised exhibition of its characteristic cyclopoietic activity. To return to our earlier metaphor: while each particularised morphology which an organism may adopt is a kind of self-contained harmony within its own set of measures—in that each is a quasi-independent instance of its constituents conforming to a particular functionally unified and robustly stable configuration—each is nonetheless fundamentally a derivative expression of the thematic overture of a *single* symphony.

Because in our account, it is the persistence of the continual activity of a structural power which serves as the metaphysical ground for the diachronic unity of an organism, and given the nature of the manifestation-types which characterise those powers we have just elucidated, a central aspect of our account can now be cast in greater relief. As we have argued, fully capturing the nature of the cyclopoietic activity of structural powers requires the conceptualisation of their morphologically multifaceted manifestation-types as the emanation of the teleological texture of their dynamics. In this way, we offer a novel type of account of the diachronic unity of organisms—one which is fundamentally *homeodynamic*, rather than merely *homeostatic*. For while our account certainly recognises the importance of the homeostatic capacity of organisms to maintain a particular morphology in the face of various perturbations of their constituents, it nonetheless views that phenomena as a kind of 'special case' of the dynamic nature of organisms: various regions in a particular morphospace will certainly be substantially weighted (in the aforementioned sense), but only relatively, and only in the context of and as a result of the dynamics of the landscape as a whole. Thus, according to our account, what "remains the same" over time in properly unified organisms their robust preservation not of a single morphological form, but rather of the dynamic specificities of their teleologically textured morphospace which encompasses the potentiality for the expression of many such forms. The diachronic preservation of that dynamic topology, we suggest, is the exhibition of the continued presence and activity of the multifaceted manifestation of a structural power and functions as the ontological ground of the unity of organisms.

5. Conclusion

As Aristotle recognised, as 'substances' *par exellance*, organisms enjoy a unique and privileged ontological status in our world—one for which our metaphysics must account. According to Aristotle, that status is conferred on them in virtue of their possessing 'principles of activity' which allow them to persist as unified beings, metaphysically unfettered from their mereological moorings. In line with this fundamental insight, we have offered a contemporary neo-Aristotelian metaphysical account of the diachronic unity of organisms according to which that unity is conferred by the diachronic possession of a special type of causal power—a structural power—whose dynamic manifestation consists in the teleologically directive impetus toward the organisation of the temporally fluctuating members of its mereological makeup according to the specifications of a multi-faceted, contextually correlative morphological repertoire. On our view, although that unity *is* exhibited in the persistent capacity of an organism to remain morphologically *static* over time, it is more fully so in the persistence of its rich dynamic capacity for a specified range of morphological *variability*. In this way, organisms are *homeodynamic* unities whose privileged position upon the ontological hierarchy is one afforded them by their possession of structural powers.[36]

Notes

1 Incorporating the ontological consequences of this phenomena into our *organism* concept is a central motivation for adopting a 'process ontology' in the philosophy of biology, a framework which is currently experiencing a slight revival. See Dupré (2013) and Jaeger and Monk (2015).

2 The notion of a 'structural power' was first introduced within a *synchronic* context by Marmodoro (2017).

3 For the application of some of the concepts discussed in the later sections of this paper to a discussion of biological 'natural kinds', see Austin (2016).

4 If it were, an organism with a divergent morphology (on account of a deformity caused by a developmental deficiency, for instance) would therefore fail to manifest its structural power and so fail to properly unite its constituent constituents into one entity—but this is false: abnormality does not a multiplicity make. Furthermore, this sort of view would also plausibly entail that the full exhibition of its causal role would amount to it ceasing to function at some point. But this cannot be, for, as Aristotle recognised, when that activity ceases, so too does the entity it belongs to; see, for instance, *On the Parts of Animals* I.1, 641a18-21, Barnes (1984).

5 These powers thus encapsulate the central sense of Aristotle's notion of 'actuality'/ '*energeia*' (literally, "being in work") in *Metaphysics* T, Barnes (1984). See Charles (2010) for an excellent recent discussion.

6 It's worth noting that Rea (2011), too, offers an account of the diachronic unity of entities which is grounded in the activities of causal powers. However, it is one fraught with substantial difficulties of various kinds—see Marmodoro (2013).

7 The term 'cyclopoietic' is meant as a nod toward Maturana and Varela's (1980) influential coinage of the term 'autopoietic', used to describe organisms as "self-building". We note in passing that Aristotle made use of this same criterion for

distinguishing between organisms and 'artefacts', claiming that only the former possessed a 'principle of its own production'—see *Physics II*.1, 192b29-34, Barnes (1984).

8 The function which structural powers perform is thus the embodiment of Aristotle's illustrative description of the distinguishing activity of 'natural beings'— their operation is like doctors doctoring themselves. See *Physics II*.8, 119b30-2, Barnes (1984) .

9 In being self-directed in this fashion, these powers are *perpetually manifesting*: their manifestation provides the sufficient stimulus conditions for their subsequent manifestations. See Marmodoro (2017) for more detail.

10 Huang and Wikswo (2006), Jaeger and Monk (2015), Bich (2016).

11 Arguably, it was this very sense of unity that Kant had in mind when discussing the intrinsic 'natural purpose' that sets organisms apart from matter in the *Critique of Judgment*; see Weber and Varela (2002) for a comprehensive historical discussion, and Walsh (2006) for a contemporary application of this Kantian schema. For more generalised contemporary discussion of 'living beings' as unified dynamically, see Ruiz-Moreno et al. (2004), Cornish-Bowden (2006), and Razeto-Barry (2012).

12 These characteristics are derived from the influential account of Sommerhof (1950). See also Nagel (1977) and, more recently, Walsh (2012).

13 Greenspan (2001), Kitano (2004), Mason (2010).

14 Edelman and Gally (2001), Whitacre and Bender (2010).

15 Occasionally the phenomenon of 'buffering', wherein a developmental system's production of a particular morphological feature is insensitive to a wide variation of alterations in some of its input values, is understood as an exhibition of generative robustness—e.g., in the segment polarity network (von Dassow et al. 2000; Ingolia 2004). However, as robustness *via* causal parameter insensitivity isn't relevant to our discussion, we have refrained from including it here.

16 Zhenglong et al. (2003), Frankel et al. (2010), MacNeil and Walhout (2011).

17 This type of 'degeneracy' is, of course, importantly distinct from the variety that holds between DNA codons and amino acids, which refers to the merely static measure which reflects the fact that the number of unique possible combinations of the former outstrip the number of unique possible types of the latter.

18 Edelman and Gally (2001), Conant and Wagner (2004), Whitacre and Bender (2010).

19 This reflects the practice of contemporary developmental systems biology, where 'robustness' is often conceptualised as an irreducibly holistic and "distributive" feature of complex systems. See Wagner (2005) and Whitacre and Bender (2010).

20 We aren't the first to characterise the intricate operations of living organisms *via* musical metaphors—see principally the famed systems biologist Denis Noble's *The Music of Life* (2006).

21 We're acutely aware that there is a lot more that both *could* and *should* be said about the philosophical implications of these ontological claims, but as a more full discussion would take the present paper too far off course, we leave their consideration and explication for future work.

22 Whitman and Agrawal (2009), Gilbert and Epel (2015).

23 Gibbs et al. (2011).

24 Laforsch and Tollrian (2004).

25 Miura (2005).

26 Pigliucci (2005), Fusco and Minelli (2010).

27 Rasskin-Gutman (2005), McGhee (2006).

28 The intimate connection between non-random regularity and 'nature' is discussed at length in Aristotle's *Physics*, Barnes (1984).

29 Schlichting and Smith (2002), Windig et al. (2004), Aubin-Horth and Renn
   (2009). For some recent insights into the complexities of the reaction-norm mapping of the aforementioned *Daphnia*, see Colbourne et al. (2011).
30 This idea that even regularly produced so-called morphological "monsters" are
   the results of an intrinsic generative 'logic' was championed in the context of
   developmental biology by Alberch (1989). We note in passing that Aristotle
   prefigured this reasoning in *On the Generation of Animals* (*IV*.4, 770b13-24,
   Barnes 1984): "Even in the case of monstrosities, whenever things contrary to
   the established order but still always in a certain way and not at random, the
   result seems to be less of a monstrosity because even that which is contrary to
   nature is in a certain sense according to nature . . . for instance, there is a vine
   which some call 'smoky'; if it bears black grapes, they do not judge it a monstrosity because it is in the habit of doing this very often. The reason is that it is in its
   nature intermediate between white and black; thus the change is not a large one
   nor, so to say, contrary to nature; at least, it is not a change into another nature".
31 This type of modelling has recently been extensively utilised in the study of the
   morphological potential of pluripotent cells—see, for instance, Bhattacharya et
   al. (2011) and Li and Wang (2013).
32 Wagner and Stadler (2003), Rasskin-Gutman (2005), McGhee (2006), Wagner
   (2014).
33 The conception of morphospace as contoured in this fashion, first proposed by
   Waddington (1957), is now a central pillar of the theoretical research project of
   evolutionary developmental biology in the form of 'developmental constraints'
   or 'generative bias' (Arthur 2002; Amundson 2005; Hallgrimsson et al. 2012;
   Brigandt 2015). For recent empirical case studies, see Young et al. (2010) and
   Rasskin-Gutman and Esteve-Altava (2014).
34 The representation of the probability measure of states *via* topological mappings
   is central to the now widely employed methodology of *dynamic systems theory*
   analyses in theoretical developmental biology (Wang et al. 2011; Huang 2012;
   Davila-Velderrain et al. 2015), within which system state robustness is often
   given a topological interpretation (Kitano 2004; Huang 2009; Huneman 2010).
35 We intentionally refer to the *manifestation* of structural powers as 'multi-
   faceted'—reflecting the fact that the context-sensitive manifestations it may
   exhibit are *facets* of a *single surface*—rather than referring to the *powers themselves* as 'multi-track' (Martin 2007; Williams 2011; Vetter 2013). As evidenced
   from the discussion below, the two conceptions are radically distinct, though we
   don't have the space to explore the distinctions here.
36 This collaborative research has been made possible by funding from the Templeton World Charity Foundation. We are also grateful for the feedback from the
   audience of our presentation of an earlier draft of this paper at Corpus Christi
   College, Oxford.

Bibliography

Alberch, P. (1989) 'The Logic of Monsters: Evidence for Internal Constraint in
   Development and Evolution', *Geobios* 22: 21–57.
Amundson, R. (2005) *The Changing Role of the Embryo in Evolutionary Thought:
   Roots of Evo-Devo* (Cambridge: Cambridge University Press).
Arthur, W. (2002) 'The Emerging Conceptual Framework of Evolutionary Develop-
   mental Biology', *Nature* 415(6873): 757–64.
Aubin-Horth, N. and Renn, S. (2009) 'Genomic Reaction Norms: Using Integrative
   Biology to Understand Molecular Mechanisms of Phenotypic Plasticity', *Molecu-
   lar Ecology*, 3763–80.

Austin, C. J. (2016) 'Aristotelian Essentialism: Essence in the Age of Evolution', *Synthese*, doi:10.1007/s11229-016-1066-4.

Barnes, J. (1984) The Complete Works of Aristotle, Volumes I and II (Princeton: Princeton University Press)

Bhattacharya, S., Zhang, Q. and Andersen, M. (2011) 'A Deterministic Map of Waddington's Epigenetic Landscape for Cell Fate Specification', *BMC Systems Biology* 5(85): 1–11.

Bich, L. (2016) 'Systems and Organizations: Theoretical Tools, Conceptual Distinctions and Epistemological Implications', in G. Minati, M. Ambram and E. Pessa (eds.) *Towards a Post-Bertalanffy Systematics* (New York: Springer): 203–9.

Brigandt, I. (2015) 'Evolutionary Developmental Biology and the Limits of Philosophical Accounts of Mechanistic Explanation', in P. A. Braillard and C. Malaterre (eds.) *Explanation in Biology: An Enquiry into the Diversity of Explanatory Patterns in the Life Sciences* (Dordrecht: Springer): 135–73.

Charles, D. (2010) 'Metaphysics Θ.7 and 8: Some Issues Concerning Actuality and Potentiality', in J. Lennox and R. Bolton (eds.) *Being, Nature, and Life in Aristotle: Essays in Honor of Allan Gotthelf* (Cambridge: Cambridge University Press): 168–97.

Colbourne et al. (2011) 'The Ecoresponsive Genome of Daphnia Pulex', *Science* 331(6017): 555–61.

Conant, G. and Wagner, A. (2004) 'Duplicate Genes and Robustness to Transient Gene Knock-Downs in Caenorhabditis Elegans', *Proceedings of the Royal Society B* 271(1534):89–96.

Cornish-Bowden, A. (2006) 'Putting the Systems Back Into Systems Biology', *Perspectives in Biology and Medicine* 49(4): 475–89.

Davila-Velderrain, J., Martinez-Garcia, J. C. and Alvarez-Buyila, E. R. (2015) Modeling the Epigenetic Attractors Landscape: Toward a Post-Genomic Mechanistic Understanding of Development', *Frontiers in Genetics* 6(160): 1–14

Dupré, J. (2013) 'Living Causes', *Proceedings of the Aristotelian Society Supplementary Volume* 87(1): 19–38.

Edelman, G. and Gally, J. (2001) 'Degeneracy and Complexity in Biological Systems', *Proceedings of the National Academy of the Sciences* 98(24): 13763–8.

Frankel, N., Davis, G., Vargas, D., Wang, S., Payre, F. and Stern, D. (2010) 'Phenotypic Robustness Conferred by Apparently Redundant Transcriptional Enhancers', *Nature* 466: 490–3.

Fusco, G. and Minelli, A. (2010) 'Phenotypic Plasticity in Development and Evolution: Facts and Concepts', *Philosophical Transactions of the Royal Society B* 365(1540): 547–56.

Gibbs, M., Wiklund, C. and Van Dyck, H. (2011) 'Phenotypic Plasticity in Butterfly Morphology in Response to Weather Conditions During Development', *Journal of Zoology* 283(3): 162–8.

Gilbert, S. F. and Epel, D. (2015) *Ecological Developmental Biology: The Environmental Regulation of Development, Health, and Evolution* (Sunderland, MA: Sinauer Associates).

Greenspan, R. (2001) 'The Flexible Genome', *Nature* 2(5): 383–7.

Hallgrimsson, B., Jamniczky, H., Young, N., Rolian, C., Schmidt-ott, U. and Marcucio, R. (2012) 'The Generation of Variation and the Develomental Basis for Evolutionary Novelty', *Journal of Experimental Zoology Part B Molecular and Developmental Evolution* 318(6): 501–17.

Huang, S. (2009) 'Reprogramming Cell Fates: Reconciling Rarity With Robustness', *Bioessays* 31(5): 546–60.

——— (2012) 'The Molecular and Mathematical Basis of Waddington's Epigenetic Landscape: A Framework for Post-Darwinian Biology?', *Bioessays* 34(2): 149–57.

Huang, S. and Wikswo, J. (2006) 'Dimensions of Systems Biology', *Reviews of Physiology: Chemistry and Pharmacology* 15: 81–104.

Huneman, P. (2010) 'Topological Explanations and Robustness in Biological Sciences', *Synthese* 177(2): 213–45.

Ingolia, N. (2004) 'Topology and Robustness in the Drosophila Segment Polarity Network', *PLoS Biology* 2(6): 0805–15.

Jaeger, J. and Monk, N. (2015) 'Everything Flows: A Process Perspective on Life', *EMBO Reports* 16(9): 1064–7.

Kitano, H. (2004) 'Biological Robustness', *Nature Reviews Genetics* 5(11): 826–37.

Laforsch, C. and Tollrian, R. (2004) 'Inducible Defenses in Multipredator Environments: Cyclomorphis in Daphnia Cacullata', *Ecology* 85(8): 2302–11.

Li, C. and Wang, J. (2013) 'Quantifying Cell Fate Decisions for Differentiation and Reprogramming of a Human Stem Cell Network: Landscape and Biological Paths', *PLoS Computational Biology* 9(8): e1003165.

MacNeil, L. and Walhout, A. (2011) 'Gene Regulatory Networks and the Role of Robustness and Stochasticity in the Control of Gene Expression', *Genome Research* 21(5): 645–57.

Marmodoro, A. (2013) 'Aristotle's Hylomorphism Without Reconditioning', *Philosophical Inquiry* 37: 5–22.

—— (2017) 'Power Mereology: Structural Powers versus Substantial Powers', in M. Paoletti and F. Orilia (eds.) *Philosophical and Scientific Perspectives on Downward Causation* (New York: Routledge).

Martin, C. (2007) *The Mind in Nature* (Oxford: Oxford University Press).

Mason, P. (2010) 'Degeneracy at Multiple Levels of Complexity', *Biological Theory* 5(3): 277–88.

Maturana, H. and Varela, F. (1980) *Autopoiesis and Cognition: The Realization of the Living* (Boston: D. Reidel).

McGhee, G. (2006) *The Geometry of Evolution: Adaptive Landscapes and Theoretical Morphospaces* (Cambridge: Cambridge University Press).

Miura, T. (2005) 'Developmental Regulation of Caste-Specific Characters in Social-Insect Polyphenism', *Evolution & Development* 7(2): 122–9.

Nagel, E. (1977) 'Goal-Directed Processes in Biology', *The Journal of Philosophy* 4(5): 261–79.

Noble, D. (2006) *The Music of Life: Biology Beyond the Genome* (Oxford: Oxford University Press).

Pigliucci, M. (2005) 'Evolution of Phenotypic Plasticity: Where Are We Going Now?', *Trends in Ecology & Evolution* 20(9): 481–6.

Rasskin-Gutman, D. (2005) 'Modularity: Jumping Forms Within Morphospace', in W. Callebaut and D. Rasskin-Gutman (eds.) *Modularity: Understanding the Development and Evolution of Natural Complex Systems* (Cambridge: MIT Press): 207–19.

Rasskin-Gutman, D. and Esteve-Altava, B. (2014) 'Connecting the Dots: Anatomical Network Analysis in Morphological EvoDevo', *Biological Theory* 9(2): 178–93.

Razeto-Barry, P. (2012) 'Autopoiesis 40 Years Later: A Review and a Reformulation', *Origins of Life and Evolution of Biospheres* 42(6): 543–67.

Rea, M. (2011) 'Hylomorphism Reconditioned', *Philosophical Perspectives* 25(1): 341–58.

Ruiz-Moreno, K., Pereto, J. and Moreno, A. (2004) 'A Universal Definition of Life: Autonomy and Open-Ended Evolution', *Origins of Life and Evolution of the Biosphere* 34(3): 323–46.

Schlichting, C. and Smith, H. (2002) 'Phenotypic Plasticity: Linking Molecular Mechanisms With Evolutionary Outcomes', *Evolutionary Ecology* 16, 189–211.

Sommerhof, G. (1950) *Analytical Biology* (Oxford: Oxford University Press).

Vetter, B. (2013) 'Multi-Track Dispositions', *The Philosophical Quarterly* 63(251): 330–52.

von Dassow, G., Meir, E., Munro, E. M. and Odell, G. M. (2000) 'The Segment Polarity Network Is a Robust Developmental Module', *Nature* 406: 188–92.

Waddington, C. H. (1957) *The Strategy of the Genes* (London: George Allen & Unwin).

Wagner, A. (2005) *Robustness and Evolvability In Living Systems* (Princeton: Princeton University Press).

Wagner, G. (2014) *Homology, Genes, and Evolutionary Innovation* (Princeton: Princeton University Press).

Wagner, G. and Stadler, P. (2003) 'Quasi-Independence, Homology and the Unity of Type: A Topological Theory of Characters', *Journal of Theoretical Biology* 220: 505–27.

Walsh, D. (2006) 'Organisms as Natural Purposes: The Contemporary Evolutionary Perspective', *Studies in History and Philosophy of Science Part C: Studies in History and Philosophy of Biological and Biomedical Sciences* 37(4): 771–91.

––––––– (2012) 'Mechanism and Purpose: A Case of Natural Teleology', *Studies in History and Philosophy of Biological Biomedical Sciences* 43, 173–81.

Wang, J., Zhang, K., Xu, L. and Wang, E. (2011) 'Quantifying the Waddington Landscape and Biological Paths for Development and Differentiation', *Proceedings of the National Academy of Sciences of the United States of America* 108(20): 8257–62.

Weber, A. and Varela, F. (2002) 'Life After Kant: Natural Purposes and the Autopoietic Foundations of Biological Individuality', *Phenomenology and the Cognitive Sciences* 1(2): 97–125.

Whitacre, J. and Bender, A. (2010) 'Networked Buffering: A Basic Mechanism for Distributed Robustness in Complex Adaptive Systems', *Theoretical Biology and Medical Modelling* 7(20): 1–20.

Whitman, D. W. and Agrawal, A. A. (2009) 'What Is Phenotypic Plasiticty and Why Is It Important?', in T. N. Ananthakrishna and D. W. Whitman (eds.) *Phenotypic Plasticity of Insects: Mechanisms and Consequences* (Enfield: Science Publishers): 1–63.

Williams, N. E. (2011) Putting Powers Back on Multi-Track', *Philosophia* 39(3): 581–95.

Windig, J., de Kovel, C. and de Jong, G. (2004) 'Genetics and Mechanics of Plasticity', in T. DeWitt and S. Scheiner (eds.) *Phenotypic Plasticity: Functional and Conceptual Approaches* (New York: Oxford University Press): 31–49.

Young, N., Wagner, G. and Hallgrimsson, B. (2010) 'Development and the Evolvability of Human Limbs', *Proceedings of the National Academy of Sciences* 107(8): 3400–5.

Zhenglong, G., Steinmetz, L., Gu, X., Scharfe, C., Davis, R. and Li, W.-H. (2003) 'Role of Duplicate Genes in Genetic Robustness Against Null Mutations', *Nature* 421(6918): 63–6.

8   A Biologically Informed Hylomorphism

Christopher J. Austin

1. Introduction

There's no denying that contemporary metaphysics is experiencing an Aristotelian revival of sorts wherein dispositions, or 'causal powers' are no longer regarded as scholastic superfluities, ideally to be explained away, but are instead being put to work in everything from theories of colour to theories of modality. But while the Aristotelian doctrine of 'potentiality' is now widely understood as being fairly innocuous and even theoretically advantageous, there has been a recent notable rise in the defenders of a much more contentious Peripatetic postulate—the doctrine of *hylomorphism*. According to the ontological principle of hylomorphism, the *natures* of entities are in some sense metaphysically, or conceptually bipartite: they have both a *material* and a *formal* aspect. Thus, fully "grasping the nature" of an entity requires understanding it as the conceptual union of *both* aspects.

The minor surge of the defence of this doctrine notwithstanding, it's certainly safe to say that hylomorphism isn't currently *en vogue*, even amongst the most ardent defenders of a neo-Aristotelian metaphysic. To my mind, there's a simple reason for this: while the contemporary defenders of this doctrine have done quite a lot of work in precisely explicating what the *conceptual* notion of 'form' amounts to, comparatively little has been done toward showing that this is a concept with *empirical* content. If we believe, as I do, that an effectual impetus to join a particular philosophical church must consist in more than simply being given a conceptual dissection of its characteristic complex metaphysical doctrine, the paucity of practitioners in the hylomorphic pews should come as little surprise.

With this in mind, this paper is a kind of altar call—its aim is to show that the hylomorphist's claim that fully grasping the nature of entities is a "two concept job" can be given firm empirical footing. To do so, I bring the conceptual focus back to its Aristotelian origin—the biological realm. My claim is that recent advances in developmental systems biology afford us an empirically tractable picture of the hylomorphic nature of biological entities by way of elucidating what the *formal* aspect of that nature consists in. The hope is that, having been enriched by an empirically informed conception of form, hylomorphism might once again be seen as good news for metaphysics.

2. Hylomorphism: A Matter of Definition

Taken generally, hylomorphism is the doctrine that fully capturing the metaphysical 'nature' of an entity requires an appeal to two distinct (though ultimately intimately interrelated) concepts—*matter* and *form*. Or, to put it another way, according to hylomorphism, any adequate metaphysical definition of an entity must be two-fold—it must encompass the nature of the entity *qua* matter and *qua* form. But what does this bipartite distinction amount to? Let us say that to define the nature of an entity *qua* matter is to define it as an organised, connected collection of discrete parts; here, 'organisation' and 'connectivity' are to be understood, at the very least, both spatially and causally (and perhaps temporally), and 'discrete' denotes their being ontologically, or existentially independent from one another. To define the nature of an entity *qua* form, on the other hand, is to define it as a holistic, dynamically directed structure; more on this momentarily.

The claim of hylomorphism is that both of these concepts must be put to use in successfully "capturing the nature" of an entity—but what is involved in this task? Clearly, "capturing the nature" of an entity is to be understood as getting a grip on *what that entity is* in some metaphysically fundamental sense. In line with the now-popular Lockean understanding advocated in contemporary metaphysics, let us say that "capturing the nature" of an entity amounts to understanding *why* and *how* that entity possesses its characteristic set of properties: getting a grip on the *nature* of a clump of gold, for instance, plausibly involves understanding *why* it has such-and-such surface-level properties (reflective surfaces, malleability, conductivity, etc.), which involves understanding *how* it comes to have them (through its molecular structure, or electron count, or etc.)—thus, Kripke's appeal to its "periodic" nature.[1] On this line of thinking, citing the *nature* of an entity affords one rich explanatory power with respect to its possession of a set of typical features—*why* those features are there (or why they *could* be there), and *how* they got there (or how they *would* have gotten there), etc.[2]

Defining the nature of an entity *qua* matter then is to cite an entity's organised, connected collection of discrete parts as explanatory with respect to its possession of a characteristic set of features.[3] I take it that this sort of definitional methodology won't be unfamiliar to the reader—it is, after all, representative of the prominent philosophical project of reductionism—and so it's probably unnecessary to spend too much time on it here. What's more important for present purposes is to flesh-out precisely what it means to define the nature of an entity *qua* form. My approach here will be to trace the Peripatetic thread as it has weaved through contemporary hylomorphic accounts by distilling a set of shared criteria for a *formal* definition present in the literature.[4] Though I've already briefly mentioned a putative description of such a definition, it's instructive to consider it in more detail. To do so, I'll distinguish three aspects of a 'formal definition'; though, as we will see, these three are in some way intertwined.

Firstly, to define the nature of an entity *qua* form is to offer an explanatory basis for its characteristic features in something "over and above" its mereological constituents. Formal definitions are often understood as demarcating *higher-order* facts *about* an entity's constituents—typically, they either pick out some privileged *relation* of those constituents,[5] or else some sort of *process* of,[6] or *metaphysical operation* on,[7] those constituents. Importantly, in virtue of referring to something appropriately higher-order, formal definitions are taken not to refer to any *extra* mereological part of those entities,[8] nor are they understood as being *reducible to* any competing material definitions which might concern those parts.[9]

Secondly, a formal definition of an entity's nature picks-out some irreducibly higher-order fact about that entity and its constituents precisely because to define that nature *qua* form is to represent the entity as an ontological *unity*—as *metaphysically one*. In contrast to a material definition, wherein appeal is made to a collection of various discrete mereological parts and pieces, a *formal* definition's explanatory prowess is grounded in a *holistic* conception of an entity.[10] The *unity* that formal definitions are meant to appeal to is understood as being importantly distinct from the "mere togetherness" that characterises the content of a material definition: to be sure, the latter cites an organised, *connected* collection of parts, but the former cites that collection *as one*.

The last aspect of this type of definition makes clear what this distinction really amounts to, as defining an entity's nature *qua* form involves an appeal to an entity as a *causally unified system*. This is typically cashed-out by the claim that a formal definition picks-out a higher-order causal activity *of the entity as a whole*,[11] or else one that is in some sense an emergent, irreducibly *cooperative activity* of an entity's constituents.[12] The causal unity implicit in a formal definition doesn't consist simply in the fact that a particular entity performs a particular higher-order activity which involves each of its parts operating in causal unison, but also that this structure orients these parts, as a whole, toward a particular causally privileged end, or ends. As one might expect of an Aristotelian account, to define the nature of entity *qua* form is to cite as explanatory (in the relevant sense) its holistic causal "directedness" toward some end-state(s).[13] In some sense, then, a formal definition represents the entity's constituents as non-autonomous participants in a singularly directed, dynamically continuous structure.[14] Thus, we see again, now more clearly, the *higher-order unity* that a formal definition is meant to capture—namely, a holistic, goal-directed activity, ontologically attributable to an entity only as a singular causal system.[15]

As I understand it then, to define the nature of an entity *qua* form is to demarcate its holistically higher-order, dynamically directed causal structure as uniquely explanatory with respect to its possession of a set of typical features. Now that we've a better grip on what a *formal* definition amounts to, the pertinent question is, given this conception, what's required in order to give a plausible defence of hylomorphism? For our purposes, as the more contested aspect of the doctrine, let us ask: what's required in order to give a plausible defence of the applicability of a *formal* definition of an entity's nature? To answer that question requires getting clearer about the nature of the defence I want to offer. As I've said, my aim is to display and defend an empirical incarnation of the conceptual framework of hylomorphism. Thus, in explicating that framework, I have focused on the doctrine's core definitional claims, rather than any of its purported ontological commitments. As it happens, precisely *what* those commitments are is widely disputed, even among its adherents. If an entity's nature admits of a formal definition, does this entail, for instance, that we must reformulate our account of mereological composition[16] or that we must countenance a novel ontological category whose members are imbued with unique, "downwardly directed" causal powers?[17] Or does such an admission merely require helping ourselves to a non-ontological free lunch, delivered simply *via* a process of abstraction?[18]

Rather than taking a particular stance on this issue, my aim is to focus on the widely accepted *definitional* project: after all, every defender of hylomorphism presumably agrees that the doctrine is committed to the claim that fully capturing the nature an entity requires an appeal the dichotomous descriptive machinery of *matter* and *form*, irrespective of whatever the ontological underpinnings or consequences of those descriptions are taken to be. With the account laid out above then, the project of this paper is to show one way in which this definitional project may be vindicated; such vindication might be taken to entail particular ontological consequences for the doctrine, and although I won't be defending them in detail here, I will briefly address them in the final section. As already mentioned, to do so, the paper will focus on the clearly more contested aspect of the doctrine—*formal* definition. For my defence to have succeeded, it will have to have shown that the concepts invoked in this type of definition have a plausible empirical instance. Importantly, with the above discussion in mind, the success of this defence requires (a) showing that form is *conceptually independent* of matter, and (b) showing that form plays a *unique explanatory role* with respect to matter: (a) is satisfied if a *formal* definition of an entity, as explicated above, can be made without explicit appeal to its *material* definition, while (b) is satisfied if such a definition is able to play an explanatory role with respect to the possession of a characteristic set of an entity's features which is uncapturable by appealing to its *material* definition.[19]

My claim is that if we focus on the biological realm, itself once the fount of Aristotelian inspiration, a contemporary defence of the principles of hylomorphism is available: recent advances in developmental systems biology have shown, or so I will argue, that fully capturing the nature of biological entities is a job which requires both matter *and* form.

3. Back to Biology: Building an Organism

Aristotle's argument that the principles of his hylomorphic metaphysic were truly *in rei* was primarily grounded in the physical principles he believed to be *in natura*—that is, in the biological realm. If you're after a robust understanding of that metaphysic, then, you'd be better off examining starfish, rather than statues.[20] Accordingly, most philosophers who've since taken up the hylomorphic mantle have placed biological entities as paradigms of that metaphysic—and rightly so. However, although few deny that the doctrine naturally dwells in the "land of the living", even fewer have taken on the project of providing a detailed account of *how*, and *in what way*, that realm is to be characterised by its metaphysical principles. Typically, at best, these philosophers merely suggestively cite practicing biologists' rather vague delineations of characteristic *phenomena of life*—homeostasis, emergence, etc.—as empirical undergirding for the doctrine's metaphysics.[21] More commonly, however, is the simple, though unexamined posit of biological entities as hylomorphic exemplars—one often finds 'humanity' atop the candidates for form, for instance.[22] In what follows, I want to offer a more empirically specific focus, by examining in detail the particularities of an important class of biological entities.[23]

Rather than taking on the "big picture" task of providing a hylomorphic account of the nature of biological entities *tout court*, I want to take up the more minute and more manageable task of providing a hylomorphic account of the nature of the biological individuals which make up biological entities. Why? One reason is practical: in my view, providing an empirically robust hylomorphic account of the nature of a biological entity—a starfish, for instance—is a complex and complicated affair, requiring a perhaps unappreciated amount of philosophical subtlety. Better, then, for the purposes of this paper, to make an attempt at the more practical task of providing such an account for the individuals which compose biological entities; ideally, the account I offer will be generalisable, "upwards" as it were, though I won't be arguing for that here.[24] Another reason is principled: the focus of my examination may not be best conceptualised as full-fledged entities in their own right, but they are most certainly biological *individuals*; more on this in a moment. And being biological individuals, it's reasonable to expect that a hylomorphic account ought to apply to them just as equally as it does to the larger individuals they compose.

That said, the individuals I want to focus on are called *developmental modules*, currently on the centre stage of research in the field of developmental systems biology. Developmental modules are discrete biological systems causally responsible for the development of particular morphological features. A foundational fact upon which the edifice of systems biology is built is that the morphological development of organisms is a rather piecemeal affair. More specifically, that an organism's development is controlled *discretely*, by individualised organismal sub-systems which initiate and direct the formation of its various body parts—eyes, legs, and the like. These sub-systems—or developmental modules—are treated as individuals in part due to their relative causal autonomy during the process of development: they are characterised equally by an extremely high causal connectivity among their constituents and an extremely low causal connectivity with other parts of the organism.[25] They are, in other words, discernible bundles of tightly knit causal loops whose activities are responsible for an organism's develop-ment of a particular trait. But developmental modules are also individuals in perhaps a stronger sense, as recent advances in evolutionary developmen-tal biology (evo-devo) have made clear: they are able to be generationally inherited and so are traceable (with modification) throughout evolutionary history,[26] a fact which may even merit them a place at the ground-floor of the elusive, proper "level of selection".[27] In a perfectly respectable sense, then, developmental modules, the organismal sub-systems causally respon-sible for the production of particular morphological traits, are biological *individuals*—and ontologically important ones at that, as it is their activities which give shape to the fully featured biological entities we're more directly acquainted with.[28]

The pertinent question then is: what is the *nature* of a developmental module? Recall that citing the nature of a thing is meant to provide rich explanatory import with respect to its characteristic feature(s). To answer that question then, we must know which feature(s) the citation of the nature of a module might purport to aid in explaining. The obvious answer seems to be that citing the *nature* of a module should help shed explanatory light upon *the development of its associated morphological trait*: it should, as I earlier put it, importantly aid in explaining the *why* and *how* of that pro-cess. Thus, providing an answer as to the *nature* of a developmental module requires some knowledge of what that process amounts to. If we consider that a fully developed morphological feature is nothing more than a particu-larised spatial configuration of various cell-types, we can get a preliminary grip on the process in question—put simply, it involves putting the correct *things* in the correct *places*. The process of "building" a morphological fea-ture is thus two-fold: it requires the creation of a certain set of cell-types particular to the feature in question, and the arrangement of this set in a particular three-dimensional configuration. More specifically, the operation of that process involves not only that the genomes of a set of cells take on particular expression profiles which determine their individual developmen-tal fates, but also that these specifically expressed cells are spatially coordi-nated in a particular configuration.

We now know that the process which begins with a collection of cells whose genomes are not in any particular expression state (i.e. pluripotent cells), known as an *imaginal disc*, which over time take on specific expres-sion profiles in a coordinated fashion, requires the activity of an entire *net-work* of genes.[29] It requires a certain set of genes that act *intra-cellularly* to produce the proteins that determine the particular cell-types which "build" the morphological feature in question *and* a set of genes whose protein products (known as *transcription factors*) act *inter-cellularly* to regulate the intra-cellular expression profiles of other genes in neighbouring cells, thereby controlling *which* genes are expressed in *which* cells throughout the disc, as well as *when* and *where* that expression takes place during the development of a morphological feature. Thus, we can model the process of the development of an imaginal disc by mapping out a *genetic regulatory network* (GRN), which includes the set of genes whose expression determines particular cell-types, the set of genes which controls their expression, and the particularities of the causal, regulatory relationships among them (activation, repression, etc.)

Understood in this way, the development of a particular morphological feature can be seen as the temporal succession of a series of expression profiles of the GRN elements in the cells which compose an imaginal disc. Importantly, this is a process governed by the "regulatory logic" of that GRN, as the expression profile of each cell within the disc evolves over time according to the particularities of its regulatory structure: if G_1 is *highly expressed* at t because it is up-regulated by G_2 at $t - 1$, then at $t + 1$, G_3 and G_4 will be *barely expressed*, due to G_1 highly down-regulating both, etc.[30] Over time, then, due to the specific regulatory logic of a particular GRN, the cells of an imaginal disc take on a controlled and continuous series of expression profiles *via* a series of patterning processes ultimately resulting in the collectively stable state of a various collection of particular cell-types arranged in a particular spatial configuration—that is, in a fully developed morphological feature.[31]

It should by now be clear that if we wish to "capture the nature" of a developmental module, we must have recourse to its associated GRN, as knowledge of its elements and the relations among them sheds explanatory light upon the development of its associated morphological trait: if we

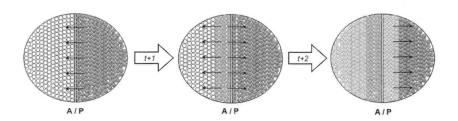

Figure 8.1 Schematic two-dimensional representation of the early developmental stages of a multi-cellular imaginal disc constituting a module: 'A/P' denotes the anterior and posterior regions of the module, distinct bubble colours represent distinct cell-types, and arrows represent the causal/regulatory influence of one cell-type upon neighbouring cell-types. Over time, the cellular constitution of a module becomes increasingly compartmentalised and spatially discrete.

want to explain the *why* and *how* of that process, we must appeal to the structural-causal mapping of its GRN. In doing so, we are citing its organised, connected collection of discrete parts as explanatory with respect to the possession of its characteristic feature. We are, in other words, showing the validity of defining the nature of a developmental module *qua* matter. But does such a definition *fully* capture the nature of a developmental module? That is, is there something yet left to account for with respect to offering the relevant explanatory utility which this definition fails to deliver? The answer, I think, is yes, for, as I argue below, the *material* definition of a developmental module leaves one *uninformed* about its nature in an important respect.

4. Modules and Morphospaces

Although I have been preliminarily modelling the causal output of a developmental module rather rigidly as a *singular*, fully specified morphological feature, a complication must now be made, as the full picture admits of rather more *flexibility*. For we now know that the morphological structure produced by a single developmental module, being underwritten by a specific genetic regulatory network, is capable of a wide variety of intra- and inter-cellular environmentally induced phenotypic variation—this is the phenomenon of *phenotypic plasticity*, attested to by the reality (read: quantifiability) of *reaction norms*.[32] As a result of "upstream" alterations consisting mainly of heterochronical and heteropical changes in inter-cellular signalling, a single developmental module is capable of producing a wide range of "downstream" qualitative alterations in its associated morphological feature with respect to its precise shape, size, pigmentation, etc.[33] Thus, the morphological feature generatively specified by a single developmental module cannot be fully characterised by a single, particularised instance with respect to these qualitative and quantitative factors, but must instead be understood as a generalised collection of various qualitative and quantitative variations on that feature—this set of possible permutations is known as the feature's *morphospace*. For this reason, capturing the generative capacity of a single developmental module with respect to its associated morphological feature must involve modelling its "variational tendencies",[34] or its set of "developmental trajectories, [correlated with] the particular set of environmental conditions to which [it] is exposed",[35] to construct an "idealised type . . . constructed from ample and acknowledged variation".[36]

With this in mind, it's clear that "fully capturing the nature of a developmental module" must involve capturing its *rich* generative capacity to produce its *entire* morphospace. The pertinent question for our purposes is: can the *material* definition we've provided accomplish this? In order to answer this, we must look again to the causal story of development. We've already seen that one can model a fully developed morphological feature as a specific spatial arrangement of a collection of cells with specific genetic expression profiles. We've also seen that the developmental process involved in generating such a feature can be modelled as the temporal succession of states of the overall expression profile of the imaginal disc (itself composed of a number of individual cells' profiles), the transitions of which are governed by the regulatory logic specified by its GRN. Of course, we have thus far only modelled a *single* developmental trajectory towards the generation of a single variant of a morphological feature, and the phenomenon of developmental plasticity shows that *many* such trajectories are possible.

However, accommodating this involves no further complication—using the same GRN and its constitutive regulatory logic, we can model each of these trajectories as the developmental consequence of its "generative rules" being applied in the context of distinct initial developmental input conditions.[37] In other words, the phenomenon of developmental plasticity reflects the fact that a single regulatory network is capable of delivering a variety of distinct morphological end-states according to a variety of distinct initial developmental conditions, as altering the initial network-state of a module has regulatory consequences (specified by the generative rules of that network) on the expression states of its cells which ripple "downwards" and "outwards" throughout an imaginal disc during the process of development.

So, modelling a module's flexibility with respect to its capacity to produce various developmental trajectories by defining it *materially*—that is, *via* its associated GRN—is easily done. However, a further complication arises when one considers that the morphospaces associated with developmental modules are not *merely* reflections of their developmental plasticity, but also of their *generative constraints*: for these systems are not *wholly* flexible, causally subject to every incoming environmental influence during the process of development, but instead reliably and repeatedly end that process within a well-demarcated range of particular states.[38] In other words, no module's morphospace consists of an ontologically exhaustive set of every possible qualitative and quantitative permutation on its associated morphological feature. Rather, the morphospace which characterises a developmental module is composed of a select set of *generatively privileged* permutations which arise within a wide range of distinct environmental (read: causal) contexts. In this way, the character of a morphospace associated with a developmental module shows that nature delights in variety without indulging in it—morphological variation is allowed, but only within certain limits.

If "fully capturing the nature of a developmental module" involves capturing its rich generative capacity to produce its entire morphospace, then any adequate definition of that nature must be explanatorily relevant with respect not only to its generation of a certain amount of morphological variation, but also with respect to the specified constraints on that variation. What we require, in other words, is not only explanatory power with respect to a module's capacity to produce various distinct developmental trajectories, but also with respect to the limitations on that capacity. Importantly, note that understanding the latter allows us to understand, for any particular module, why *these* morphologies are *privileged*, and *why* they are so—something that cannot be achieved by simply appealing to any *single* developmental trajectory, nor to the entire *set* of privileged trajectories. Capturing this fact, I suggest, is crucial to capturing the *nature* of a developmental module.

Accomplishing this, as I will show, requires conceptualising these organismal sub-systems in a radically novel fashion, *via* the conceptual framework of *dynamic systems theory* (DST). Indeed, in doing so, it requires, as I argue below, that we conceptualise developmental modules *holistically*, as *higher-order*, *dynamically directed* systems.

5. Dynamic Systems Theory: A Formal Science

The desire to more fully understand the *developmental constraints* of organismal systems was perhaps the founding motivation for the development of DST, a project begun in spirit by Waddington's posit of an 'epigenetic landscape',[39] and subsequently fleshed-out with insights from Kaufmann's Boolean modelling[40] of GRNs.[41] DST, as a novel modelling technique of such systems, has afforded researchers a set of unique conceptual resources with which to understand the process of development, and is now rather widely applied[42] in analyses of everything from sub-organismal cell-fate[43] to the evolvability of organism populations.[44]

In order to show the utility of DST in this respect, and in application to our current project, let us take stock. We have seen that the developmental process involved in a module's generation of a morphological feature can be modelled as the temporal succession of states of the overall expression profile of the imaginal disc (itself composed of a number of individual cells' profiles), the transitions of which are governed by the regulatory logic specified by its GRN. This fact forms the foundation of DST modelling, and the thought is: if we construct an abstract multi-dimensional state-space whose individual points represent particular disc-wide expression profiles (where each specifies the expression-state of each GRN element within each cell in the disc), arranged continuously (according to cellular expression values) on axes which represent a particular cell-type in a particular spatial region, we can model a particular instance of the development of a morphological feature as a temporal trajectory through this state-space, ending in the expression-state representing that feature; the figure below illustrates this type of model with respect to a simplified GRN, represented on a two-dimensional state-space.[45]

Accordingly, utilising the data derived from experimental evidence of the phenomenon of phenotypic plasticity, we can represent the generative progression of a variety of the module's possible developmental routes by tracing-out distinct trajectories through a single multi-dimensional state-space. The resulting picture provides a representation of the multiple

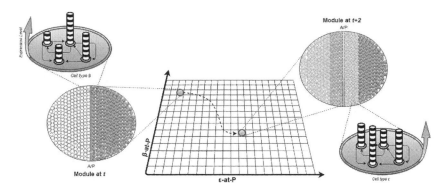

Figure 8.2 Schematic representation of a single developmental trajectory of a module through a (truncated) abstract state-space, in reference to Figure 8.1.
On either side, the 'module at *t*' and 'module at *t* + 2' depict the spatial
arrangement of two cell types (*β* and *ε*) within the imaginal disc with
respect to its anterior (A) and posterior (P) regions. Each cell-type is represented as consisting of the module's GRN elements (depicted as elliptical bases), their regulatory connections (depicted by arrows), and their
particular expression levels (depicted as stacked elliptical elements). In the
middle of the figure, the temporal transition of the spatial arrangement
of *β* and *ε* with respect to P is modelled as a trajectory through a two-
dimensional plane whose edges represent unique disc-wide cellular GRN-
expression states, arranged such that the distance between any two edges
reflects quantitative similarity with respect to spatially specific cellular
expression. The 'module at *t* + 2' here represents the expression levels
of the module's GRN which constitute its developmental end-state.

developmental pathways, each defined by distinct trajectories through GRN
expression-value space, which are responsible for the production of the various morphological permutations which comprise the morphospace of a particular module.

As theoretically interesting as this model may be, it yet fails to offer us
a comprehensive understanding of the structural limitations on a module's
capacity to produce these permutations. In other words, as I earlier put it,
this representative framework doesn't offer any elucidation with respect to
why *these* permutations are privileged, or *why* they are so. I think it's clear
that examining more closely any *single* trajectory corresponding to such
a permutation isn't going to do the requisite work, but nor will a similar
scrutiny of the entire *set*—in the end, we're still left in the dark as to what
singles *these* trajectories out from among many possible ones, and thus, *this*
collection of disc-wide GRN expression values from among many possible
multi-cellular expression configurations. However, a natural way forward
should suggest itself: if we want to see why *these* pathways are privileged,
we ought to compare them to a set of less developmentally fortunate ones.

Given the representative machinery of our multi-dimensional state-space, we can do just that, as mapping out a trajectory on this space only requires our picking a state (a disc-wide cellular GRN expression profile) and iteratively applying the associated GRN regulatory logic to derive its temporally successive states. In other words, "determining the next move" of a developmental trajectory within state-space from any state requires a simple *conditionalising* process: for any particular regulatory network, by plugging in a specific set of expression values for the members of that network, and applying the activities of the causal connectives which constitute its regulatory logic, we can derive its members' subsequent expression values. Thus, because the regulatory logic of a GRN effectively acts to assign a Boolean function to each state within this state-space, we can vectorise any single state and trace the directionality of temporally successive states within that space.[46] We can, in other words, plot any possible developmental trajectory for a particular imaginal disc.[47]

If we do so, after a significant number of iterations, we find that the collection of these trajectories exhibit interesting properties. Firstly, we find that localised collections of trajectories follow similar curvatures through state-space: they appear to "stick together", bending around similar regions of that space. Secondly, we find that multiple trajectories end in the same general areas in state-space: these regions appear to "attract" trajectories from various originating points within that space. As one may have guessed, these regions correspond to the disc-wide expression states that define the various morphological permutations which comprise the module's morphospace.

Notice that taking a "bigger picture" look at the characteristics of this state-space reveals precisely the features we were interested in, for here we see *privileged* permutations *qua* attractor-regions (e.g., φ in *Figure 8.3*) and *constraints* on possible permutations *qua* curvature structures on that space. What we want to know then is: what explains this *shaping* of state-space? We've seen that the developmental transition from any particular point in state-space to the next is determined by a kind of Boolean function which utilises the GRN's regulatory logic operating on the particular expression profile of the GRN elements which define that state. However, the transitions between states in this space is not reflective of merely simple analytical operations—for note that the transition-function in question is a *regulatory* one, and so each step within a single trajectory is a step toward disc-wide regulatory *stability*. In other words, although the state-to-state transitions within that space take place according to the aforementioned Boolean model, each step throughout developmental time is in fact a transition from a *less* stable disc-wide expression profile to a *more* stable one, given the relevant regulatory structure. So, from any origination point within that space, the subsequent state-transitions which comprise its trajectory follow the multi-cellular expression profile of the disc's "search" for *regulatory stability*, where the relevant GRN elements' expressions "even-out" in such

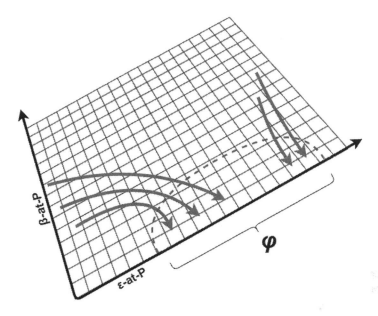

Figure 8.3 Schematic representation of a simplified, two-dimensional state-space depicting a small selection of a module's developmental trajectories. This truncated state-space represents the disc-wide cellular expression levels of the module with respect to two cell types (β and ε) in a particular spatial region (posterior, P). Multiple individual trajectories (depicted as arrows) from distinct initial conditions converge on a general region (φ) of developmental end-states with quantitatively similar, spatially specific, disc-wide cellular expression values (with respect to ε and P).

a way that their collected values no longer cause further significant inter-network expression alterations.

With this in mind, we can add another aspect to our state-space: each state can be given a *stability measure* which specifies the GRN elements' expression values tendency to substantially shift (given the relevant regulatory logic) to a subsequent state; in effect, in this process, we are properly vectorising the state-space, in that the arrows we earlier assigned to each state now have a direction *and* a kind of magnitude.[48] In DST modelling, this aspect is represented by assigning each state a particular *elevation* value (along another dimension), where the *higher* the elevation value, the relatively higher level of expression *instability* of the state—i.e. the more likely the disc-wide expression values of its GRN elements will shift (again, given the relevant regulatory relations in operation).[49] Once we have done so, our abstract state-space is now a structured *topology* complete with high hills and low-lying basins with various gradient measures connecting them.

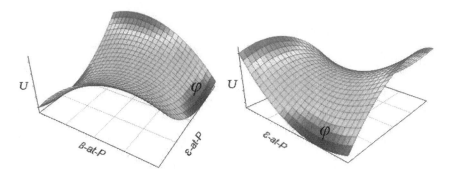

Figure 8.4 Schematic topological representation of the state-space from Figure 8.3. The third dimension (U) reflects the elevation level of any particular disc-wide spatially-specific expression profile for any specific coordinate, itself a measure of the relative regulatory stability; here, a higher U-value and warmer colouration are inversely correlated with regulatory stability. φ, denoting a set of quantitatively similar developmental end-states with respect to ε-type expression profiles within the posterior compartment of a module (P), is shown as a low-lying basin within state-space. NB. although representing a complete such topology for a particular module would require a rather complex, multi-dimensional state-space, the same principles at play in this schematic would apply.

With this stability-based topological mapping of our state-space in hand, we can now understand the process of the development of a particular module in a novel fashion: if we depict the state of the module as a kind of frictionless orb, we can model the temporal succession of various distinct states of the module throughout the process of development as the dynamic trajectory of that orb through a pathway geometrically constrained by the topological ridges and valleys of the system's Boolean regulatory configuration. This novel modelling puts us in the position to understand more clearly why, for the corresponding morphospace of a particular module, *these* morphological permutations are privileged, and *why* they are so: they represent disc-wide patterns of *regulatory stability* with respect to intra-module cellular expression states which "carve out" wide, low-lying basins in the topology of state-space, and their privilege consists in the fact that the dynamics of the process of development is shaped and constrained by the geometric curvature of that topology.[50] In this way, the framework of DST affords us a more complete picture of the rich generative capacity of a developmental module—for it not only allows us to understand a module's ability to produce the varied morphological permutations which comprise its associated morphospace, but, importantly, also the causal-*cum*-structural "shape" of that capacity with respect to both the developmental privileging of and constraint on those permutations. Thus, by utilising the conceptual resources of DST we are able to more fully "capture the nature" of a developmental module, having been equipped with the explanatory resources necessary to account for the multi-faceted character of the developmental process of its associated morphological trait.

Importantly, however, note that in order to have this rich understanding of the nature of a developmental module, we have had to *abstract away* from its compositional particularities and their mechanistic interactions in an appeal to a *higher-order structure* which is neither a compositional part of the module, nor strictly reducible to any such part (or set of such parts).[51] Of course, this process of abstraction required an initial appeal to its compositional elements and their mechanistic arrangement in order to define a network and its associated regulatory logic, but the resulting topology from which we have drawn the aforementioned explanatory prowess (a) is itself constructed purely from a set of functionally defined, weighted Boolean connectives which (b) form a continuous mapping over an exhaustive set of various iterations on the values of those compositional elements and their causal connectives. In as much as functionally defined operators are unable to qualify as "proper parts" of a biological system, (a) entails that this topology cannot be strictly understood as a contributing to the constitution of a module. Furthermore, given that a highly abstract, functional mapping which plots the interrelation of every possible configuration of an entire system is incapable of being bijectively assigned to the set of elements which compose that system, (b) illustrates the irreducibility of a topology to such a set; here, you might say, 'the possible' outstrips 'the actual'.

Note further that, in utilising the explanatory resources afforded by our topological understanding of a developmental module, we have had to conceptualise it as a higher-order, *dynamically holistic* system: these are resources granted to us only by modelling the system's causal activity as an iterative operation on a continuous, integrative mapping of its entire collection of possible system-wide state-values. Indeed, each point in the collection that comprises a complete state-space is intimately connected to its neighbouring points to form a smooth gradient contour so that the resulting geometry of that topology—and thus, its dynamic "flow"—cannot be attributed to any particular GRN element, nor the entire GRN, but only to the system *as a whole*, by taking into account its exhaustive set of possible disc-wide expressions states.[52] For within that topology, each individual vector is merged into a *holistic* dynamic structure, and it is this integrated flow (and not the specification of any underlying operating mechanisms) which plays an explanatory role with respect to the multi-faceted developmental of the module's characteristic morphological feature *via* system-wide stability measures and their resulting topological curvature.

What's more, the flow which characterises this higher-order structure doesn't just represent the dynamic activity of the system acting *as a whole* (as "one"), but as a whole with respect to *its directedness toward certain states*: the flow of the system, characterised by its vector-summed stability measures, presents a topology whose geometrical configuration *directs* a module's process of development *toward* certain morphological end-states *qua* disc-wide expression patterns of regulatory stability.[53] For the system's causal progression, represented as the temporal traversal of that process through the two-dimensional state-space of disc-wide expression profiles, is no random walk—it is *guided by* (and *restricted by*) the three-dimensional contours of its holistically defined topology *toward* certain developmentally *privileged* morphologies. Importantly, this "goal-directedness" which bestows explanatory utility with respect to morphological development is attributable only to that topology, and thus to the system *as a whole*—as we have seen, the stability-measure which defines that topology cannot be gleaned from the mere specification of the module's GRN, or any single iteration of that GRN within a possible disc-wide expression state, or even any particular developmental trajectory guided by the regulatory strictures of that GRN.

6. Hylomorphic Modules: Explanation and Ontology

With all of the above in mind, the point I wish to make ought to be clear: in order to have a sufficiently rich understanding of the nature of a developmental module and its associated generative capacity, we have had to appeal to a *holistic* conception of its system-wide *causal structure* in which its various possible developmental trajectories *toward* particular morphological end-states are *dynamically united*.

Importantly, although this higher-order causal structure to which we must appeal is in an intimate way metaphysically tied-up with the mereological makeup of a developmental module, as its constitutional elements specify the module's possible expression profile (which define its corresponding state-space) and the regulatory logic which governs the temporal transitions between them, the preceding discussion has strongly indicated that this abstracted causal structure is importantly *conceptually independent* of that make-up, in that each state which comprises its space is defined *functionally* (as a weighted Boolean function), and the resulting topological structure, *qua* functional mapping, is conceptually independent of the mechanistic particularities of the activities of the module's GRN elements. This is further evidenced by the fact that a wide variety of permutations in the mereological makeup of a module which are nonetheless causally connected by the same regulatory architecture will result in that system's higher-order, topological structure being unchanged:[54] thus, a particular *geometrical-cum-dynamical mapping* cannot be conceptually wed to any particular set of constitutional elements.[55] Indeed, the now popular evo-devo project of individuating homologue-specifying developmental modules *via* processual definitions, itself grounded in the overwhelming evidence that *distinct* GRNs have underwritten the *same* developmental modules over time,[56] depends upon this fact.[57]

Furthermore, although this higher-order structure is conceptually distinguishable from the diverse array of its mereological underpinnings, it cannot for that reason be regarded as a mere heuristic artefact, as an appeal to its nature licences unique *explanatory* and *predictive* power with respect to the causal structure of the process of morphological development.[58] As we have seen, understanding the process of the development of a particular morphological feature as a dynamic traversal through a topological mapping of expression stability affords a novel, non-mechanistic explanation of the *shape* and *structure* of a module's developmental capacity: this is an explanatory oblation purchased by an appeal to a module as a higher-order, dynamically integrated system, rather than by its mere characterisation as a specific set of "entities and activities".[59] But this understanding also provides novel, non-mechanistic *predictive* power with respect to that process, for the particularities of the higher-order, pseudo-kinetic curvature of the system's stability topology licences inductive inferences regarding both the probability of the module following particular developmental trajectories (under certain conditions, and more generally) and the probability of the module producing particular morphological permutations (under certain conditions, and more generally). This prowess is exhibited perhaps most prominently in cutting-edge cell biology, where the regulation and re-programmability of cell fate is analysed *via* the higher-order topological dynamics of stem cells,[60] but it is present (and increasingly so) in the study of everything from plant morphology[61] to carcinogenesis.[62]

Thus, in satisfaction of the twinned goal I earlier introduced, I have shown not only (a) that a higher-order, holistically dynamic, goal-directed structure can be conceptually distinguished from the particular vagaries of a developmental module's mereological underpinnings, but also (b) that by appealing to this structure, one is afforded a wealth of unique explanatory resources with respect to the generative capacity of that module and its associated morphospace. In other words, to return to our original formulation, I've shown that fully "capturing the nature" of a developmental module requires not only having a grip on its specific constitutive collection of genetic elements and the particular arrangement of their causal connectives, but also on the dynamically directed topology of its higher-order causal structure. Or, to put it yet another way: it is a job which requires an appeal to both matter *and* form.

While providing a plausible, empirically informed vindication of the Lockean definitional project of hylomorphism in the biological realm—which has been the sole aim of this paper—is no trivial task, one might yet wonder what the metaphysical worth of this toil is: what does a successful defence of (a) and (b) tell us, for instance, about the *ontology* of organisms? In line with the purpose of this paper, as stated in §1, I have intentionally remained silent on this issue in the hope that the results of the discussion might be of applicable value to a wide variety of specific accounts (of the kind earlier mentioned), and not stand or fall on the posits of any particular ontology. And although for that reason, I have refrained from giving those results any ontological gloss, I think it's instructive to end by briefly more explicitly noting the ways in which they aren't in any way inimical to, and in fact offer conceptual support to, the typical ontological claims of contemporary hylomorphism.

Note first that showing that (a) is true is a prerequisite for attempting to defend the truth of the central claim of hylomorphism—that fully capturing the nature of an entity requires an appeal to both matter and form: whatever your particular ontological commitments, if the nature of entities cannot be shown to be at the very least *conceptually* bipartite, that claim is clearly off the table. Of course, (a) being true only secures the conceptual independence of form from matter, and one might reasonably expect a project which aims to aid the cause of hylomorphic ontologies to do better: wouldn't showing that form is also *existentially independent* from matter be of more use? In this instance, the answer is no. For although hylomorphism conceptualises entities as ontological *unities* of form and matter, this is a unity which is not taken to be established by metaphysically tying together—either through composition or connection[63]—two existentially separate sub-entities. And because hylomorphism denies the very possibility of the existence of *uninformed* matter, or *immaterial* form, a call for the truth of something more robust than (a) betrays a fundamental misunderstanding of the doctrine.

That said, however, vindicating the conceptual independence of form aids in supporting the ontological claims of hylomorphism in only a limited fashion—namely, by securing a metaphysical foundation for them. Showing that (b) is true, on the other hand, may go some way further in that task. If (b) is true, and the higher-order dynamic structure of developmental modules licenses *irreducibly novel* explanatory power with respect to the ontogenic processes of its mereological makeup, then, plausibly, given that explanation often traces causation, we may have *prima facie* reason for thinking that structure possesses irreducibly novel *causal power*. Importantly, while this sort of move is certainly defeasible, any proposed annulment of it on the grounds that "existential dependence *entails* causal ineffectuality" ought to be dismissed.[64] Not only would this sort of objection beg the question against hylomorphism, but as its defenders have been at pains to point out,[65] the *emergent* properties of entities which are typically acknowledged to existentially depend upon their 'realisation bases' are often assigned causal roles, and treated with ontological sincerity—a practice now widely adopted in contemporary developmental biology.[66]

If the holistically higher-order dynamic structure of developmental modules can be understood as a *causal* structure then, in line with the 'Eleatic Principle' ("to *be* is to *be powerful*")—widely adopted among neo-Aristotelians in the defence of dispositional realism—we have good reason for thinking it represents a fact about the *ontology* of those modules.[67] Indeed, the recent surge in support for adopting a Whiteheadean 'process ontology' in the philosophy of biology[68] can be seen as a reflection of the growing consensus that such mechanistically irreducible, higher-order causal structures must be understood as genuinely "carving at the joints" of organisms.[69]

Putting particular ontologies aside however, the more general lesson I wish to draw from the preceding discussion is that both (a) and (b) being true not only reflects the assumption in contemporary developmental biology that this *formal* structure is no mere metaphor, or philosophical phantasm, but also functions as the conceptual soil in which a neo-Aristotelian hylomorphic ontology might flourish. That said, though the further question as to whether and to what extent any of the ontologies currently on offer bear philosophical fruit is no doubt an important one, it is an enquiry I leave for another time.

7. Conclusion

Though the neo-Aristotelian congregation has grown considerably in recent years, most of its members have hesitantly refrained from adopting a doctrine historically central to its metaphysical catechism, and understandably so—for while many have demonstrated its theoretical plausibility, few have offered a compelling account of its empirical viability. Throughout this paper, by focusing on the biological realm, and appealing to recent theoretical advances therein, I have attempted to do just that. To that end, I've argued that the hylomorphic claim that fully "capturing the nature" of a biological individual requires an appeal both to it *qua* an organised, connected collection of discrete parts and *qua* a dynamically directed higher-order holistic structure can be given empirical content. In doing so, I've focused on a particularly important class of biological sub-systems with the hope that, given their role as developmental building blocks, the account can eventually be generalised to a higher-level hylomorphic account of organisms.[70] While that crucial work yet lies ahead, the hope is that this paper has shown it a task worth its toil by making a compelling case that the hylomorphic creed is one worthy of contemporary conviction.[71]

Notes

1 Kripke (1980; *cf.* Putnam 1975).
2 In his *Essay Concerning Human Understanding*, Locke referred to this as the dependence of an entity's 'nominal' essence upon its 'real' essence (Woolsey 1964). For an instance of this in the context of contemporary hylomorphism, see Oderberg (2011).
3 This contemporary notion of 'matter' is closest to what commentators have called 'functional matter' in Aristotle—see Lewis (1994). Notably, this contemporary formulation doesn't place any particular emphasis, as Aristotle did, on matter's definition as *pure potentiality* and its subsequent role in underlying accidental property-change.
4 Note that this won't involve any careful exegesis of Aristotle—the reader is free to think of these aspects of a formal definition as *neo-Aristotelian*.

5 Fine (1999), Johnston (2006).
6 Koons (2014).
7 Marmodoro (2013).
8 Johnston (2006), Rea (2011), Marmodoro (2013). The exception to this rule is Koslicki (2008), who views formal definitions as picking out some further "non-material", though mereological, part of an entity. However, as this isn't widely held, and as Aristotle himself expressly argued against this type of position—see *Metaphysics* Z 3-6 and 17 , Barnes (1984) -, I haven't considered her view in any detail here.
9 Robinson (2014), Jaworski (2016).
10 Johnston (2006), Oderberg (2007), Rea (2011), Marmodoro (2013).
11 Jaworski (2012).
12 Rea (2011).
13 Oderberg (2007), Rea (2011), Jaworski (2012), Marmodoro (2013).
14 Marmodoro (2013) refers to this phenomenon as the "re-identification" of an entity's constituents with respect to the function of its 'substantial form'.
15 Jaworski (2016).
16 Fine (1999), Johnston (2006), Koslicki (2008).
17 Oderberg (2007), Rea (2011), Jaworski (2012).
18 Marmodoro (2013).
19 The requirement that *form* provides novel *explanatory* power with respect to an entity's constituents, rather than a *causal* power *over them*, is explicitly defended by Rea (2011) and Jaworski (2012).
20 The choice of creature here was no accident—Aristotle was quite interested in sea creatures (in *History of Animals*, Barnes (1984)), and sea urchin mouths are now known as 'Aristotle's Lanterns'.
21 As in Jaworski (2012).
22 As in Rea (2011).
23 None of this is meant to suggest that these philosophers haven't dressed the doctrine with interesting and elucidating metaphysical flourishes—they certainly have. The point is simply that their doing so is often largely independent of any examination of the finer biological details. A notable exemption is Walsh (2006), and to a lesser extent, Boulter (2012).
24 Aristotle argues (in *Ethics* I.7, Barnes (1984)) that if the *parts* of a thing (a human eye, for instance) are understood as teleological—that is, having a *form*—so too must the *whole* thing (the human as an entire organism, in this case).
25 Raff and Sly (2000), Erwin and Davidson (2009).
26 Hall (2003), Davidson and Erwin (2006), Wagner (2014).
27 Brigandt (2007), Brakefield (2011), McCune and Schimenti (2012).
28 The case is even stronger if one thinks, as Clarke (2013) suggests, that any bits of our biological ontology upon which natural selection operates have the right to be called biological *individuals*.
29 Gurdon and Bourillot (2001), Tabata (2001), Mann and Carroll (2002).
30 For more on the regulatory "logic" found in GRNs, see Yeger-Lotem et al. (2004) and Alon (2007).
31 Salazar-Ciudad et al. (2003).
32 Pigliucci (2001), West-Eberhard (2003), Gilbert and Epel (2015).
33 Schlichting and Smith (2002), Aubin-Horth and Renn (2009).
34 Von Dassow and Munro (1999: 316).
35 Pigliucci et al. (1996: 81).
36 Love (2009: 57).
37 Gurdon and Bourillot (2001), Tabata (2001), Mann and Carroll (2002), Müller (2008).

38 Rasskin-Gutman (2005), Newman and Müller (2006), McGhee (2006), Wagner (2014).
39 Waddington (1957).
40 Kauffman (1969).
41 See Wang et al. (2011) and Huang (2012) for more details on the conceptual union between Waddington and Kaufmann in DST.
42 There are now a number of specialist journals which focus on holistic treatments of developmental phenomena—see, for instance, *Molecular Systems Biology* and *BMC Systems Biology*.
43 Bhattacharya et al. (2011), Verd et al. (2014).
44 Striedter (1998), Jaeger and Monk (2014).
45 For a (relatively) accessible introduction to how this mapping is done, both theoretically and with the aid of empirical data, see Huang (2009) and Wang et al. (2011).
46 Wang et al. (2011), Davila-Velderrain et al. (2015).
47 This is, of course, a rather complex task, given that performing it requires taking into account *multiple* cells, their spatial *arrangement*, and both *intra-* and *intercellular* regulatory interactions.
48 Kim and Wang (2007), Bhattacharya et al. (2011).
49 Technically, assigning an elevation value involves stochastic simulation of *groups* of cells, etc.—but I pass over this complication here. See Bhattacharya et al. (2011) for the finer details.
50 Kitano (2004), Huang (2009), Huneman (2010).
51 See Levy and Bechtel (2013) for a good discussion of this general sort of abstraction process in biological modelling.
52 Jaeger and Monk (2015).
53 Interestingly, Von Dassow and Munro (1999: 310) briefly note in passing the conceptual similarity between the causal privileging of end-state morphologies in DST models and an Aristotelian form of "goal-directedness".
54 Gilbert and Bolker (2001), Jaeger and Monk (2015).
55 Thus, in accord with the classic Aristotelian picture, 'form' will be *multiply realisable*—the "one over many"—in at least an explanatory sense. See Mitchell (2012) for a comprehensive look at the phenomenon's various incarnations in contemporary biology.
56 Rieppel (2005), Brigandt (2007), Love (2009), Wagner (2014).
57 For an account which more explicitly defines homologous morphological structures in the framework of DST, see Striedter (1998).
58 Even if the explanatory virtues provided by higher-order, dynamic models must ultimately somehow "bottom out" in the activity of mechanisms (as Kaplan and Craver 2011) argue), it's not clear that this detracts from their having genuinely novel explanatory power (Brigandt 2015); for an opposing view, see Kaplan (2015). I discuss these issues in Austin (2016b).
59 Cf. Huneman (2010: 214–19).
60 Bhattacharya et al. (2011), Li and Wang (2013).
61 Álvarez-Buylla et al. (2008).
62 Kaneko (2011).
63 As Aristotle makes clear in *Metaphysics* E:6, Barnes (1984).
64 There are, of course, other, independent reasons one might have for rejecting that move—see Robinson (2014) for a recent critique.
65 Rea (2011), Koons (2014), Jaworski (2016).
66 Boogerd et al. (2005), Mitchell (2012), Salazar-Ciudad and Jernvall (2013).
67 The principle originated in Plato's *Sophist* and was reintroduced into contemporary debates by Armstrong (1997).

68 Henning and Scarfe (2013), Dupré (2013), Jaeger and Monk (2015).
69 Waddington (1969) himself, the progenitor of the 'epigenetic landscape' concept, professed to being deeply influenced by Whitehead, as Gilbert and Bolker (2001) note. More recently, Hall (2013) has characterised the contemporary topological models of DST as having a natural home within a Whiteheadean ontology.
70 The conceptual resources utilised here may even be applicable to a hylomorphic account of biological 'natural kinds', the first steps of which are undertaken in Austin (2016a).
71 With thanks to audiences at both Oxford and Cambridge, and to the editors of this volume for their invaluable feedback on earlier versions of this paper.

Bibliography

Alon, U. (2007) 'Network Motifs: Theory and Experimental Approaches', *Nature Reviews Genetics*, 450–61.
Álvarez-Buylla, E., Chaos, A., Aldana, M., Benitez, M., Cortes-Poza, Y., Espinosa-Soto, C. et al. (2008) 'Floral Morphogenesis: Stochastic Explorations of a Gene Network Epigenetic Landscape', *PLoS ONE* 3(11): e3626.
Armstrong, D. (1997) *A World of States of Affairs* (Cambridge: Cambridge University Press).
Aubin-Horth, N. and Renn, S. (2009) 'Genomic Reaction Norms: Using Integrative Biology to Understand Molecular Mechanisms of Phenotypic Plasticity', *Molecular Ecology*, 3763–80.
Austin, C. J. (2016a) 'Aristotelian Essentialism: Essence in the Age of Evolution', *Synthese*, doi:10.1007/s11229-016-1066-4.
——— (2016b) 'The Ontology of Organisms: Mechanistic Modules or Patterned Processes?', *Biology & Philosophy* 31(5): 639–62.
Barnes, J. (1984) The Complete Works of Aristotle, Volumes I and II (Princeton: Princeton University Press)
Bhattacharya, S., Zhang, Q. and Andersen, M. (2011) 'A Deterministic Map of Waddington's Epigenetic Landscape for Cell Fate Specification', *BMC Systems Biology*, 1–11.
Boogerd, F. C., Bruggeman, F. J., Richardson, R. C., Stephan, A. and Westerhoff, H. V. (2005) 'Emergence and Its Place in Nature: A Case Study of Biochemical Networks', *Synthese* 145: 131–5.
Boulter, S. J. (2012) 'Can Evolutionary Biology Do Without Aristotelian Essentialism?', *Royal Institute of Philosophy Supplement*, 83–103.
Brakefield, P. (2011) 'Evo-Devo and Accounting for Darwin's Endless Forms', *Philosophical Transactions of the Royal Society*, 2069–75.
Brigandt, I. (2007) 'Typology Now: Homology and Developmental Constraints Explain Evolvability', *Biology & Philosophy*, 709–25.
Brigandt, I. (2015). 'Evolutionary developmental biology and the limits of philosophical accounts of mechanistic explanation', In P. A. Braillard & C. Malaterre (Eds.), *Explanation in biology: an enquiry into the diversity of explanatory patterns in the life sciences,* (pp. 135–173). Dordrecht: Springer.
Clarke, E. (2013) 'The Multiple Realizability of Biological Individuals', *Journal of Philosophy*, 413–35.
Davidson, E. and Erwin, D. (2006) 'Gene Regulatory Networks and the Evolution of Animal Body Plans', *Science*, 796–800.
Davila-Velderrain, J., Martinez-Garcia, J. C., and Alvarez-Buyila, E. R. (2015) 'Modeling the Epigenetic Attractors Landscape: Toward a Post-Genomic Mechanistic Understanding of Development', *Frontiers in Genetics* 6(160): 1–14

Dupré, J. (2013) 'Living Causes', *Proceedings of the Aristotelian Society Supplementary Volume*, 19–38.

Erwin, D. and Davidson, E. (2009) 'The Evolution of Hierarchical Gene Regulatory Networks', *Nature Reviews Genetics*, 141–8.

Fine, K. (1999) 'Things and Their Parts', *Midwest Studies in Philosophy*, 61–74.

Gilbert, S. and Bolker, J. (2001) 'Homologies of Process and Modular Elements of Embryonic Construction', *Journal of Experimental Zoology*, 1–12.

Gilbert, S. F. and Epel, D. (2015) *Ecological Developmental Biology: The Environmental Regulation of Development, Health, and Evolution* (Sunderland, MA: Sinauer Associates).

Gurdon, J. and Bourillot, P. (2001) 'Morphogen Gradient Interpretation', *Nature* 413: 797–803.

Hall, B. (2003). 'Evo-Devo: Evolutionary Developmental Mechanisms', *International Journal of Developmental Biology*, 491–5.

Hall, B. (2013). 'Epigenesis, Epigenetics, and the Epigenotype: Toward An Inclusive Concept of Development and Evolution', In B. Henning, & A. Scarfe (Eds.), *Beyond Mechanism: Putting Life Back Into Biology*, (pp. 345–368). Plymouth: Lexington Books.

Henning, B., & Scarfe, A. (Eds.). (2013). *Beyond Mechanism: Putting Life Back Into Biology*. (Plymouth: Lexington Books).

Huang, S. (2009) 'Reprogramming Cell Fates: Reconciling Rarity With Robustness', *Bioessays*, 546–60.

—— (2012) 'The Molecular and Mathematical Basis of Waddington's Epigenetic Landscape: A Framework for Post-Darwinian Biology?', *Bioessays*, 149–57.

Huneman, P. (2010) 'Topological Explanations and Robustness in Biological Sciences', *Synthese*, 213–45.

Jaeger, J. and Monk, N. (2014) 'Bioattractors: Dynamical Systems Theory and the Evolution of Regulatory Processes', *Journal of Physiology*, 2267–81.

—— (2015) 'Everything Flows: A Process Perspective on Life', *EMBO Reports*, 1064–7.

Jaworski, W. (2012) 'Hylomorphism: What It Is and What It Isn't', *Proceedings of the American Catholic Philosophical Association*, 173–87.

—— (2016) 'Hylomorphism: Emergent Properties Without Emergentism', in M. Garcia-Valdecasas and N. Barrett (eds.) *Biology and Subjectivity: Philosophical Contributions to Non-Reductive Neuroscience* (Springer).

Johnston, M. (2006) 'Hylomorphism', *Journal of Philosophy*, 652–98.

Kaneko, K. (2011) 'Characterization of Stem Cells and Cancer Cells on the Basis of Gene Expression Profile Stability, Plasticity, and Robustness', *Bioessays* 33(6): 403–13.

Kaplan, D. (2015) 'Moving Parts: The Natural Alliance Between Dynamical and Mechanistic Modeling Approaches', *Biology & Philosophy*, 757–86.

Kaplan, D. and Craver, C. (2011) 'The Explanatory Force of Dynamical and Mathematical Models in Neuroscience: A Mechanistic Perspective', *Philosophy of Science*, 601–27.

Kauffman, S. A. (1969) 'Metabolic Stability and Epigenesis in Randomly Constructed Nets', *Journal of Theoretical Biology*, 437–67.

Kim, K. and Wang, J. (2007) 'Potential Energy Landscape and Robustness of a Gene Regulatory Network: Toggle Switch', *PLoS Computational Biology*, 0565–77.

Kitano, H. (2004) 'Biological Robustness', *Nature Reviews Genetics*, 826–37.

Koons, R. (2014) 'Staunch vs. Faint-hearted Hylomorphism: Toward an Aristotelian Account of Composition', *Res Philosophica*, 151–77.

Koslicki, K. (2008) *The Structure of Objects* (Oxford: Oxford University Press).

Kripke, S. (1980) *Naming and Necessity* (Cambridge, MA: Harvard University Press).

Levy, A. and Bechtel, W. (2013) 'Abstraction and the Organization of Mechanisms', *Philosophy of Science*, 241–61.

Lewis, F. (1994) 'Aristotle on the Relation Between a Thing and Its Matter', in T. Scaltsas, D. Charles and M. L. Gill (eds.) *Unity, Identity, and Explanation in Aristotle's Metaphysics* (Oxford: Oxford University Press): 247–77.

Li, C. and Wang, J. (2013) 'Quantifying Cell Fate Decisions for Differentiation and Reprogramming of a Human Stem Cell Network: Landscape and Biological Paths', *PLoS Computational Biology* 9(8): e1003165.

Love, A. (2009) 'Typology Reconfigured: From the Metaphysics of Essentialism to the Epistemology of Representation', *Acta Biotheoretica*, 51–75.

Mann, R. and Carroll, B. (2002) 'Molecular Mechanics of Selector Gene Function and Evolution', *Current Opinion in Genetics & Development*, 592–600.

Marmodoro, A. (2013) 'Aristotle's Hylomorphism Without Reconditioning', *Philosophical Inquiry*, 5–22.

McCune, A. and Schimenti, J. (2012) 'Using Genomic Networks and Homology to Understand the Evolution of Phenotypic Traits', *Current Genomics*, 74–84.

McGhee, G. (2006) *The Geometry of Evolution: Adaptive Landscapes and Theoretical Morphospaces* (Cambridge: Cambridge University Press).

Mitchell, S. (2012) 'Emergence: Logical, Functional and Dynamical', *Synthese*, 171–86.

Müller, G. (2008) 'Evo-Devo as a Discipline', in A. Minelli and G. Fusco (eds.) *Evolving Pathways: Key Themes in Evolutionary Developmental Biology* (Cambridge: Cambridge University Press): 3–29.

Newman, S. and Müller, G. (2006) 'Genes and Form: Inherency in the Evolution of Developmental Mechanisms', in E. Neumann-Held and C. Rehmann-Sutter (eds.) *Genes in Development: Re-Reading the Molecular Paradigm* (Durham: Duke University Press): 38–77.

Oderberg, D. (2007) *Real Essentialism* (New York: Routledge).

——— (2011) 'Essence and Properties', *Erkenntnis*, 85–111.

Pigliucci, M. (2001) *Phenotypic Plasticity: Beyond Nature and Nurture* (Baltimore: Johns Hopkins University Press).

Pigliucci, M., Schlichting, C., Jones, C. and Schwenk, K. (1996) 'Developmental Reaction Norms: The Interactions Among Allometry, Ontogeny and Plasticity', *Plant Species Biology*, 69–85.

Putnam, H. (1975) 'The Meaning of "Meaning"', *Minnesota Studies in the Philosophy of Science*, 131–93.

Raff, R. and Sly, B. (2000) 'Modularity and Dissociation in the Evolution of Gene Expression Territories in Development', *Evolution and Development*, 102–13.

Rasskin-Gutman, D. (2005) 'Modularity: Jumping Forms Within Morphospace', in W. Callebaut and D. Rasskin-Gutman (eds.) *Modularity: Understanding the Development and Evolution of Natural Complex Systems* (Cambridge: MIT Press): 207–19.

Rea, M. (2011) 'Hylomorphism Reconditioned', *Philosophical Perspectives*, 341–58.

Rieppel, O. (2005) 'Modules, Kinds, and Homology', *Journal of Experimental Zoology*, 18–27.

Robinson, H. (2014) 'Modern Hylomorphism and the Reality and Causal Power of Structure: A Skeptical Investigation', *Res Philosophica*, 203–14.

Salazar-Ciudad, I. and Jernvall, J. (2013) 'The Causality Horizon and the Developmental Bases of Morphological Evolution', *Biological Theory*, 286–92.

Salazar-Ciudad, I., Jernvall, J. and Newman, S. (2003) 'Mechanisms of Pattern Formation in Development and Evolution', *Development*, 2027–37.

Schlichting, C. and Smith, H. (2002) 'Phenotypic Plasticity: Linking Molecular Mechanisms With Evolutionary Outcomes', *Evolutionary Ecology*, 189–211.

Striedter, G. (1998) 'Stepping Into the Same River Twice: Homologues as Recurring Attractors in Epigenetic Landscapes', *Brain, Behavior and Evolution*, 218–31.

Tabata, T. (2001) 'Genetics of Morphogen Gradients', *Nature*, 620–30.

Verd, B., Crombach, A. and Jaeger, J. (2014) 'Classification of Transient Behaviours in a Time-Dependent Toggle Switch Model', *BMC Systems Biology*, 1–19.

Von Dassow, G. and Munro, E. (1999) 'Modularity in Animal Development and Evolution: Elements of a Conceptual Framework for Evo Devo', *Journal of Experimental Zoology*, 307–25.

Waddington, C. H. (1957) *The Strategy of the Genes* (London: George Allen & Unwin).

Waddington, C. H. (1969). 'The Practical Consequences of Metaphysical Beliefs on a Biologist's Work: An Autobiographical Note', In C. H. Waddington (Ed.), *Towards a Theoretical Biology 2: Sketches* (pp. 72–81). Edinburgh: Edinburgh University Press.

Wagner, G. (2014) *Homology, Genes, and Evolutionary Innovation* (Princeton: Princeton University Press).

Walsh, D. (2006) 'Evolutionary Essentialism', *British Journal of the Philosophy of Science*, 425–48.

Wang, J., Zhang, K., Xu, L. and Wang, E. (2011) 'Quantifying the Waddington Landscape and Biological Paths for Development and Differentiation', *Proceedings of the National Academy of Sciences of the United States of America*, 8257–62.

West-Eberhard, M. (2003) *Developmental Plasticity and Evolution* (New York: Oxford University Press).

Woolsey, A.D. (1964) John Locke: An Essay Concerning Human Understanding, (London: Fontana Library)

Yeger-Lotem, E., Sattath, S., Kashtan, N., Itzkovitz, S., Milo, R., Pinter, R. Y., Alon, U. and Margalit, H. (2004) 'Network Motifs in Integrated Cellular Networks of Transcription-Regulation and Protein-Protein Interaction', *Proceedings of the National Academy of Sciences*, 5934–9.

9 The Great Unifier

Form and the Unity of the Organism

David S. Oderberg

1. Introduction: The Unity Problem

In his monumental treatise *On Growth and Form*, the famous mathematical biologist D'Arcy Wentworth Thompson commented as follows: 'The biologist, as well as the philosopher, learns to recognise that the whole is not merely the sum of its parts. It is this, and much more than this. For it is not a bundle of parts but an organisation of parts, of parts in their mutual arrangement, fitting one with another, in what Aristotle calls "a single and indivisible principle of unity"; and this is no merely metaphysical conception, but is in biology [a] fundamental truth . . .'.[1]

The kind of unity to which Thompson is referring—the organisational harmony of interacting parts—is a phenomenon we find both within and without biology. The unity of chemical compounds, of an atom, a molecule, of a lump of iron or uranium, is also a subject of wonder, a phenomenon asking for an explanation. In biology, however, as Thompson makes clear, following Aristotle before him, there is a special kind of unity. Terence Irwin puts it thus when commenting on Aristotle's discussion of animal souls:[2] '. . . a collection of flesh and bones constitutes a single living organism in so far as it is teleologically organized; the activities of the single organism are the final cause of the movements of the different parts'.[3]

For the many philosophers who reflexively recoil at talk of teleology and final causes, the idea can be put in a different yet familiar way: organisms act for their own sustenance, maintenance, and development. Their parts all serve the overall goal of the organism's flourishing. The organism, unless it has reason, does not set itself this goal; and even rational animals such as ourselves do not set every element of our goal of flourishing as human beings: much of what we do is no more than what happens to us or consists of the processes we inevitably undergo for our own sustenance, maintenance, and development. Yet the goal is there, however we got it and however any organism of any kind got it. Using more traditional terminology, I claim that organisms display *immanent causation*: causation that originates with an agent and terminates in that agent for the sake of its *self-perfection*. By 'self-perfection' I do not mean that there is some *ideal type* that every organism strives to reach. The idea is far more modest—namely that every organism aims, whether consciously or not, at the fulfilment of its potentialities such that it achieves a good state of being, indeed the best state it can reach given the limitations of its kind and its environment. Immanent causation is a kind of teleology, but metaphysically distinctive in what it involves. It is not just action for a purpose, but for the agent's *own* purpose, where 'own purpose' means not merely that the agent acts for a purpose it possesses, but that it acts for a purpose it possesses such that fulfilment of the purpose contributes to the agent's self-perfection.

This unity cries out for explanation. There is, as I have already implied, a 'unity problem' for all substances, organic and inorganic:[4] to put it crudely, what holds their essences together?[5] This has also been called the 'problem of complex essences', in other words, the question of 'the linkage of inherently separable components into a single kind-essence'.[6] A typical example is the electron: it shares its unit negative charge with the tau lepton but not its mass; it shares its mass with the positron but not its charge. The same applies to organisms: the flying squirrel (tribe *Pteromyini*) and sugar glider (species *Petaurus breviceps*) share a gliding membrane but not a pouch; the latter shares a pouch with the kangaroo (genus *Macropus*) but not a gliding membrane. The properties I have mentioned are all essential:[7] they are partially definitive of the things to which they belong. Yet they are also really distinct: they are separately instantiated in different kinds of thing yet co-instantiated in others. When they are found together, what holds them together? In the case of organisms the unity problem is even more acute than in the inorganic case, since not only are there distinct yet co-instantiated essential properties, but these properties all subserve the organism's overall flourishing, and the organism itself seeks to bring about its overall flourishing by employing its own parts, powers, and other characteristics. In other words, the organism, by engaging in immanent causation, displays a further kind of unity beyond the harmony and integration of its parts.

Now the immediate objection one is likely to raise is that there is no unity problem, only a pseudo-problem. What could it mean for essential properties to be 'held' together other than that, in a given case, they are properties belonging to the same essence? This seems to be the view of Jonathan Lowe, for whom the particular but regular combination of powers and liabilities in members of a given kind consists in the fact that, precisely, the objects in question *are* members of a given kind and that kinds are real universals[8]. Although this account, as Lowe points out, reduces the number of brute facts that must be countenanced compared to nominalism, brute facts there are nevertheless. It might be thought I am unfair to Lowe here, since he also holds that kinds are governed by laws linking them with their attributes, which adds some depth, as it were, to the bruteness.[9] But Lowe's official view of laws is that they *consist in* the characterisation of kinds by attributes,[10] so no new information is added to the account. What if one had a different view of laws, say that they involved some sort of metaphysical determination or production;

on such a view, one would hold that the essential behaviour or operation of members of a kind was governed by the essence of the kind. One might say that kangaroos nurse their young in a pouch because this is metaphysically necessitated by their essence. But even so, this cannot account for all essential properties. Since having a pouch and having a flying membrane can come apart, there cannot be a *law* uniting them if the law involves necessitation. But if the law is contingent, what kind of law is it? If it is a law of biology, what law? If metaphysical, how can it be contingent? Further, kangaroos are essentially mammals, but it is hardly a *law* that they are mammals in any sense beyond that being a mammal is part of the essence of being a kangaroo. Similarly, it is not a law that electrons have negative unit charge: it's just part of what it is to *be* an electron.[11]

Attempting to explain unity in terms of laws is bound to fail. Taking it to be a brute fact is also unacceptable. There is a difference, or so it seems, between an organism and an organ, on the one hand, and on the other between an organism and a collective of which it is a member, such as a colony.[12] That is why appealing simply to immanent causation is insufficient to mark out organisms as a unique category of living thing. Organs, too, work for their own self-perfection: consider homeostasis within the organism, self-repair, intake of nutrients, and so on. An organ[13] has a similar unity to the organism of which it is a part—call it, for now, *tight*. Yet the organ is *subservient* to the organism in a way that the organism is not subservient to anything. Again, many collectives—consider ant and bee colonies, among many others—also work for their self-perfection. Unless we merely stipulate—which seems ad hoc—that immanent causation excludes this kind of colonial collective agency, we should accept that living collectives, too, display immanent causation. By contrast with the organ, however, a collective has a similar unity to the organism that is a member of it inasmuch as neither are subservient to anything in the way the organ is subservient to the organism. But the collective's unity is *loose*. So the organ's unity is tight but subservient, the collective's is loose but not subservient, and the organism's is tight but not subservient. What is the metaphysical explanation of these differences?

In what follows, I explore and defend the traditional distinction between organs, organisms, and collectives by utilising Aristotelian conceptual tools that have, for one reason or another, fallen out of favour. Along the way I will consider a number (though by no means all) of the hard cases that have been raised in the literature as a possible threat to this tripartite metaphysical distinction. The pivotal concept for clarifying and defending it is that of substantial form, which is where I begin the analysis.

2. Form as a Unifying Principle

The Aristotelian hylemorphist claims that the differences are to be accounted for in terms of *form*, more precisely *substantial* form. Forms are universal determining principles whereby things are endowed with substantial natures and accidental characteristics. Are forms just what we call universals? It depends what one means by 'universal'. There are a number of things that can be said here, but for my purpose, the main point is that there are universal forms, whether of substance or of accident, but the former should not be thought of as *kinds* along the lines of, say, Lowe's four-category ontology,[14] except by way of synecdoche, inasmuch as having a substantial form entails membership of a substantial kind. There is the form of the Eastern Gray Kangaroo (*Macropus giganteus*), which determines the animals that have it to be in a corresponding substantial kind. Being in a substantial kind is, so to speak, part of what it is to have a substantial form. But the form is not the kind, as seen by the fact that on the hylemorphic theory there is only one substantial form per substance—the famous doctrine of the *unicity* of form—whereas every substance instantiates more than one kind. Boxer the Easter Gray Kangaroo has the single form of the Eastern Gray Kangaroo but instantiates numerous kinds, such as that of, once again, *Eastern Gray Kangaroo* but also the kind *marsupial* and the kind *mammal*. Membership in all the higher metaphysical genera is explained by the substantial form inasmuch as there are real features possessed by Boxer, in virtue of his form, that are shared by marsupials and mammals that are not Eastern Gray Kangaroos. But these various groupings of features are *abstractions* from Boxer's form and from the form of any other Eastern Gray Kangaroo, among other infima or lowest species, again to use the hylemorphic terminology. There is no space to defend the unicity of form here;[15] I raise it only to clarify the difference between form and kind.

As well as substantial forms there are *accidental* forms, such as being cloven-hoofed, and these too are universals. Kinds of substances, being universals, have multiple instances—individual substances such as mammals. Accidental forms have particular accidents as their instances, now called tropes but traditionally called modes; cases of cloven-hoofedness are an example. These universals exist both in the mind and in reality, the difference being that in the mind the universal exists as a single, unified idea, whereas outside the mind the universal exists as *multiplied*: it exists *in* its instances or, to use Lewis's well-known and apt description, it is 'wholly present wherever and whenever it is instantiated'.[16] In other words, although the *selfsame* universal is wholly present in each instance, it is *multiplied* in the sense of having multiple instances.[17] The same goes for substantial forms: they are grasped *as* unified ideas by the mind, but in reality, they are *multiplied* in their instances, which are *particularised* forms. The form of Skippy the Eastern Gray Kangaroo is the same as that of Boxer the Eastern Gray Kangaroo inasmuch as they share the universal form of that species, which explains their belonging to the correlative universal kind. Whereas substantial kinds have substances as instances, however, substantial forms have *particular forms* as instances. Skippy and Boxer have different *particular* forms inasmuch as each possesses its own principle of unity, which is also the principle of its specificity as a certain kind of kangaroo, and this because, and as surely as, they are each a distinct, individual kangaroo. We can see now that for the hylemorphist, while in many respects forms—henceforth, I mean *substantial* forms, unless otherwise indicated—are on a par with other universals, the former nevertheless do their own metaphysical work.

Our concern is with form's work as the *unifier* of an organism—an individual living substance. The first thing that needs to be appreciated is that this is not strictly a scientific but a metaphysical matter. Form is not a scientific postulate but a metaphysical one. One way of thinking of it as an *organising principle*, where by 'principle' I mean, as the scholastic philosophers did in this context, a real, objective *cause* of something's being the kind of thing it is, what Aristotelians call a 'formal cause'. Not every cause is efficient, on this picture of reality: form as organising principle is a cause in the sense of being metaphysically responsible for something's having a certain nature. As such, form—the formal cause—is not the sort of thing a biologist or any other natural scientist could ever discover. What they discover are the *kinds* of things there are, to be sure, but they do not discover that form is responsible for the essential unity of any kind of thing, either as a kind or as an instance of a kind.

Biologists have not and could not discover the existence of form any more than a physicist could discover, or ever did discover, the existence of *matter*. It is through properly philosophical reflection that we know such things must exist. Without going into detail here, matter is known to us as the metaphysical principle of change and potentiality. What we, either as ordinary observers or scientists, know are the particular material objects that exist. What we know *philosophically* is that that they have something in common that is the permanent substrate, to put it tendentiously, of their change and powers. So matter, as understood in purely metaphysical terms, is not the everyday matter we bump into when we interact with different kinds of thing. It is a metaphysical *posit* without which, claims the Aristotelian, insoluble philosophical problems arise, and which *underlies* the everyday matter of our common experience. The same goes for form. We know *philosophically* that substances have something in common that is responsible for their unity and specificity, but we know through observation, whether ordinary or scientific, the particular forms of substances that exist. So on this score the by-now stale derision of substantial forms that we have inherited from Galileo, Descartes, Spinoza, Hume et al. can be seen to be far less compelling than most philosophers, raised in these post-scholastic world views, have thought. Put another way—and quite gesturally, I accept[18]—the early modern rejection of substantial forms owes more to anti-scholastic prejudice, in my view, than to irresistible philosophical critique.[19]

We should, however, only postulate form as a metaphysical principle if it can do work in explaining the unity that needs explaining. What I want to focus on here is that aspect of unity whereby an organism is clearly neither an organ nor a collective of substances, be they other organisms or anything else. I take this to be intuitively clear even if the boundaries between organs, organisms, and collectives are hard to draw. One might object that we have no pre-theoretical intuition as to the special metaphysical status of the organism; I take it that John Dupré would agree, espousing as he does a pluralistic account of the ways in which cells 'combine to form integrated biological wholes'.[20] Multicellularity, he goes on to say, comes in many varieties, and we should not think of the organism in a 'naïve and static way' as a 'living individual'. It is, rather, a 'process' or 'life cycle'.[21]

Yet is the intuition so easy to dismiss? It cannot be a scientific discovery that the intuition does or does not latch onto reality: who made the discovery one way or the other? Biologists have, to be sure, discovered all kinds of unicellular entities as well as multicellular organisation, but no biologist discovered—or *could* discover—that there is no difference in kind between organisms as living individual substances, organs that are material parts subservient to organisms, and aggregates of which organisms are members. The difference between parts of substances, substances, and aggregates of which substances are members cuts across the entire ontological realm: it is not special to biology but reflects the way the world in general is organised. And here we have the first philosophical argument in favour of the intuition: that if these categories are instantiated universally outside biology, we should expect them to be found within biology as well. Moreover, given the special teleologically loaded unity of the living world, we should expect the division to be even more pronounced than in the world of the inorganic. We should be able to identify parts by the *service* they render to the whole, the whole by its integral teleology and hence the service rendered to *it* by its parts, and the aggregate by the service rendered to *it* by its substantial members.

The second philosophical argument for the intuition is very simple: it just *seems*, from both common and scientific observation, that there are many instances of organs, organisms, and collectives or aggregates[22] and that the categories are mutually exclusive for those instances. And, to paraphrase Richard Swinburne's Principle of Credulity, things usually are the way they appear to be.[23] Now perhaps the most common way of rejecting the intuition is by appeal to vagueness, which is effectively what a pluralist such as Dupré is doing when he points out, quite correctly, the multifarious ways in which multicellularity presents itself. The vagueness, to be of any power against the intuition, needs to be ontic. But assuming it is, there are two responses one might be tempted to make and yet which I recommend resisting in the case of the organic world even if, as I am happy to accept, they can be deployed in the inorganic case. One is to point out that vagueness is everywhere and appealing to it is often a cheap shot. We all want to solve the problems of vagueness. We learn nothing special about biology when we see that we can soritify biological predicates like '. . . is an organism' as much as any others. Another response is the Johnsonian one:[24] the existence of twilight does not mean we cannot distinguish between day and night. If there are clear ontic intermediaries between organs, organisms, and collectives, then to insist upon the importance of these three while ignoring the in-between cases is to treat the latter as second-class biological citizens—curious departures from nature's most important paradigms. This is to introduce unwarranted metaphysical (and perhaps methodological) bias into what should be a dispassionate allocation of ontological status.

Instead, I propose that we resist the idea that any vagueness in this domain is ontic. There is no promise here of an a priori, knockdown argument (which is not to say there mightn't be one). I prefer an appeal to ignorance, as it were: what *kind* of thing could there *be* lying between the organ, the organism, and/or the collective? What would be its essential features? What kind of teleology would it manifest? Since, however, the goal of this paper is more to justify form rather than to prove the existence of what it explains, the questions will be left hanging, with the burden, in my view, on the opponent of my threefold taxonomy. Moreover, any knockdown argument, if there is one, should be consistent with what we find in nature, but showing this requires detailed, case-by-case evaluation.

As an example, Dupré cites *lichens* as an case of 'multispecies organisms . . . symbiotic associations of photosynthetic algae or bacteria with a fungus'. He notes the anomalousness of such an object from the perspective of the 'traditional dichotomy between unicellular organisms and monogenomic multicellular organisms', adding that it is 'quite unproblematic' when we approach multicellularity from a more 'comprehensive' perspective.[25] Now a relatively innocent reading of this passage reflects commitment to the special category of organism, but only a broadening of the category to include multigenomic organisms. A less innocent reading, which I adopt given the entire context of the chapter, is that it downgrades the category itself: after all, if there are even multispecies organisms, what is so special about the term 'organism'? Aren't lichens an example of the vast heterogeneity we find in living systems? Even if there are paradigm cases of good old-fashioned organisms such as cats, and of organs such as hearts and livers, and lichens are quite unlike either of these, this no more elevates the paradigms above our scientific and other interests than, recalling the earlier dictum, the existence of twilight brings out anything *ontologically* special about day and night. Let's talk about organisms, to be sure; but let us not pretend we are carving nature at a privileged joint.

In reply, lichens are not nearly as worrisome for my view as they might seem. They are sometimes called 'dual organisms' because of the symbiotic relation between the mycobiont (fungus) and the photobiont (green algae or cyanobacterium); but a dual organism is no more an organism than a dual carriageway is a carriageway. That a lichen behaves differently from its component organisms does not make it an organism since the same is true of any collective and it would be question-begging merely to claim that collectives are all organisms for the same reason lichens are. Although there are still many gaps in our knowledge, a lichen is usually and best regarded as a 'miniature ecosystem'[26] consisting of two kinds of individual organism, a fungus and an alga or colony of bacteria (acting as photosynthesising agents), working in extremely close symbiosis. As far as we know, most lichen-forming fungi, as too their photobionts, also occur in a free-living state in nature *or* can be cultivated in a laboratory (albeit with generally less success than as lichen components, as one would expect).[27] The fact that either partner *can* exist free-living distinguishes them metaphysically from a true *organ*, such as an arm, leg, or eye, that has no free-living state. For an organ to exist separately from its natural body, similar or identical bodily conditions have to be *simulated*.

The other distinguishing feature is that lichen symbionts *both* reproduce. Hence to speak of lichen reproduction without qualification is misleading. There is genuine sexual reproduction, wherein the germinating mycobiont (fungal) spores have to find a suitable photobiont in the environment in order for successful symbiosis to arise again. Here it is the *fungus* that reproduces, not the lichen as a whole. There is also the *propagation* (rather than reproduction) of the lichen as a whole, fungus and photobionts, through the asexual reproduction of the fungus—the breaking off of a part of the fungus (a propagule) but containing photobionts within it. The fungus and the photobionts continue to grow as did their respective parents once a suitable environment is found. In both the sexual and asexual cases, the mycobiont (fungus) has its own mode of reproduction, and the photobiont continues to reproduce, always asexually as far as we know, in the mycobiont with which it is simultaneously propagated or which captures it following sexual reproduction. Now although many individual organisms have more than one mode of reproduction, none reproduces itself twice over at the same time, or more precisely none engages in two distinct processes of partial reproduction. I take this not to be a truth of biology but of metaphysics.[28] Moreover, there would be something strange going on biologically were it to be that in the sexual case the fungus reproduces, not the lichen, whereas in the asexual case the lichen reproduces, not the fungus. It is far more economical, and less bizarre, to take the asexual case to be one of lichen *propagation* via fungal and photobiont reproduction. Many collectives, such as various kinds of bee or ant colony, propagate by splitting or budding. But if the components are themselves engaging in reproduction, via their own identifiable processes, it is both biologically obfuscatory and metaphysically quite dubious to say that the whole collective is itself reproducing. The very idea of reproduction itself is called into question.

Returning to form, the central idea is that only an organism has a substantial form *simpliciter*: organs and collectives have them only *secundum quid*, or in a manner of speaking. It would be useful to have distinguishing terminology here, so I will stipulate that an organism *has* or *possesses* a substantial form, a collective *contains* one or more substantial forms, and an organ—to use an unfortunate neologism—*abtains* a substantial form. An organism *has* or possesses a substantial form inasmuch as this is its unifying principle as an individual substance of its essential kind. The blue whale (*Balaenoptera musculus*) has precisely the substantial form in virtue of which it is a member of that biological species. A collective *contains* one or more substantial forms inasmuch as it consists, inter alia, of one or more individual organisms in some systemic combination. Examples include ant colonies, bacterial colonies, forests, obligate colonies such as corals and facultative ones such as carpenter bees (*Xylocopa pubescens*).[29]

The *abtaining* of a substantial form by an organ is a trickier concept to grasp, but it goes a long way towards showing how substantial form acts as a unifying principle. I now proceed to elaborate the concept of abtaining, which will clarify how, ontologically, an organ is to be distinguished from an organism.

3. Organs and Organisms

An organ—a term I am, to reiterate, using stipulatively to denote any biologically identifiable part of an organism that subserves the whole—does not have its *own* substantial form; for if it did, it would be a substance. But it is not a substance since substances are ontologically independent. Now there is an important literature on ontological independence, but I have no space or need to enter into a technical discussion of its definition.[30] For present purposes, it is enough to say that a substance has existence in itself and by virtue of itself as an ultimate distinct subject of being. This definition encompasses several notions. Substance has existence in itself in the sense that it is not in anything else, not a modification of, a part of, an aspect of, some other thing. It exists by virtue of itself since its continued existence does not require it to be a product or projection of something else. As a distinct and ultimate subject of being, it is the bearer of qualities, but nothing bears it or is a subject of it.

An organ is clearly not encompassed by this concept of substance. It is a part that serves the whole, and cannot be constituted as an object with its own identity in a way that is metaphysically independent of that whole. To be more precise, we should say that something is not an organ unless either: (i) it *is* serving the whole; or (ii) is in some way able to carry out its functions *as if* it were serving a whole, such as when a heart is kept warm, pumping and oxygenated outside the body; or (iii) is kept in a state whereby its *powers* of serving the whole are preserved, such as when a heart is kept on ice before transplant. The first condition corresponds to Aristotle's *second* actuality— the actual exercise of powers; the second to a *simulated* second actuality; the third to *first* actuality—having powers but not exercising them.[31] I cannot see how first actuality could be simulated. In all three cases, the identity of the organ is still constituted by the function it performs with respect to the whole organism to which it belongs. As such, following Aristotle's famous homonymy principle,[32] an object that fulfils none of the above three conditions is not genuinely an organ, no matter how much it resembles one.

With this in view, what it means for the organ to *abtain* the substantial form of the whole is as follows. Although the organ does not have its own substantial form in any condition, when actually subserving an organism the organ is united to the whole *by* the substantial form. Here the organ is in its normal, natural state. It is thoroughly *permeated* by the substantial form in the sense that every part and property of the organ is co-opted to the service of the whole (barring damage or disease). The organ has *no life of its own*: it is the metaphysical slave of the whole, forming just one part, however important, of the organism's total organisation, which is dictated by the substantial form.

We still have to explain, though, what is going on when the organ is in condition (ii) or (iii) above, which we can call *simulation* and *dormancy*, respectively. Hoffman and Rosenkrantz think that both cases demonstrate the existence of 'organic living entities' that are neither organisms nor parts of organisms.[33] This seems to me wrong, for it misconstrues part-hood by overemphasising the property of being *joined* to the whole. In fact, Rosenkrantz, on his own account of whole organisms in suspended animation, allow that opponents of intermittent existence might plausibly construe suspended animation as involving the continued existence of the organism as constituted by the preservation of its potentiality for metabolic activity in a sense weaker than what he calls 'capability'.[34] This looks just like the distinction between Aristotelian first actuality ('second potentiality') and second actuality referred to above. If he thinks it can apply to whole organisms, so he should also allow it for organs. Ad hominem aside, the point is that just as an organism can retain its essential powers in a state of dormancy or suspended animation, so too can an organ retain its essential powers in a similar state although detached from the whole. It is the retention of essential powers that is crucial, not the state of being joined. This is what unites the dormancy case to the simulation case, the only difference being that in simulation the organ actually exercises some or all of its powers, albeit in a way that merely simulates subservience to a real organism.

Yet how is retention of power to be reconciled with the organ's essentially subservient nature? Why not count it as a substance in its own right, one that happens in the normal case to reside within another substance, namely a whole organism, but that also may not? Yet this is precisely to reject the organ's essentially subservient nature, not to reconcile it with retention of power. The organ is a dependent entity, its very identity defined by that of which it is an organ. Simulation cannot be understood independently of the genuine case, and there is no a priori reason to consider dormancy a privilege of substances. I claim that the most satisfying, perhaps the only plausible, explanation of retention of power must appeal to the organ's abtaining the substantial form of the whole. It does not have its *own* substantial form, but it does have many forms—accidental forms, as the scholastics put it. The organ is a discrete, identifiable, biologically significant portion of matter possessed of many accidents, the essential ones of which—i.e., the *propria* or properties in the strict sense of which—have a certain *organisation* by which the organ is defined as the kind of organ it is for the kind of species to whose members it normally belongs. What unifies those accidents is the very substantial form of the organism to which it is subservient. But given that the substantial form is *not* present in the organ, the organ must somehow *derive* or *borrow* the unifying power of the substantial form in one of the following ways.

(i) The organ actually *belongs* to its connatural organism, this being the normal case. We might include as a deviant sub-type of (i) the case where an organ is grown inside the organism's body using the organism's own tissue, such as when a person's nose is rebuilt and grown on their forehead.[35] One might legitimately wonder, though, whether such an entity fails to satisfy any of the three necessary conditions for being an organ stated earlier, and hence does not merit being called a nose at all until it is moved to its proper place. (ii) The organ once *belonged* to its connatural organism at some prior time. An organ removed from an organism and still satisfying the dormancy or simulation conditions would be a typical case of (ii). (iii) The organ has come into existence via a causal process that *began* with a distinct organ satisfying (i) or (ii). Case (iii) covers organs synthesised from other organs (such as cells—recall my stipulative use of the term 'organ') belonging or having belonged to a connatural organism. An example of (iii) would be currently typical organ synthesis, where the organ is grown outside the organism's body using detached cells from the latter, say, a liver cultured in a lab from the organism's stem cells. In case (iii), the organ still abtains its unity from the organism, but only indirectly via some other organ that abtains it directly.

Now to many, this way of explaining the difference between an organ and an organism will seem to partake of the kind of metaphysics that gives metaphysics a bad name. This is unfortunate and short-sighted. It is unfortunate because it bespeaks a refusal to engage seriously with hylemorphic metaphysics, free of the anti-scholastic prejudices of a bygone age. It is short-sighted because it reflects a preoccupation with surface illusions rather than the depth of the position. On the surface, the idea that an organ might 'abtain' a substantial form looks like 'spooky metaphysics' involving 'occult qualities' akin to the 'dormitive virtues' mocked by Molière. Yet one also might wonder how a universal can be wholly present wherever and whenever it is instantiated, how any parts can compose a whole, how real causal influence can be transmitted from one thing to another; and the list goes on. All such things have been wondered about, and many more; if the wonder seems insuperable, one adopts the appropriate position, whether nominalism, or compositional nihilism, or regularity theory, and so on. Quite why the scholastic framework or any of its key posits should be treated as of special concern because of its particularly 'spooky' nature is itself a matter for wonder.[36]

The real focus needs to be on whether a given posit can explain something that needs to be explained, in a way that is not ad hoc or incoherent. I don't see either criticism applying here, but that is where the opponent needs to concentrate their energy. Substantial form as unifier is clearly a metaphysical posit: I claim that we must acknowledge its existence if we are to explain the unity of the organism and its different metaphysical status from the organ and from the collective. It would be quite mistaken to think of form as a kind of suprasensible metaphysical 'glue' that holds the organism together, something we need special scholastic spectacles to 'see'. We do not see substantial form, and we do not 'see' it either. On my position, we know that it must exist or we have no explanation of unity. But to postulate form as no more than *that which unifies* would be ad hoc: claiming simply that we need a unifier to explain unity borders on the tautologous, but this is something no scholastic has ever said. Substantial form not only unifies, it *determines* the identity of a substance, it *actualises* matter, it is a principle of identity and stability in a substance, and more. In other words, it is a key element in an entire metaphysical picture. Needless to say, devotees of Quinean 'desert landscapes' will have no time for form, but they will have no time for universals, substances, powers, and much else besides. To single out form as having a special mystery about it is the opposite of special pleading, what we might call special prosecution.

Having defended, by appeal to form, the distinction between organs and organisms, I move now to the other part of the tripartite distinction—between organisms and collectives. Here the issues seem to me less subtle, though there are still important empirical challenges to the sharpness of the distinction. Once again, the appeal to form will show that the distinction is both plausible and clear cut.

4. Organisms and Collectives

Although collectives come in many kinds, as noted above, what they have in common—at least in biology—is their consisting of one or more organisms in some sort of systemic relation (whether including or excluding non-substances such as an ecological niche or organismically produced tools or habitats). This, I argued earlier, is how we should understand lichens.

Yet the idea of a 'superorganism' has found its way into the literature, and as Michael Ghiselin notes, it has become a recent fad.[37] He himself has done much to counter some of the excesses found in employment of the concept.[38] Unreflective comparisons between, say, an organism's eyes and the combined eyes of an insect colony's members, or between an organism's skin and a colony's nest, do not withstand scrutiny.[39] The argument that collectives must be organisms because natural selection works on organisms as well as on collectives is an elementary fallacy.[40] As far as the historical debate goes, I am on the side of those who regard the idea that collectives are literally a kind of organism as 'bad metaphysics'[41] or as consisting of 'poetic metaphors in scientific guise'.[42]

That said, for my purposes the main idea to keep in focus is that collectives *contain* substantial forms but do not *possess* them. It is not that we can appeal to substantial forms to *demonstrate* that there is a difference between organisms and collectives, but that given the distinction, only forms can explain it. Pointing to a list of disanalogies will not do the job: superorganism theory has long had a 'magnetic appeal'[43] precisely because of the many *analogies* that can be found between the two kinds, in terms of such phenomena as selection (at individual and group level) and division of labour. Even if the disanalogies overwhelmed the analogies, we would still not have arrived at an *explanation*. For, we should ask, why the disanalogies? Disanalogies point to a significant difference; they do not explain it. For an explanation, we must engage metaphysically: organisms are unified by a single substantial form that constitutes them as an individual substance. An ant colony's behaviour, for all its beguiling resemblances to multicellular co-operation in an organism, does not reflect the existence of a unifying form, since the colony *contains* substantial forms already—those found in the individual ants. But if the ants have their own substantial forms, this *excludes* a further, superorganismal form. Why? Couldn't one object that the superorganismal form can be superimposed on the individual ant forms, such that the ants each have their own form and the *plurality itself* has a colonial substantial form?

The objection fails. For assume the existence of the individual forms and the superorganismal, colonial substantial form. The colony would itself be a substance and the ants would be its parts. But the parts of an organism—the organs, as I am calling them—are, as set out earlier, essentially subservient, ontologically dependent entities whose very *identity* is defined by the organism of which they are parts. Putting it loosely, there is no such thing as a *heart* simpliciter, only the heart of a lion, of a man, or of a reptile. Organs are defined by the organisms they subserve. So the ants would be defined by the superorganism they subserve. But they cannot be, since—ex hypothesi—they already have their own substantial forms as ants; and they are defined by these.

They certainly don't have an *extra* substantial form, namely the colonial form, since this would absurdly imply that each ant was a colony. The colonial form is supposed to be the form of the plurality of ants yet also function so as to *define* each individual ant as an organ of the colony. In other words, the individual ant is supposed both to possess its own substantial form *and* abtain another substantial form. It is, then, supposed to be both ontologically independent and ontologically dependent; and this is a contradiction.

So, on the substantial forms view, we either have to deny that the ant is a substance or deny that the colony is a substance. It is clear what we must do. Denying that the ant is an individual substance is a non-starter, unless we deny that there are any individual substances at all—not a position I am questioning here. If anything is an individual substance an ant is, since it displays all the hallmarks of ontological independence. *What* it is, its quiddity, is in no way defined by reference to colonies in general or any colony in particular. An account of how ants *behave*, of course, must make reference to necessary colony formation (albeit, note, not for facultatively colonial organisms)[44], but this is not the same as defining an ant as a kind of organism with its distinctive and independent physiology, anatomy, developmental processes, and so on—all of which constitute its immanent causal behaviour, and none of which require defining it in terms of its colonial behaviour, whatever the causal connections between the two.

We are faced, then, with the obvious and only remaining move, which is to deny that the colony is a substance. It is ontologically dependent inasmuch as the kind of thing it is depends, metaphysically, on the kinds of organism that belong to it, each having its own independent physiology, anatomy, developmental processes, and so on. This is quite different from the substantial members of the aggregate, whose identities as the kinds of thing they are do not depend metaphysically on their membership of the aggregate. This is so even if it is essential to a given substance to *be* a member of an aggregate. Now it is not for the metaphysician to say what aggregates there are any more than what substances there are: this is for the biologist, and hence is a wholly a posteriori matter. What the metaphysician can say, however, is that we should expect there to be a sharp division between substances and aggregates because they are distinct metaphysical categories; so-called borderline cases should reflect our investigative limitations rather than a blurriness of the categories.

We see this sharpness, I submit, in the supposed borderline cases of biofilms and slime moulds. A biofilm, at its most general, is a colony of bacteria adhering to a surface and often to each other, producing an extracellular polymeric substance (EPS) that acts as a matrix holding the colony together, protecting it against predators and toxins, helping to digest and pass on nutrients to the colonists, and facilitating communication between them.[45] Slime moulds, on the other hand, are a paraphyletic class of protists that form masses of protoplasm containing one or more nuclei and reproduce by means of spores emanating from sporangia, much like plants and many fungi.[46] There is much that we do not know about biofilms and slime moulds, making interpretation of their natures a difficult matter. One might wonder, however, whether they give credence to the idea of a superorganism constituted by other organisms. There is far too much to be said about slime moulds and biofilms to be anything other than cursory here, but a few remarks are in order.

It is difficult to see biofilms as anything but extraordinarily close-knit colonies of individual organisms. All biologists, as far as I can tell, refer to them as such[47] or as 'communities'[48] or with similar terminology. Virtually all bacteria can form biofilms,[49] and yet all live quite happily in the planktonic state unless either stressed by lack of nutrients or the presence of toxins, or attracted by a suitable surface. The bacteria within the biofilm maintain their cellular integrity. The biofilm forms precisely through the colonisation of a surface by free-living bacteria. It grows by the reproduction of the bacteria themselves and the addition of new bacteria. The biofilm does not itself reproduce, it *disperses* by the detachment of the bacteria from the colony, their movement to a new location, and reattachment to a new substrate. The bacteria themselves produce enzymes that degrade the EPS or the substrate. Some biofilms undergo 'seeding dispersal', whereby hollow cavities in the matrix fill with planktonic bacteria that then breach the colony wall and emerge to form new colonies.[50]

Nevertheless, Ereshefsky and Pedroso argue that biofilms are 'individuals and not merely communities', by which they mean individual organisms.[51] Without going into the detail of their own account, note for example their contrast between biofilms and 'symbiotic complexes, such as the symbiotic relation between ants and acacias . . . Bacteria in a biofilm exchange genetic content; ant/acacia symbionts do not'.[52] This lateral gene transfer is taken by them to be one of many markers of individuality since it facilitates communication between the bacteria and hence development of the colony, especially the EPS as a boundary between the colony and its environment. Yet although they make the contrast with lack of LGT in ant/acacia symbiosis they also refer to a supposed lack of LGT in aphid/bacteria symbionts. In support of the latter, they cite a paper in which the authors find no evidence of LGT between the pea aphid (*Acyrthosiphon pisum*) and its obligate mutualist Gammaproteobacteria (*Buchnera aphidicola*). Yet these same authors *do* explicitly suggest that there is evidence of functional gene transfer to *A. pisum* from *prior* rickettsial endosymbionts, given that Rickettsiales symbionts are found in some aphids. (Albeit the evidence is consistent also with transfer by bacterial infection rather than endosymbiosis.)[53] So the claim that biofilms are 'better candidates for biological individuals than aphid-symbiont combinations'[54] in virtue of lack of LGT is not warranted by the evidence.

Again, getting less technical, Ereshefsky and Pedroso are impressed by the way the biofilm EPS defends the bacteria, digests nutrients and passes them on to the bacteria, and facilitates communication between them. Yet a bee hive is not far away from this: it protects the bees, facilitates communication, and although it doesn't digest nutrients, it stores them (which is just as important as digestion). I am not denying the EPS is perhaps more complex than a bee hive (how do we measure complexity?), but what I am suggesting is that being impressed by this or that marker or group of markers can cause one to lose the wood for the trees. A hylemorphist will look at all the characteristics of the entity under consideration, forming a holistic judgment as to whether the unity provided by substantial form is present. Anything less is bound to risk being skewed or arbitrary.

The same, I submit, applies to slime moulds. In fact they come in two quite different kinds, the plasmodial or 'true' slime moulds and the cellular slime moulds. The cellular slime moulds have been described by the world's expert on them as 'no more than a bag of amoebae encased in a thin slime sheath', albeit capable of the most remarkable behaviour 'equal to those of animals who possess muscles and nerves with ganglia, that is, simple brains'.[55] The amoebae in a cellular slime mould retain their cell membranes. They feed and reproduce normally as individual amoebae as long as food is plentiful. When they begin to starve, they aggregate into pseudoplasmodia that produce fruiting bodies inside which are spores that germinate as free-living amoeba that drift to other locations and will remain free-living as long as conditions are favourable. Cellular slime moulds, then, do not strictly reproduce: they have a fungus-*like* property of sporulation that produces yet more free-living amoebae, but fungal spores are juvenile fungal parts that germinate into the vegetative mycelium or fungal body.

For all their amazing behaviour, cellular slime moulds are rightly designated as not true slime moulds. They are, in my view, tightly integrated colonies, much like biofilms. Their motility, chemotaxis, direction-finding behaviour, and the like, can all be found in colonies and swarms constituted by birds, bees, fish, and other animals.

Plasmodial slime moulds, the *true* slime moulds (*Myxomycetes*), do by contrast appear to be genuine individual organisms. The constituent amoebae do not retain their individual membranes, but instead mass into a single-celled, multinucleate plasmodium. Reproduction is also typically by sporulation, but here the spores are either haploid gametes that later fuse with suitable gamete partners to form the juvenile plasmodium via a zygote, or (less commonly) the diploid zygotes themselves. The zygotes grow into plasmodia through repeated nuclear (but not cellular) division. The entire life cycle seems to instantiate the growth and reproductive pattern of an individual organism, with no entities within the single-celled plasmodium that correspond to potential free-living amoebae.

There is much more to say, and much more to be known, about the various kinds of entity that might challenge the idea of a sharp boundary between organisms, organs, and collectives. So far I have made out a positive case for the distinction's plausibility, utilising substantial form. I want, however, to return to more foundational issues by means of a negative argument. If we grant that the unity of the organism is what marks it out from the organ on one side and the collective on the other, why must we go so far as to posit substantial form? Mightn't a lesser principle of unity be sufficient for the task? If some of the leading candidates for such a principle are not successful we will have further, indirect reason at least to think that only a principle as strong as substantial form can do the required work.

4. Lesser Unity Principles Will Not Work

An objection to substantial form that no doubt arises in many minds is that they seem redundant. Why couldn't I say everything I have about organs and collectives, contrasting them with organisms, without even mentioning substantial forms? Why not simply talk about substances and non-substances? Aren't substantial forms an ontological spare wheel? In reply, as I have already argued, there must be a principle of unity for substances— that which unifies otherwise disparate elements into a whole. If that principle is to have ontological reality, it must be either a substantial form or something else. So to underscore the reality of form, we should consider

alternative unity principles to see if they can do the required work. I want
to look briefly at two alternatives, and will argue that both are found wanting. The first is that of Hoffman and Rosenkrantz,[56] who have a complex
account of organismic unity in terms of functional organisation, one which
also requires the existence of a 'master part' that controls the processes of all
the other parts. Details of their specific account of functional organisation
aside, their basic, quite Aristotelian idea is that organisms are functionally
united in such a way that their parts subserve or contribute to the typical
life processes of growth, metabolism, development, and/or reproduction. As
such, however, there is no distinction between organs, organisms, and collectives, given especially that they do not propose these basic life processes
to be necessary and sufficient for any living thing. Hence organisms cannot
be singled out by some privileged set of life processes.

What does single out organisms, according to Hoffman and Rosenkrantz,
is that they must have a 'master part' that controls and regulates their processes. They, like I do, take it as a datum that organisms are not proper parts
of other living things whereas organs are parts of organisms, but they seek
to explain this terms of the organism's processes' being controlled by its
master part, which excludes the organism's being functionally subordinate
to another living entity. A part of an organism, such as a liver cell, is functionally subordinate to the organism and its life processes are controlled by
the organism's master part, either directly or indirectly. Hence the nucleus
of the liver cell is not its master part.

The implied argument for the master part thesis is precisely that anything
less won't make the right distinctions: mere functional unity, however complex, is too coarse grained to do the work they require.[57] Although they don't
put it this way, it would be a miracle were there to be a special kind of functional unity that applied to all and only organisms. The explicit argument,
however, is one by induction from observed cases of paradigmatic organisms.
Eukaryotic single-celled organisms seem to have the nucleus as a 'highly centralised regulatory system', and prokaryotes have the system of their DNA
and mRNA molecules. If a proteinoid microsphere were an organism, its protein chains would constitute its master part.[58] The (central) nervous system of
an 'adult vertebrate' is its master part, as are the roots, stem and leaves of a
'typical mature plant',[59] as well as the nerve net of a jellyfish, the nucleoid of
a bacterium, and the 'nuclear system' of a plasmodial slime mould.[60]

This approach seems to me wrongheaded, for it leads to Rosenkrantz's
view that the master part of a plant is all of the plant minus its sap, since the
sap appears to be the only part that does not control the plant's life functions.
He calls the plant minus its sap a 'decentralized' master part without showing how to distinguish between centralised and decentralised master parts.
Yet I submit this is to abuse the term 'part', since for biological purposes
the plant minus its sap is no more a proper part of the plant that is Tibbles
minus his tail. Moreover, before a vertebrate embryo gastrulates,[61] there is
nothing resembling a nervous system to serve as master part, yet there is still

an organism. Furthermore, what on their master-part theory of organisms is to be said about non-vascular plants such as algae and bryophytes, all of which lack sap? Are we to say that they are their own master parts? After all, if we want to justify the idea that, say, the 'nuclear system' of a plasmodial slime mould is a master part simply because it is a proper part according to extensional mereology (not that Rosenkrantz and Hoffman say that, but it is an obvious move), why not hold that non-vascular plants are their own improper master parts? All in all, it seems that the master-part theory is unacceptably hostage to empirical fortune when it comes to defining the organism.

The other unity principle I briefly want to mention is *structure*, as exemplified in the work of Kathrin Koslicki.[62] She holds that we need something *like* substantial form to act as a scientifically respectable analogue of a metaphysically discreditable idea. So she proposes structure—a perfectly natural and understandable thought. Moreover, instead of the prime matter that accompanies substantial form on the hylemorphic theory, she proposed *content* as that on which structure operates to produce a structured whole. I have criticised her view elsewhere,[63] primarily on the ground of its being subject to what I call the 'content-fixing problem', which concerns the impossibility in principle of choosing any particular content as *the* content on which structure is imposed to constitute a substance. What, for instance, is *the* content on which equine structure operates to constitute a horse? For different contents, whether it be flesh and bones, cells, atoms, quarks, there will be a different structure, and yet there is no reason in principle to choose one over the other.

There is, however, another problem for structure-based hylemorphism, which I call the 'qualitative problem'. Related to the content-fixing problem, and in a way more fundamental and also explanatory of why the first problem arises, the qualitative problem is that most if not all the candidate contents for a structural account come *too late* in the metaphysical analysis to be viable partners for any given structure. Certainly, if we restrict ourselves to *biological* contents, whether flesh and bones, organs, biological systems, genes, or DNA itself, they are all *already* defined by the organism to which they belong *before* structure even comes on the scene to organise them into a substantial whole. No structure imposed on, say, equine organs can explain any unity beyond the unity of those specific organs. The unity *within* each organ is left unexplained: and yet each organ is already defined as an equine organ *before*[64] the proposed equine structure does any work. The unity problem, however, is a problem concerning the *whole* substance, including *all* of its biological parts: in virtue of what are they all united into an organic substantial whole? Structural hylemorphism cannot answer this question.

5. Conclusion

For traditional hylemorphism, only substantial form penetrates, in its unifying power, to every element of the organic substantial whole. It leaves no aspect of unity in need of further explanation. This is why the traditional hylemorphist holds, for example, that there is as much of the substantial form of humanity in my little finger as there is in me as an individual human being. The difference, however, is that as an organ my little finger *abtains* my substantial form, whereas I *possess* it. The unifying power of the form nevertheless descends to my finger as much as it does to any other part of me and as much as it determines me, as a whole, as the kind of thing that I am.

To reiterate, substantial form is a metaphysical posit, not the subject of an empirical hypothesis. It is not 'something we know not what' that is vacuously postulated to explain unity. It is what we *must* have if unity is to be explained. If this is hard for biologists and/or philosophers to swallow, it is because both have abandoned the quest for a genuine *philosophy of nature* that combines scientifically informed metaphysics with metaphysically informed science. An adequate philosophy of nature will resist the wild flights of fancy found in superorganism theory, which undermines organismic unity from both ends—blurring the boundaries both between organisms and collectives and between organisms and their parts. If D'Arcy Wentworth Thompson was right that the unity of the organism is a fundamental truth requiring explanation, then only substantial form can claim to be the 'great unifier' that does the work that needs to be done. With this, I can cite the agreement of another Thompson—William R. Thompson (1887–1972), one-time research scientist at the Imperial Institute of Entomology and Fellow of the Royal Society—who said:

> In the unification of a multiplicity, with reference to a specific end, the organism resembles a machine; but it is not, like the machine, unified by the participated activity of a separate mover. It moves itself, and what we call "physico-chemical properties" or "cytological activities" are simply the living unit envisaged in abstracto at various levels. They are not true nonliving or nonorganismal agglomerates unified and moved by something higher. There is nothing to move them but the thing they constitute; in other words, nothing to move them but themselves.[65]

By this, Thompson means that the unity of an organism's parts does not derive from anything external to the organism. The organism moves its parts, but only because it moves itself. And it moves itself because of its unity as a single substance. I happily side with both Thompsons in regarding *form* as the only thing that can make this possible.[66]

Notes

1 Thompson (1945: 1019).
2 *De Anima* I.5, 411a24-b30, Ross (1931).
3 Irwin (1988: 288).
4 I set out the unity problem in Oderberg (2007), and much of the subsequent defence of hylemorphism is designed to solve it.

5 A non-essentialist version is: what holds their natures together? In the latter case, there is no implication about the modal status of a thing's nature. I presume essentialism in what follows, without requiring it as part of my solution to the unity problem.

6 Dumsday (2010: 620).

7 Strictly, there is a difference between the constituents of a thing's essence and the essential properties ('necessary accidents') that flow from that essence, but for the present discussion I treat the two kinds similarly and call them all 'essential properties'. For more, see Oderberg (2011).

8 Lowe (2006: 161).

9 Ibid.

10 Ibid.: 141.

11 Here I side with Bird (2007: 208–9) against Lowe (2006: 154).

12 As Clarke (2010: 316) puts it, in the context of the 'problem of biological individuality', the task is to focus 'on what properties separate living individuals from living parts and from living groups, while taking the property of life itself for granted'.

13 By which I mean any biologically identifiable part of an organism rather than a random or gerrymandered hunk of tissue or other organic material.

14 Lowe (2006).

15 I defend the doctrine in Oderberg (2007: ch. 4.2).

16 Lewis (1986: 202).

17 For more on this, see Oderberg (2007: ch. 4.5).

18 For the detailed response to sceptical worries about substantial form, and an equally detailed defence, see Oderberg (2007).

19 The interested reader may also like to consult Oderberg (2012) for a somewhat unorthodox exposure of Hume's prejudice against the scholastic theory of substance.

20 Dupré (2012: 88).

21 Ibid.: 99.

22 I will use the terms interchangeably.

23 Swinburne (2004: 303).

24 Allegedly. Johnson is quoted as having said this by Anscombe (1961: 60) but I have not been able to find it in his works. Edmund Burke made a somewhat similar remark: 'though no man can draw a stroke between the confines of day and night, yet light and darkness are upon the whole tolerably distinguishable' (*Thoughts on the Cause of the Present Discontents*, 1770; thanks to Robert Koons for alerting me to this).

25 Dupré (2012: 90).

26 Nash (2008b: 7–8); 'minute ecosystems' (Tuovinen et al. 2015: 130).

27 Putting together Tuovinen et al. (2015: 130), Friedl and Büdel (2008: 9), Honegger (2008: 29–30), and Nash (2008b: 3), this seems to be the natural inference albeit the evidence is not conclusive. In other words, as far as we know, the majority of lichen photobionts are capable of existing in the free-living state, whether found in nature (even if only rarely) or cultured in the laboratory (even if only with difficulty and limited success).

28 For more on lichen reproduction, see: Paulsrud (2001), Friedl and Büdel (2008), Sarma (2013: 351–2), Australian National Herbarium (2012).

29 For carpenter bees' facultative sociality, see Dunn and Richards (2003).

30 See Chisholm (1994), Lowe (1994), Fine (1995), Hoffman and Rosenkrantz (1997).

31 For Aristotle on first and second actuality, see *De Anima* II.5, 417a21-35, Ross (1931).

32 *Metaphysics* Z:10, 1035b23 and elsewhere, Ross (1928).

33 Hoffman and Rosenkrantz (1999: 87, n.13).
34 Rosenkrantz (2013: 95).
35 See www.rhinoplastyinseattle.com/blog/updates-and-analysis/post/can-a-plastic-surgeon-really-grow-a-nose-on-a-forehead.
36 In Oderberg (2007) I give an extensive defence of the same viewpoint in the context of biology (and some physics).
37 Ghiselin (2011).
38 Such as in Hölldobler and Wilson (2009).
39 Ghiselin (2011: 163).
40 I agree with Ghiselin (2011: 153) that this is the 'basic thrust' of the complex argument of Wilson and Sober (1989).
41 Ghiselin (2011: 153).
42 Wilson and Sober (1989: 338), speaking of many versions of the superorganism theory.
43 Ibid.: 338.
44 Should the superorganism theorist say that facultative colonies are not superorganisms whereas obligate colonies are? Why? After all, the colonial behaviour is the same, yet in facultative colonies the organisms are quite liable to disperse and live their own *substantial* lives. To maintain in the face of this that the obligate colonies are still superorganisms looks like a manoeuvre with no independent motivation.
45 For the details, see Romeo (ed.) (2008)
46 For more on slime moulds, see Stephenson and Stempen (1994) [plasmodial] and Bonner (2009) [cellular].
47 E.g., Romeo (ed.) (2008) *passim*, Karatan and Watnick (2009).
48 Kaplan (2010).
49 Karatan and Watnick (2009).
50 See Kaplan (2010) for the details.
51 Ereshefsky and Pedroso (2016).
52 Ibid.: 109.
53 Nikoh et al. (2010).
54 Ereshefsky and Pedroso (2016: 108).
55 Bonner (2009, Kindle edition): ch. 5, 'Behavior of Multicellular Slugs'.
56 Hoffman and Rosenkrantz (1997, 1999), Rosenkrantz (2013).
57 As Clarke (2010: 316) perceptively puts it in the context of 'functional integration' accounts of individuality: 'The trouble is that pretty much everything is organized, in some sense . . . [T]here are many respects and degrees to which functional integration is evident in systems we clearly don't want to describe as biological individuals'.
58 Hoffman and Rosenkrantz (1997: 126–7).
59 Hoffman and Rosenkrantz (1999: 96).
60 Rosenkrantz (2013: 97–8). Gratifyingly, they come to the same view as I do about slime moulds, namely that plasmodial slime moulds are organisms but cellular slime moulds are ' a collective of "social" unicellular organisms'.
61 Gastrulation is the reorganisation of the hollow sphere of cells (blastula) to a three-layered gastrula. In human embryos, this occurs at around 16 days.
62 Koslicki (2008).
63 Oderberg (2014).
64 In the metaphysical sense of 'before', of course, not the temporal.
65 Thompson (1947: 154).
66 I am grateful for comments and feedback on a version of this paper to colleagues and graduate students at the University of Reading, and to participants in the conference on Biological Identity organised by Anne Sophie Meincke and John Dupré at Senate House, London, in June 2016.

Bibliography

Anscombe, G. E. M. (1961) 'War and Murder', in W. Stein (ed.) *Nuclear Weapons: A Catholic Response* (London: Sheed and Ward): 45–62.

Australian National Herbarium (2012) 'Australia's Lichens: Reproduction and Dispersal', at www.anbg.gov.au/lichen/reproduction-dispersal.html [last accessed 22.4.16].

Bird, A. (2007) *Nature's Metaphysics* (Oxford: Clarendon Press).

Bonner, J. T. (2009) *The Social Amoebae: The Biology of Cellular Slime Moulds* (Princeton: Princeton University Press).

Chisholm, R. (1994) 'Ontologically Dependent Entities', *Philosophy and Phenomenological Research* 54: 499–507.

Clarke, E. (2010) 'The Problem of Biological Individuality', *Biological Theory* 5: 312–25.

Dumsday, T. (2010) 'Natural Kinds and the Problem of Complex Essences', *Australasian Journal of Philosophy* 88: 619–34.

Dunn, T. and Richards, M. H. (2003) 'When to Bee Social: Interactions Among Environmental Constraints, Incentives, Guarding, and Relatedness in a Facultatively Social Carpenter Bee', *Behavioral Ecology* 14: 417–24.

Dupré, J. (2012) *Processes of Life: Essays in the Philosophy of Biology* (Oxford: Oxford University Press).

Ereshefsky, M. and Pedroso, M. (2016) 'What Biofilms Can Teach Us About Individuality', in A. Guay and T. Pradeau (eds.) *Individuals Across the Sciences* (Oxford: Oxford University Press): 103–21.

Fine, K. (1995) 'Ontological Dependence', *Proceedings of the Aristotelian Society* 95: 269–90.

Friedl, T. and Büdel, B. (2008) 'Photobionts', in T. H. Nash (ed.) *Lichen Biology* (Cambridge: Cambridge University Press): ch.2.

Ghiselin, M. T. (2011) 'A Consumer's Guide to Superorganisms', *Perspectives in Biology and Medicine* 54: 152–67.

Hoffman, J. and Rosenkrantz, G. S. (1997) *Substance: Its Nature and Existence* (London: Routledge).

——— (1999) 'On the Unity of Compound Things: Living and Non-Living', in D. S. Oderberg (ed.) (1999) *Form and Matter: Themes in Contemporary Metaphysics* (Oxford: Blackwell): 76–102.

Hölldobler, B. and Wilson, E. O. (2009) *The Superorganism: The Beauty, Elegance, and Strangeness of Insect Societies* (New York: W.W. Norton & Co.).

Honegger, R. (2008) 'Mycobionts', in T. H. Nash (ed.) *Lichen Biology* (Cambridge: Cambridge University Press): ch.3.

Irwin, T. (1988) *Aristotle's First Principles* (Oxford: Clarendon Press).

Kaplan, J. B. (2010) 'Biofilm Dispersal: Mechanisms, Clinical Implications, and Potential Therapeutic Uses', *Journal of Dental Research* 89: 205–18.

Karatan, E. and Watnick, P. (2009) 'Signals, Regulatory Networks, and Materials That Build and Break Bacterial Biofilms', *Microbiology and Molecular Biology Reviews* 73: 310–47.

Koslicki, K. (2008) *The Structure of Objects* (Oxford: Oxford University Press).

Lewis, D. (1986) *On the Plurality of Worlds* (Oxford: Blackwell).

Lowe, E. J. (2006) *The Four-Category Ontology* (Oxford: Clarendon Press).

——— (1994) 'Ontological Dependency', *Philosophical Papers* 23: 31–48.

Nash, T. H. (ed.) (2008a) *Lichen Biology* (Cambridge: Cambridge University Press).

——— (2008b) 'Introduction', in T. H. Nash (ed.) *Lichen Biology* (Cambridge: Cambridge University Press): ch.1.

Nikoh, N., McCutcheon, J. P., Kudo, T., Miyagishima, S., Moran, N.A., and Nakabachi, A. (2010) 'Bacterial Genes in the Aphid Genome: Absence of Functional Gene

Transfer from *Buchnera* to Its Host', *PLoS Genetics* 6: 1–21, at http://journals.plos.org/plosgenetics/article?id=10.1371/journal.pgen.1000827 [last accessed 4.5.16].

Oderberg, D. S. (2007) *Real Essentialism* (London: Routledge).

———— (2011) 'Essence and Properties', *Erkenntnis* 75: 85–111.

———— (2012) 'Hume, the Occult, and the Substance of the School', *Metaphysica* 13(2012): 155–74.

———— (2014) 'Is Form Structure?', in D. D. Novotný and L. Novák (eds.) *Neo-Aristotelian Perspectives in Metaphysics* (London: Routledge): 164–80.

Paulsrud, P. (2001) *The Nostoc Symbiont of Lichens* (Uppsala: Acta Universitatis Upsaliensis; PhD dissertation).

Romeo, T. (ed.) (2008) *Bacterial Biofilms* (Berlin: Springer).

Rosenkrantz, G. (2013) 'Animate Beings: Their Nature and Identity', in D. S. Oderberg (ed.) *Classifying Reality* (Oxford: Blackwell): 79–99.

Ross, W. D. (1928) *Aristotle: Metaphysics*, 2nd ed, Vol. VIII of *The Works of Aristotle* (Oxford: Clarendon Press).

———— (1931) *Aristotle: De Anima*, Vol. III of *The Works of Aristotle* (Oxford: Clarendon Press).

Sarma, T. A. (2013) *Handbook of Cyanobacteria* (Boca Raton: CRC Press/Taylor and Francis).

Stephenson, S. L. and Stempen, H. (1994) *Myxomycetes: A Handbook of Slime Molds* (Portland, OR: Timber Press).

Swinburne, R. (2004) *The Existence of God* (Oxford: Clarendon Press; 2nd edition).

Thompson, D. C. (1945) *On Growth and Form* (Cambridge: Cambridge University Press; reprint of 1942 edition).

Thompson, W. R. (1947) 'The Unity of the Organism', *The Modern Schoolman* 24: 125–57.

Tuovinen, V., Svensson, M., Kubartová, A., Ottosson, E., Stenlid, J., Thor, G., and Dahlberg, A. (2015) 'No Support for Occurrence of Free-living Cladonia Mycobionts in Dead Wood', *Fungal Ecology* 14: 130–2.

Wilson, D. S. and Sober, E. (1989) 'Reviving the Superorganism', *Journal of Theoretical Biology* 136: 337–56.

10 Action, Animacy, and Substance Causation

Janice Chik Breidenbach

1. Introduction

Perhaps Macbeth was justified in doubting that Birnam Wood could ever come to Dunsinane, despite the witches' prophecy. Barring supernatural intervention, we normally have no reason to suspect that trees in a forest might spontaneously sprout legs and walk: indeed, when the prophecy is fulfilled, it is fulfilled (however unexpectedly!) by means of *human* agency and artifice. Philosophical thought on the subject of action has frequently produced the suggestion that actions are unique among the forms of movement in the natural world: the advancement of Malcolm's army, shielded with the leafy boughs of Birnam to secure Macbeth's doom, seems absolutely unlike the advancement of an electrical storm through a forest or the inevitable fall of dry timber. It may even be suggested that action contains a hint of the divine: for agents are the ultimate and free originators of states of affairs in the world, and are therefore like gods. Agency, it is thus thought, is exceptional among causal phenomena in the natural world. In a world of material causes, action is thought to represent an exceptional causal type.

This chapter pursues the claim that such causal exceptionalism for human agency is false. The idea that the causation of action is unique among other causal phenomena poses an obstacle to our understanding of agents and their doings as belonging to a 'naturalistic' explanatory framework, i.e., one that assumes that all causal phenomena, including agency, may be explained without invoking *non-natural* or even *divine* causal elements.[1] For action theorists attracted to naturalistic thinking about mind and action, this latter claim will come as a welcome proposal. For others, perhaps, any mention of a naturalistic program arouses suspicions of a rather predictable, and yet understandable sort. Most significantly, there is a tendency to associate naturalistic explanations of agency with a reductive materialist account of agents and their actions, as well as the worry that a denial of causal exceptionalism for agency involves a denial of human exceptionalism *tout court*. Yet it is unacceptable to think that these latter conclusions must be embraced, having exhausted all other possibilities for a more sympathetic naturalism. For one may reject causal exceptionalism with regard to agency without likewise rejecting the concept of an agent cause. Nor indeed, for that matter, must one jettison the reasonable judgment that human beings are truly exceptional within the natural world.

I argue that we can and should retain both ideas: firstly, that agents are themselves causes, and agency an irreducible type of causation; secondly, that human beings are exceptional within the animal kingdom, in the variety of powers exhibited and activities pursued. Moreover, I think it is possible to retain both of these ideas within a naturalistic programme of a certain sort: specifically, one committed to the concept of action as irreducible to its fundamentally physical constituent parts. This commitment, obviously, requires the support of metaphysical argument. The main point that I wish to press about agents and agency is that the causation (or varieties of causation) concerning them should not be understood as being of an exceptional type within the natural world, even while one grants that human beings are surely a most exceptional subclass of animal beings. Indeed, I think it is impossible to sufficiently appreciate this latter fact while insisting on an absolute or more radical conception of human alterity over animal nature.

2. Are We Animal Agents?

Are we *animal* agents? That is, when we think of ourselves as agents, is it in light of 'animacy', i.e. the concept of an animal, that we understand agency? Or is there something else, not included within nor entailed by the idea of animacy, on account of which we take ourselves to be agents? If we affirm the latter, then we cast doubt on the plausibility that non-human animals might also be agents, at least in a sense univocally understood with respect to our being agents. If we affirm the former, then we must be prepared to admit that we are not the only agents in the world.

Thinking that we are not the only agents challenges the assumption that the concept of 'action' is one pertaining uniquely to human beings. The reasons for excluding non-human animals from the class of agents have tended to centre on complex notions such as rationality, freedom, self-recognition, and moral judgment. It is thought that these powers, and others related to them, are possessed by human beings. I think we should have no trouble in granting that human beings possess these powers, certainly to the greatest observable extent within the animal kingdom, and with an avowedly remarkable range of activities. Yet it is a questionable further step to infer from the latter observation that, in light of our exceptional abilities, we therefore are the *only* agents there are: that the concept of being an agent is one that necessarily excludes all non-human animals.

Donald Davidson, in 'Rational Animals' (1982), offers an argument against the possibility of non-rational agents. Davidson argues that any creature with the capacity of truly possessing beliefs must also possess the capacity of exercising propositional speech. Thinking requires the concept of a thought, claims Davidson: in order to have a belief, I must recognize my belief *qua* belief, i.e. as dependent on an entire network of other beliefs. And this is a recognition that is supplied only by language, or propositional speech concerning the correctness or incorrectness of my beliefs.[2] Since dumb animals are, by definition, incapable of expressing propositional speech, they must be incapable of truly possessing beliefs. And since belief possession is a requirement for rational action, according to the causalist's basic definition, Davidson concludes that non-human animals are therefore incapable of performing actions, which by his lights represent events of the *rational* type.[3] Rational animals are the only agents there are.[4]

Davidson's argument in 'Rational Animals' has been widely discussed, and in particular, much debate arises over his claim that mute animals lack the particular sort of belief possession that he holds relevant for agency.[5] However, I would like to dispute his more general premise that agency requires this kind of belief possession in the first place. For if it turns out that agency does not require it, then it remains an open question whether or not non-rational animals ought to be considered agents, *regardless* of whether they have Davidsonian beliefs.

The deeper mischief with Davidson's argument, I suggest, lies in his causalist approach to the concept of action. Causalists argue that an agent *S*'s *A*-ing is an action if and only if *A*-ing was caused in the right way by appropriate mental states or events.[6] This view has been criticized on the basis that the causal relation is placed 'between events or states of affairs or facts of which [the agent] is *a constituent*' (Broadie 2013: 574, italics mine).[7] The agent is understood merely as a logical constituent of the causal relata *actually* productive of action: i.e., the events of *S*'s intending to *A* and *S*'s *A*-ing, where *A* is a set of movements belonging to an agent *S*. On many causalist accounts, this shift from the agent to the causal relations within her implies an ontological division between events or states of the mental and physical types; this latter division may be expressed via an explicitly Cartesian-style dualism, or by a more moderate account, e.g., non-reductive materialism.

The causalist account states that the cause of action is not the agent herself, but rather mental occurrences, such as beliefs and desires. Moreover, causalism posits a break that occurs, whenever there is an action, between the causal relata cited by the causalist: i.e., between agential intentionality and its bodily effects. The causalist account involves recognition of certain *types* of entities that qualify as such relata in the causation of action. This qualification based on types includes, whether explicitly or implicitly, the Cartesian thought that the causal relata can be sorted as mental (or intentional) and physical (or bodily), even if these binary concepts are not understood in contemporary accounts as corresponding precisely to Descartes's original idea.

The worry about a strictly Cartesian ontological division is a familiar one. Seeing the agent not as causes but rather constituents of diverse types of events is clearly related to the problem of how to explain the efficacy of mental events in physical domains, known in the philosophy of mind as the problem of causal exclusion. The requirement of causal closure states that every physical effect must have a complete physical cause. Thus, a bodily movement must have a physical cause, but the mental items required by the causal theory of action lack, by themselves, the kind of efficacy presumed to generate physical causes. If one views intentions or reasons as metaphysically 'determined' or 'realized' by corresponding physical states, such as neurophysiological occurrences in the brain, then causal closure can be satisfied by positing that the bodily movement is caused by the physical realizers of mental events. But this kind of solution has the cost of literally 'excluding' mental events, such as intentions or reasons, from the causal story. They are rendered epiphenomenal, or causally inefficacious, and may be dismissed as mere artifacts of 'folk psychology'. But this is a cost that relatively few causalists are willing to accept.[8]

Davidson rejected the traditional Cartesian account, although he agreed with Descartes's earlier analysis against non-linguistic animal agents. Clearly, there is no *necessary* argumentative connection between a Cartesian-style dualism and the exclusion of mute animals from the class of agents. Yet it is of more than merely historical interest to recall that the original Cartesian argument against animal agents not only gave the same argument from the capacity for speech as a condition for mind and action, but it also essentially linked this argument to an event-type dualism, characterized by fundamentally disparate realms demarcated by *res cogitans*, the thinking or mental substance, and *res extensa*, the extended or physical substance. For Descartes, the connection between substance dualism and a creature's propensity for propositional speech was one of simple logical entailment: it followed from his view of physical substance as being, for the most part, separately bounded from mind. Declarative or propositional speech, he wrote, 'is the only certain sign of thought hidden in a body' (1649/1991: 244–5). Although non-human animals exceed human beings in certain abilities, Descartes argued, their lack of inventiveness with respect to declarative speech

> proves that they do not have a mind, and that it is nature that acts in them, according to the disposition of their organs—just as one sees that a clock made only of wheels and springs can count the hours and measure time more accurately than we can with all our powers of reflective deliberation.
>
> (1637/1998: 59)[9]

Davidsonian-style causalists may wish, for obvious reasons, to resist the conclusion that non-human animals lack minds altogether, or the assertion that they are constituted much like the mechanistic assembly of Descartes's clock. They may point out that whereas Descartes draws the stronger conclusion that non-human animals' inability to speak entails the absence of mind, Davidson claims that such inability merely implies the absence of belief formation and possession. It may be that Davidson's account does not outright deny some very rudimentary notion of mind to non-human animals, but what would be the significance of retaining such a concept if—as Davidson claims—it does not involve the capacity for thought? Nor is the significance of this question one that pertains only to mute animals. If the concept of agency is one from which non-human animals are excluded, this exclusion—likewise, of animacy from the concept of action—is one that cannot fail to have significant implications for our idea of human action. These implications, I will argue, are undesirable beyond an acceptable point: therefore, we should prefer a concept of agency that includes the possibility of animal agents, both rational and non-rational, linguistic and mute.

Causalists do not deny that non-human animals engage in some kind of *movement*, even if such movement falls short of qualifying as action. However, according to the Davidsonian causal theory, the movement of non-human animals fails to be intentional, because of the absence of propositional beliefs. This means that, for a movement to be intentional (i.e., for the movement to count as an action), it must be caused, *inter alia*, by a psychological event or state with certain propositional contents. The account of human agency that necessarily follows distinguishes *mere movement*, where such movement is uncaused by such a cognitive state or event, from true action (intentional movement) where a proper cause is present.

To see why this picture of human agency is problematic, it is helpful to revisit a passage from the *Philosophical Investigations* that has been treated over the last decades as an important starting point for philosophical thinking about action. Wittgenstein asks:

> Let us not forget this: when 'I raise my arm', my arm goes up. And the problem arises: what is left over if I subtract the fact that my arm goes up from the fact that I raise my arm?
>
> (2001: 136)

The question is often presumed to ask for an explanation of the difference between what agents do and what merely happens to them. There may be cases in which one's arm goes up without the agent having raised it: perhaps the agent suffers from 'alien' or 'anarchic' hand syndrome, a neurological disorder (usually caused by brain injury or surgical separation of left and right brain hemispheres) that causes autonomous, 'rogue' limb behavior. These cases are obviously distinct from cases in which the agent, having full and healthy control over her limbs, raises her arm, and yet it seems that they also have something in common: in both cases, there is a bodily movement. They may even look and feel the same: imagine a scenario in which a neurophysiologist has disabled my ability to control my limb movements and connected them artificially so that, when my arm is made to go up, the sensation is identical to one in which I have absolute control over my body. The causalist's explanation of what distinguishes them lies solely in the causal etiology leading to the behavioral event. As earlier considerations showed,

Davidson's version of the causal theory insisted that such etiology include cognitive states or events with specifically propositional or declarative content. Davidson's causal picture suggests that human activity is comprised of many movements, but only those that have a suitable causal etiology count as intentional and agential. The first difficulty with this picture lies, again, in its implausibility. This account, observes Helen Steward,

> has always faced a particular sort of embarrassment. The embarrassment is that a great deal of our purposive activity does not really seem to be preceded at all by the sorts of mental events and states that figure in the story. Often, I just seem to *act*, without any prior deliberation, without first deciding or choosing to do anything, and without at any stage consciously forming an intention to act in the way that I do, or being conscious of the existence of such a prior intention.
>
> (2012: 66)

In reply to the causalist, Steward suggests that we take notice of all the bodily movements that occur in us in the space of just ten minutes. These may include crossing and uncrossing of the legs, touching one's face, stretching, etc., not to mention the 'scratchings, shufflings, twiddlings, and jigglings that together constitute quite a large sub-class of our bodily movings—[that] present the most obvious counterexamples to the claim that to act is to have a bodily movement caused by something like a prior intention, decision, or choice' (2012: 66). Causalists will assert that these counterexamples are not conclusive. Steward counters that such replies are inevitably 'tortuous'; and, in any case, 'what exactly is the reason for insisting on preserving [the claim that actions are always caused by some prior intention, decision, or choice]?' (2012: 67). Such insistence on preserving the basic causalist idea of action's specific mental antecedents 'must be a reason for suspicion that the . . . model . . . is basically Cartesian. The agent is identified with certain of her mental states and events, and her settling how her body will move is then thought of as *their* settling how her body will move; that is to say, in what looks from this point of view as though it ought to be the best case, as their deterministically causing the wanted bodily movements. But we humans are not purely mental beings, we are embodied ones' (2012: 67). Davidson's causalist account *is* basically Cartesian,[10] particularly in its solution to Wittgenstein's infamous question. For Descartes before him gave essentially the same explanation of action in the Sixth Meditation:

> I might regard a man's body as a kind of mechanism that is outfitted with and composed of bones, nerves, muscles, veins, blood and skin in such a way that, even if no mind existed in it, the man's body would still exhibit all the same motions that are in it now except for those motions that proceed either from a command of the will or, consequently, from the mind.
>
> (1641/1998: 101)

The causalist need not assert such metaphysical independence of mind from the body, in order to agree with the Cartesian view of matter belonging to the agent (where 'matter' includes not merely the body but also the bodily motions and behavioral events exhibited by that body) as a kind of underlying substratum for intentional movement. For the causalist, what differentiates action from mere movement is simply the addition of a mental antecedent, 'a command of the will or . . . the mind', which plays a role in the etiology of a behavioral event.[11]

These philosophical considerations against causalism should not escape the notice of the naturalist, for whom the exclusion of non-human animals from the class of 'true' agents cannot possibly go unchallenged.[12] Nor can the philosopher prescind from the more basic forms of inquiry about animal behavior, learning, and preferences: namely, in our observations from the natural sciences, which may go even farther than the philosophical in revealing a creature's potential status as an agent.

An obvious problem with excluding non-human animals from the class of agents is how implausible such a claim seems in the face of current knowledge about animal behavior and learning.[13] Observation of the group hunting patterns exhibited by *Orcinus orca*, the largest species of oceanic dolphin known commonly as the orca or 'killer' whale, suggests that learnt, collaborative strategy is crucial in species survival.[14] The evidence also suggests that hunting and dietary tactics are not universal to all varieties of *Orcinus orca*: different groups, or 'pods', specialize in different hunting strategies and hold distinct preferences regarding available prey. Such group-specific specialization suggests that the behavioral patterns are learnt over generations of the group, rather than being simply mechanistic expressions of an innate feature of the species. (These patterns include cooperative hunting behavior, known as 'wave-washing',[15] evidence of butchering or 'meticulous postmortem prey processing', and 'intentional stranding hunting techniques'.)[16] Similar examples of complex social, goal-directed behavior exist across a variety of animal species.[17] In light of the evidence, it seems truly implausible to claim that *all* of these instances fail to count as examples of real agency.

My discussion so far does not amount to a *definitive* argument for the claim that the concept of action is necessarily based on the concept of being an animal. In this part, I have sought merely to establish the insufficiency of the causalist's claim to the contrary. There are no good reasons, philosophical or scientific, for excluding animals from the class of agents, and the behavioral evidence seems to suggest that they *are* agents. So we humans, too, are animal agents: this description is apt insofar as our conception of human action necessarily includes features involving our own animality. Of course, such a conclusion does not entail that we are nothing but animal agents, or nothing but animals *tout court*. The point simply is that if animals other than us can be agents, then we must revise our concept of agency to take account of animacy as integral to that concept. Expectedly, this latter point has some rather far-ranging consequences, to be considered in the remainder of this chapter.

3. The Causal Structure of Action

3.1 Agency as Substance Causation

In *A Metaphysics for Freedom*, Helen Steward argues that the concept of agency is best grasped via considerations of the concept of an animal. She begins her inquiry by rejecting the proposition that non-rational animals are 'free' in the same sense or to the same extent that human beings are. Yet it remains the case that we are nonetheless animals: 'It would be surprising if anyone thought the freedoms available to a shark or a horse or even to a chimpanzee, in and of themselves, were really terribly desirable from the point of view of a human being, were "worth wanting". But for all that, it is evident that in order to be a human being, one has to be an animal' (2012: 5).[18] Much of Steward's book is a defense against 'the idea that it is natural-istically incredible that there should be any such thing as I have alleged an animal must be: an object, a substance, that things can genuinely be up to' (2012: 248). Moreover, she seeks to redress the widespread, post-Cartesian picture of 'all non-human animals as deterministic automata' (2012: 81).

It is important to clear up some points of misunderstanding that poten-tially arise from an initial canvassing. First, Steward's account does not entail that all animals are agents. Nor does it entail that all agents are ani-mals. However, there is 'strong evidence for the view that the concept of agency is an outgrowth of the concept of *animacy*', i.e., the concept of an animal. The central notion defining what it is to be an animal (agent) is that of 'settling': as Steward explains, 'it is natural to think of such animals [as cows and sheep] as the *settlers* of various matters that concern the move-ment through time and space of their own bodies' (2012: 75). The concept of an agent, on her account, has the following additional features in addi-tion to the idea of 'settler of matters' (2012: 71–2):

(i)   an agent can move the whole, or at least some parts, of something we are inclined to think of as *its* body;
(ii) an agent is a centre of some form of subjectivity;
(iii) an agent is something to which at least some rudimentary types of inten-tional state (e.g., trying, wanting, perceiving) may be properly attributed;
(iv) an agent is a settler of matters concerning certain of the movements of its own body in roughly the sense [that] . . . the actions by means of which those movements are effected cannot be regarded merely as the inevitable consequences of what has gone before.

There is much to say about each of these features. My aim in this sec-tion is to elaborate on (iv) above, by way of the following proposal: 'The crucial determinant . . . of whether a creature truly can be said genuinely to be a self-mover has to do with whether there is any irreducible role to be played in the explanation of that organism's motor activity by a certain kind of *integration* which I believe is part and parcel of the functioning of most animals of a certain degree of complexity' (2012: 16). The focus of the remainder of this chapter will be on how such 'integration' might be coherently understood as playing an explanatory role in the concept of agency.

It is clear that, for Steward, 'integration' is a metaphysical concept. Further elaboration reveals that the agent, as a 'settler of matters', exerts a kind of top-down causation. Steward's argument posits a way of organizing the higher-level and lower-level parts and processes that exist in whole living organisms or certain biological systems. With regards to agency in particular, top-down causation entails that 'the whole of an organized and integrated living system is able to affect the intuitively lower level processes that go on in its parts' (2012: 114). Her approach to substance causation distinguishes itself from other 'top-down' approaches. Many of these other approaches are committed to the notion of 'emergent properties', which possess independent causal efficacy over the lower-level properties from which they are argued to emerge (Humphreys 1996, 1997a, b; O'Connor 2000; Lowe 2000). Steward's argument, however, is exclusively in terms of the causation of wholes versus their parts. Emergent properties are normally properties of a complex whole, but causation by emergent properties is not necessarily identical to the causation by the complex whole itself (2012: 225). Moreover, an important aspect of Steward's argument, as will become clear, is her avoidance of the presumption that causation between properties is primary to causation by substances (inanimate things or animals), an aspect that further distances her approach from that of emergent properties.[19]

Steward illustrates her case for substance causation by considering the phenomenon of a whirlpool. A whirlpool is a system that involves the unfixed constitution of individual molecules of water, which are at one moment caught up in the forces of the whirlpool 'system' and at other times left out. Reductionists argue that the diachronic persistence of the whirlpool can be understood entirely in terms of the lower-level molecular arrangement, which changes from moment to moment. In other words, a 'supervenience base', or certain set of 'basal conditions', is entirely sufficient for explanations of apparently higher-level phenomena such as whirlpools. But whirlpools are generated by specific macro-level forces, i.e., the flow of opposite running currents within a body of water, and these forces are what sustain the phenomenon of a whirlpool, regardless of which individual molecules of water are constituting the system from one moment to the next. Steward suggests, 'to understand these forces and how they work, we do not look to each momentary individual supervenience base and consider how it generates the next. The persistence of the whirlpool is a phenomenon entirely blind to the details of individual molecules. It may indeed be a

complete accident that any given individual molecule is part of the supervenience base of the whirlpool at any given moment' (2012: 241).

The whirlpool is 'a phenomenon entirely blind to the details' of its lower-
level parts, much as my walking home is a process in which I am typically
unconscious of any details concerning the physiological, neurological, and
microphysical processes that constitute my journey. Nonetheless, the 'complete description' of a whirlpool or a walk necessarily includes reference to
the lower-level constituents or basal conditions that exist from moment to
moment. Steward raises a puzzle of coincidence with regard to the 'assumption that each momentary individual supervenience base necessitates the
next' (2012: 241). Nothing in the basal conditions themselves, she argues,
in the complex arrangement of molecules and other lower-level occurrences,
reveals how or why these conditions have aggregated just so, with striking
coordination, as to produce a whirlpool. She asks: 'What explains the coordination, the collocation? What we need to know is how it has come about
that all these conditions have managed to obtain so fortuitously together'
(2012: 238). Referencing the temporally prior basal set that generates the
supervenience base of the whirlpool—such as the molecular conditions that
initially occurred when two opposing water currents were forced together
at the formation of the whirlpool—does not solve the puzzle of how these
prior molecular conditions are produced together, of how they are coordinated in such a way by themselves alone. Steward writes, 'No matter how
far back we go along the chain, if we never raise our eyes from these lower
level events, we will lack what is requisite to explain why these conditions
are (at any stage of the chain) produced *together*. And yet this is what is
extraordinary; it is what needs to be explained' (2012: 238).

This puzzle of coincidence with regard to the whirlpool is not primarily a puzzle about the natural history of molecular particles, or why such
particles co-exist at all. Rather, the puzzle arises out of a need to explain
the metaphysical arrangement of microphysical parts with respect to their
wholes. Hence, the question is not simply a request for an explanation of
the co-location of the microphysical elements in a particular space and at
a particular time, but instead: why should such elements exist *qua* parts,
with spatial and temporal inter-relations, belonging to a functional, acting,
macrophysical whole?

A key aspect to understanding Steward's puzzle of coincidence (of these
lower-level events) is that the availability of a *sufficient condition* (or set
of conditions) for certain phenomena does not by itself entail *necessitating*
microphysical conditions or prior circumstances. The issue of sufficiency
has at least two relevant implications for the analysis of top-down causation. First, we must ask what the sufficient condition is a condition *for*. It
may well be the case that there is a sufficient condition, e.g., a prior microphysical movement at t1, for a particle x to be in just the right place at the
right time: that is, within the formation and flow of a whirlpool at t2. Even
so, the existence of *this* sufficient condition does not by itself entail that

there exists an entire set of sufficient conditions for a vast microphysical collection of particles to be co-located altogether, as a whirlpool, at t2. Even if one were to identify a sufficient condition for each particle within the whirlpool at t2, such a set of conditions would fall short of explaining how it is that the movement of each particle obtains together. Their simultaneous co-location is a coincidence for which there is no sufficient condition or set of conditions, at least, not as far as one can see from the perspective of lower-level (microphysical) events alone.

The second implication related to the issue of sufficiency concerns the concept of a 'sufficient condition' itself. One may believe that 'sufficient condition' means conditions that are *necessitating*: such that, granting such conditions for the occurrence of an event or phenomenon, none other could have occurred. Here, Steward cites Anscombe: 'For "sufficient condition" sounds like: "enough". And one certainly *can* ask: "May there not be *enough* to have made something happen—and yet it not have happened?"' (2012: 240). Steward argues in the affirmative:

> We cannot just help ourselves to the assumption that everything that happens is inexorably necessitated by some prior state of the world. We *do* know, of course, that in another sense of the word 'sufficient', anything that actually happens must have had causally sufficient conditions, i.e. conditions that were 'enough' to allow for its occurrence, which is to say that nothing strictly requisite for the occurrence of anything that actually occurs could have been lacking. But that is different from there being conditions in place such that nothing else could then *possibly* have happened.
>
> (2012: 240)

Weakening the assumption that 'sufficient' entails 'necessitating' is particularly important for Steward's account of top-down causation. For if the coincidental microphysical circumstances of a whirlpool are to be understood entirely in terms of necessitating conditions, all in temporal succession generating the next set of microphysical circumstances, then there is literally no metaphysical space for the causal workings of higher-level entities, such as whirlpools or water currents. Steward comments, 'Here one can indeed see no gap into which a phenomenon like top-down causation might be fitted' (2012: 240).

Steward's puzzle of coincidence is thus exacerbated by the assumption that 'sufficient' entails 'necessitating' condition.[20] Her solution to the puzzle is, firstly, to recognize that the presence of certain sufficient conditions for phenomena such as whirlpools does not *ipso facto* eliminate the availability of other sufficient conditions—or, indeed, the possibility that no satisfactory answer to a particular event or phenomenon exists at all. Her point is aptly illustrated via Aristotle's example of a man who visits a well at the 'wrong' time and is murdered by thieves (*Meta.* VI, 3). To explain what happened to the victim, we have at our disposal a variety of sufficient conditions for his being at the well at just the same time as his assailants. 'But the mere existence of a sufficient condition does not by itself imply that the question how it was that these phenomena obtained *together* has a satisfactory answer or any answer at all', points out Steward. 'However, in the case of the complex phenomena of which our world consists . . . we surely have the right to demand that there be such an answer' (2012: 239).

The kind of answer she has in mind brings us to the second part of the solution to the puzzle of coincidence. The challenge that the case of the whirlpool poses specifically for the concept of action is, as Steward puts it, 'to understand how on earth it can be that the animal [agent] has any real, independent efficacy *of its own*: any efficacy that does not merely reduce to the efficacy of its various parts' (2012: 227). The argument that the animal agent *does* have an irreducible causal efficacy, just as the whirlpool exists 'entirely blind to the details of individual molecules', purports to offer *the* answer to the question of how the microphysical phenomena all happen to obtain together in an apparent coincidence—wherever there are actions like one's walking or complex non-agential occurrences like whirlpools. The provision of a higher-level perspective counts not merely as an explanatory point in favor of (higher-level) substance causation, but is part of the 'relevant *metaphysics* of causation' (2012: 239). Agents are causes of their own movements, operating independently of—or irreducibly with respect to— the causal succession of microphysical parts constituting action and activity.

This concept of agent cause treats causation by animal agents as having 'independent' efficacy; it is fair to ask precisely what such independence entails. That substances exist 'entirely blind' to microphysical details might be glossed to mean that agent causes do not depend at all on the causal powers of their constituent parts. This would be a problematic conclusion for an account that explicitly aims ultimately to secure the 'integrity' of parts with respect to their wholes. Top-down causation admits of the constitutive necessity of the bottom-level, microphysical parts that constitute actions such as walking and systems such as whirlpools. The foregoing objection to the idea of sufficient conditions as necessitating sought to establish that, although there exists such dependence on microphysical circumstances as sufficient conditions, the latter conditions do not by themselves determine (i.e., causally necessitate) or explain how it is that causal powers at the higher level are possible. The causal efficacy of agents is 'independent', therefore, *not* in the sense that there is no dependence upon the causal powers possessed by an agents' parts: the existence of such dependence is simply not also 'upwardly' determinative of the agent's movements. In arguing for the 'independent' causal efficacy of the animal agent, Steward thus avoids a traditional charge leveled against top-down causationists: that the higher-level causal efficacy of agents, unmoored from the reality of its lower-level supervenience base, results in a kind of causal power so independent from the latter as to evoke something 'ethereal'. Steward argues:

Nothing problematically ethereal is . . . involved; just as a whirlpool could have no effects were it not for the molecules of which it is composed having effects, so the way is clear for the straightforward admission that an animal could do nothing unless its parts simultaneously did things. The action of an animal is indeed constituted in this way by the actions of its parts. But as with the whirlpool, the key to seeing how the action of the animal is more than the sum of the actions of its parts is . . . likely to be the idea that higher-level processes can dominate and dictate the evolution and distribution of certain lower-level ones, so that, as far as the important causal metaphysics is concerned, the explanation of how a certain complex neural state of affairs has come to be depends upon higher-level processes and ontologies. And as with the whirlpool, we avoid the puzzle about how there could be anything more to be said than what is given by the low-level neural story (or indeed than by what is given by stories told at much lower levels than the neural) by simply refusing the assumption that each momentary lower-level supervenience base necessitates the next.

(2012: 244)

The above passage seems aimed at assuaging physicalist worries that Steward's variety of substance causation does not adequately account for microphysical processes and events: she offers assurances that the ontological independence of an agent's causal powers is not incompatible with a naturalistic account. So far, however, the analysis has neglected to mention how the causal efficacy of agents prevails ontologically in the face of a kind of reduction distinct from that involving causal powers (to the causal efficacy of an animals' lower-level parts): namely, reduction in terms of agent-neurological or microphysical events. For let us momentarily assume the standpoint of one who resists the foregoing argument, i.e., that the higher-level causal powers of animal agents are irreducible to the efficacy of their lower-level parts (presumably since the account of top-down causation is deemed uncompelling): such an opponent may well resist Steward's main argument based on the idea that the phenomenon of higher level substance causation ultimately reduces (both metaphysically and explanatorily) to causation between events at a lower level, i.e., the physiological, neural, or microphysical.

The argument for the independent causal efficacy of animal agents, particularly in the face of this latter threat of reduction, ultimately appears to rest in Steward's account of causal pluralism. Some philosophers have denied that events can be causes at all, thereby asserting *only* the causal powers of agents or substances (Lowe 2008; Ayers 1968). Steward eschews this route, instead adopting the view that 'causation' is a category of diverse concepts, under which we may accept not only a causation by events, which Steward calls 'makers-happen', but also substances or collections of substances[21], which she calls 'movers'. Her causal pluralism also includes causation by facts, which she calls 'matterers'. Each of these causal types is ontologically independent from and thus also irreducible to one another. This is not to say that the causal types may not be related to one another in important ways, but Steward insists that

> we must not get the categories *mixed up* and assimilate them wrongly to one another in the service of a chimerical uniformity [. . .] In particular, it is absolutely essential to recognize that some types of cause (both the movers and the makers-happen) are proper spatiotemporal *particulars* while others (the matterers) relate to *general* factors (properties, features, aspects, etc.) and we only make a horrendous hash of our causal thinking if we fail to recognize that we are interested, when looking for causes, in *both* things that are particular *and* things that are general . . . [for] causal thinking involves a concern *both* with particularity *and* generality.
>
> (2012: 211)

Steward's distinction between spatiotemporal particulars and general factors highlights the fact that particulars actually *do* things, i.e., they enact changes (or perform actions). This fact is suggestive of a stronger claim, that substance causation is not just another causal kind in addition to events and facts, but indeed, a category that holds a sort of ontological primacy over all other categories. Sketching the case for this primacy of substance causation, and in particular the agency of animals, will be the task of the next section. Establishing the primacy of substance causation is one way to assert the implausibility of reduction, both explanatory and ontological: from the level of the substance to the causation by events involving that substance, or a causation between relevant facts. It may be, of course, that Steward's argument for the irreducible causal efficacy of animal agents already settles this question. The aim of this section, after all, was to sketch her case for the claim that agents, as settlers of matters[22] and initiators of their own movements, can be causes of a kind also found pervasively in other natural but non-agential phenomena. The argument for top-down causation purports to provide this explanation and to establish the irreducibility of animal agency to causation by the animal's parts.

Having laid out an account of agency as substance causation, I turn next to some of its central difficulties.

3.2 *The Problem of Synchronic Dependency*

Many theories of top-down or downward causation presume emergentism, which posits that emergent properties exert special, or 'novel', causal influence over the basal properties from which they emerge. These emergentist interpretations of top-down causation are susceptible to a number of objections, most notably Kim's (1999, 2006). Although Steward's account rejects emergent causal powers,[23] it may be thought that the variety of top-down causation defended by Steward faces these problems as well, so it is worth reviewing them. Kim's criticisms are directed mainly at the emergentist thesis that downward causation by higher-level properties is dependent, whether synchronically or diachronically, on basal physical conditions that serve as micro-constituents (again, whether synchronically or diachronically).[24] The synchronic dependency is 'viciously circular', he argues, in that the emergentist claims that what produces higher level mentality, the basal set of physical properties, is simultaneously altered by the very mental properties that the basal set produces, and so on.

It may be thought that the variety of top-down causation defended by Steward escapes the brunt of these criticisms, primarily because her view is in terms of parts and wholes rather than emergent properties. But Kim's objection—that the basal conditions (or the bottom level parts) are causally responsible for the whole that the parts constitute—may also apply to Steward's version of top-down causation. Whether emergence is assumed or not, a similar problem seems to arise for the mereological version of downward causation: for the whole is causally dependent on the parts that constitute it as a whole, and simultaneously, claims the substance causationist, the parts somehow causally depend on the whole of which they are parts. Vicious circularity arguably results, as in the emergentist variation. For it seems absurd to think that the parts of a whole cause it to exist, while simultaneously the whole that depends on these parts also causes *them* to exist, as parts of the whole. Reduction is a tempting solution, in the face of this circularity. Should we not admit that a thing's parts fundamentally determine the thing itself, and that this 'upward determination' is more foundational than any form of causation one might otherwise introduce via 'downward' causation?

The substance causationist may respond in a variety of ways to this objection. First, she could deny that the parts are constitutive of the whole in such a way that they causally determine the latter. A whirlpool is not upwardly determined by individual water molecules, since from one moment to the next, the whirlpool is not constituted by the same collection of water molecules that happen to be taken up into the whirlpool's current. That any particular arrangement of molecules happens to be a part of the supervenience base of the whirlpool could be, Steward points out, 'a complete accident', especially insofar as we grant that 'we have no right to the assumption that each momentary individual supervenience base necessitates the next' (2012: 241). Without this assumption, causal determination from below, or the upward causation of lower-level parts on wholes, has no conceptual foothold: how is the determination of a constant (in this case, the whirlpool) possible when the assumed bottom-level determinants themselves are absolutely variable? So at least in the case of a whirlpool, the notion of 'upward determination' of parts on the whole can be set aside.

In the meantime, the question remains: does the rejection of lower-level causal determination exclude *all* 'upward' causal influence? An extreme variation on Steward's rejection of upward determination goes so far as to deny attributions of any causality exerted by the lower-level parts exerted on the whole. The whirlpool surges on regardless of which and how many water molecules happen to be taken up into it as parts: its existence is wholly independent of the latter, entirely blind to them both explanatorily and ontologically. Perhaps one way to understand this approach is to think of a whirlpool as the kind of entity or system that simply happens to involve water molecules at each instantaneous time slice: no specific molecules of water are 'necessary' in the sense that the whirlpool would be conceived as causally dependent on those particular molecules of water. According to this extreme view, then, the whirlpool is a sort of thing whose parts exert no causal influence on it whatsoever, even as temporary constituents.

This view is unsatisfactory for the obvious reason that water molecules are not merely accidental features of whirlpools. There is at least one further problem with a view like this, however: it appears to downgrade the sense of parthood, in that it begins to sound incoherent to speak of parts and wholes at all. As mere accidental accompaniments to the whirlpool's path, the water molecules that from one moment to the next happen to be taken up into the whirlpool apparently lack the status that parts or constitutive elements of a whole characteristically have. The obvious reply to this extreme position is that the molecular constituents of a whirlpool are important, *causally* important, for the higher-level substance itself. Even when they are not strictly identified with the whole that persists through time—for the whirlpool is constantly exchanging its molecular parts for new ones, and artifacts such as wheels, although somewhat more stable, have parts that one by one may also gradually be replaced, as was the whole ship of Theseus—even then, the molecular constituents may be understood to have some kind of causal influence on the fact that the whole is what it is. For example, the fact that the molecular constituents of a whirlpool are *water* molecules, even though they undergo constant change, explains why the whirlpool ceases to exist when temperatures rise above water's boiling point or falls below its freezing point. One could say that constituent parts, qua parts of a whole, are causes of another kind.

The more moderate approach to the objection of upward causal determination, however, still faces the earlier question posed by Kim, concerning synchronic circularity. The problem is that the substance causationist cannot maintain causal priority of the whole, while simultaneously claiming that its parts exert upward causal influence on the whole. The only way to solve this problem, I think, is to take seriously the idea of causal pluralism invoked earlier in Steward's argument for substance causation. Her explanation is that substances, or 'movers', are causes in an entirely different way from facts, or 'matterers', and events, 'makers-happen'. That is, there are certain kinds of effects that simply cannot be attributed to facts, and other kinds of effects that cannot be attributed to substances.[25] In order to deflect Kim's criticisms, however, we must add at least another causal type to Steward's list.[26] Specifically, we should consider the category of 'constituent parts' or the 'stuff' that makes up a whole: whatever we call it, these do not cause the whole to exist in the same way that the whole causally organizes or spatially arranges *them*, in a very precise way, specifically as parts.[27] Constituent parts are causes insofar as they make up a whole, but they make it up only insofar as the whole takes them in and spatially arranges them as such.

Kim's criticism holds weight only if we presume that the causal priority of the whole over its parts is the very same kind of concept or causal relation as the upward causal exertion that constituents have on the wholes of which they are parts. Substance causation claims that the whole causally governs its parts—as a cat slinks towards a mouse, or as water funnels continuously off the coast of Suriname—in a way entirely distinct from the causal relation that constituent parts have with regard to their whole matter. There is no problem of vicious circularity for these cases because circularity assumes a directional conflict between identical ontological types.[28] The circularity problem relies on the assumption that it is the parts *simpliciter* that both causes and is caused by the whole, and likewise the whole *simpliciter* that both causes and is caused by its parts. But these forms of causation are ontologically diverse not merely because wholes obviously are not their parts. Rather, causation by the whole on its parts results in a precise spatial arrangement of these material constituents, with related causal powers that are context-dependent (i.e., dependent on the spatial arrangement produced by the whole substance). Upwards causation by these material constituents on the whole of which they are a part is radically different: it is not the context-dependent causal powers of these parts (which are caused by the whole substance) that cause the existence of the whole, but rather their generic compositional powers that sustain the existence of the whole as a material thing.[29]

3.3 The Problem of Explanatory Redundancy

The more moderate reading of substance causation does not, however, immediately escape the potential challenge that such causation is explanatorily redundant. Steward's earlier argument concerning the puzzle of coincidence was intended to meet the challenge apropos constituent parts: i.e., that higher level substance causation is theoretically redundant as a proposed conception of 'cause', given that the lower level microphysical constituents and their efficacy constitute all that is needed to explain the movement of whirlpools and the agency of animals. Even if Steward's reply eases the trouble as regards the 'sufficiency' of constituent parts, the complaint of redundancy remains given the availability of explanations involving event causation. The popularity of Davidsonian-type causalism in particular has ensured that higher-level substance or agent causation is received with skepticism[30] or viewed as explanatorily (and ontologically) superfluous. The Davidsonian reading, instead, encourages us to view events as the prime causal candidates. If it is true that my noticing a sign for a scenic detour causes me to impulsively take that detour, why must it also be said that *I* caused a significant alteration to the planned journey? Presuming that there is no issue concerning my responsibility for subsequently arriving late to my appointment, then citing agent causation seems like an added and unnecessary feature of what happened. Contrary to criticisms advanced by 'disappearing agency' theories (Velleman 1992), proponents of event causation need not be opposed to the idea of agency or agents, for there is no diminishment of the sense that *I* took the detour path; it signals no conflict with my status as agent, in attributing the cause of my action to the event of my noticing the detour sign.

One way to counter the challenge of redundancy from the direction of event causalism is to deny that events are causes of actions at all (Lowe 2008; Ayers 1968). When it comes to actually getting something done, events simply do not seem to be the sort of things that do causal work. But, as Steward rightly points out, this is an extreme way to respond to the present challenge of redundancy (2012: 209–10). The event of my noticing a sign for a scenic detour does seem to figure causally in my subsequent action of adopting the detour route: the event of my noticing what I did explains why I went that route instead of another. If we assume the view that the 'causal' is broadly understood in terms of what explains or enlightens, an understanding that should have particular resonance for Aristotelians, then events as well can be causes of some kind.

Granting the latter need not amount to conceding that events are *the* causes of actions (even when it is specified, with sensitivity to the concepts of agents and agency, that such actions are *my* actions). Just as one potentially concedes too much in saying that the molecular constituents of an animal cause *it* to exist, one likewise grants too much status to events in allowing that they are the causes of actions. Events are better understood analogously: as the kind of causes that constituent parts and facts are. Facts are causes only insofar that they may play an explanatory role for actions. Constituent parts, in turn, are causes only insofar as they help to make up the whole that organizes them as parts. Constituent parts are therefore causes only in a limited sense, limited relative to the paradigm case of (higher-level) substance causation. We can see the limitation of constituents especially in the case of living organisms. For attempts to explain the causation of action by way of parts or properties involving the brain and body fall below the level of the animal agent as a whole: they result in explications of action in terms of sub-agential, even fragmented elements. Explication of action in terms of facts and events achieves a similarly limited result, from either direction: facts and events involving molecular, neurological, or merely physiological changes in the brain and body result in a similarly sub-agential picture of action, while facts and events involving the agent's wider circumstances may sometimes result in a viewpoint so far above and outside the agent that one fails to see how action involves an agent at all.

Events are sometimes indistinguishable from facts, as when 'the plant being wilted on Saturday morning' is put forth as an event.[31] Even more straightforward events, e.g., 'the sun coming out after the rain', 'the convocation of the new academic year', 'the child's turning five', etc., can be understood abstractly and recast as facts.[32] However they are conceived, the critical issue here is whether such entities are responsible for necessitating action.[33] They are not. Indeed, an important feature of the agent causal view under consideration is that we should deny that actions are *ever* 'necessitated' by entities even other than events, such as agents themselves. Given the pervasive equivocation between 'cause', 'necessitate', and 'determine', considerations thus far have suggested that we cannot unproblematically claim that agents cause or necessitate their actions, even on an agent causationist view. The agent causal view reviewed in the preceding section has, instead, advanced the idea that animal agents are causes of their bodily movements, not as necessitating or determining causes, but simply as 'settlers' of matters.[34] As such, agents are causes par excellence. For the varieties of the concept of cause apropos action, and the limited influence that the sub-agential (or extra-agential) forms have in bringing it about, generate the demand for a higher level whole that explanatorily 'organizes' the different causal types of event, fact, constituent part, and property; its organization of these is such that they may count as playing causal roles (especially in the explanatory sense) in action, and without the ontological gaps from which causal deviance arises. This organizing higher-level whole for action, of course, is the animal agent herself. The threat of reduction, raised in the previous section, is overcome once it is understood that the sub-agential (or extra-agential) causal features of agency could not be intelligible, qua such features, without higher-level substance causation, specifically of the agential kind. Why, then, should we assume that event or fact causation, or indeed causation by properties or molecular and other microphysical elements, are the starting points for explicating action?

Rejection of the latter assumption is problematic, however. Steward's arguments involving the example of the whirlpool made the case against reduction, specifically of the mereological kind: i.e., against the challenge that the whirlpool as constituted by water molecules can be understood entirely in molecular terms, so that the whirlpool itself is recognized as, explanatorily and ontologically, a superfluous entity. Her response to the challenge, as reviewed before, was to argue that the organization of the lower-level elements do not make sense without an 'organizing' cause from the higher-level whole. She states: 'It is quite true that the lower-level arrangements (once we have them) are "all we need", but the crucial question is how the requisite causally proximal lower-level arrangements are to be provided for in the first place' (2012: 236). The success of reductionist accounts depend upon the presumption that the entities such as the agent's microphysical parts (as well as the events involving the agent's brain and body, related facts and properties, etc.) are entities or elements that are taken for granted: it is assumed that 'we already have them', and as such, they are entirely adequate to explain agential phenomena without Steward's question, of how they are 'provided for in the first place', having to ever arise.

Steward's argument is that her question *does* arise: for what dictates that the microphysical elements are parts of the whirlpool, if not the higher level forces of the whirlpool itself? The argument must succeed, however, not just against the challenge of mereological reduction, but also over the other explanatory routes favored by causalists. For that, it must be shown that an agential cause has both ontological and explanatory primacy to generate the kind of causation involving these other sub-agential or extra-agential features. Just as mereological priority is attributed to a system such as a whirlpool, agent causation should garner explanatory first place among its related causal explanations. In the case of actions, it is natural for us to countenance first of all those explanations concerning the agent as cause, e.g., 'The cat pursued the mouse'. An analysis in terms of events possibly follows from the latter explanation, as e.g., 'The event of the cat's catching sight of the mouse caused its subsequent pursuit of it'. But analysis in terms of events or properties cannot possibly be explanatorily prior to explanations involving the agent as cause: what I mean by this is that we could not see the reductive analyses as true, without first acknowledging the agent's causal work. For the events in which 'the cat catches sight of the mouse' and 'its subsequent pursuit' both take for granted that the cat does something, i.e., that an agent causal explanation is already in place. According to this line of argument, the explanation of agents as causes is irreducible because the supposedly reductive analyses rely on *it* for their own intelligibility.[35]

The claim that agential or substance causes are ontologically and explanatorily prior to all other causal categories is not explicitly countenanced in Steward's considerations. It may be that my latter remarks in this section signal a stronger thesis with which her account disagrees. I am arguing, however, that the success of her account relies on this stronger thesis. It is not enough to simply posit that agential causes are one of many irreducible causal kinds.

There may be doubt that Steward's puzzle of coincidence proves what it purports to prove: that a higher-level substance cause is necessary for the lower level microphysical elements to come together, in the first place, as an arrangement recognizable as a supervenience base. What does the puzzle of coincidence show, but—at most—that the arrangement of microphysical entities requires *some* explanation: *any* plausible explanation, which may or may not cite higher level substance causation?[36] Perhaps her arguments against the main alternative, i.e., the necessitated succession of microphysical elements from one moment to the next, are compelling enough for one to assume that there could be no other explanation (for the phenomenon of coincidence) but that of top-down causation. But one may doubt that the latter represents a positive case for substance causation, let alone establishes it as the explanatorily primary situation from which all other varieties of causation are derived.[37]

Notes

1 Though this may seem a far-fetched prospect in contemporary philosophy of action, the concept of agent causation frequently faces the criticism that its proposed causal relation is metaphysically mysterious, even supernatural. Consider such agent causationists as Chisholm (1964) or O'Connor (2000), whose characterization of agents as 'unmoved movers'—a term applied in medieval philosophy to divine being—evokes a notion of causation for agency that is unlike any other mode of causation in the natural world: indeed, an exceptional cause.

2 'One belief demands many beliefs, and beliefs demand other basic attitudes such as intentions, desires, and if I am right, the gift of tongues' (Davidson 1982: 318).

3 McDowell offers a similar argument, on Kantian grounds (1996).

4 Davidson resists any strict identification between 'rational animals' and 'human beings'. His purpose is not to attribute the concept of an agent universally to any particular group, but rather, simply, to pursue the question of 'what makes an animal (or anything else, if one wants) rational' (318). It follows from his view that human beings may not be the only rational animals there are, and also that not every human being is necessarily a rational animal, by his definition of that term.

5 See Tim Crane's 'What is Distinctive About Human Thought?' (Inaugural Lecture, Knightbridge Professorship at University of Cambridge): 'In particular, the premise that one can only have beliefs if one has the concept of belief is crucially unsupported, and without that, there is no reason to accept his conclusion, and no reason to deny thought to non-linguistic animals. In the relevant sense, a belief can be a simple representational state, which Ramsey's chicken can have. We can call the chicken's belief a belief *that chickens [sic] are poisonous* if we like, but this does not require that we attribute to the chicken the "concept" of poison. Calling this a belief is just a way of indicating that the chicken represents the world in a way that guides its actions, and in way [sic] that can be correct or incorrect' (11).

6 'States and dispositions are not events, but the onslaught of a state or disposition is' ('Actions, Reasons, and Causes', 1963: 12).

7 Broadie's use of the term 'constituent' apparently expresses a logical relation between agent and event, comparable to the way that 'the vase is on the table' may be said to have the logical constituents of 'the vase', the 'being on' relation, and 'the table'. Perhaps this is an overly loose way of describing the relation between agent and events (or states of affairs, or facts), but it should be clear that the constituent relation employed here bears no significant similarities (except analogously) to metaphysical parthood, e.g., as when the legs of the table are understood as its constituent parts, among others.

8 Even if one excludes determinism from the causal picture, it is possible to construe lower-level states and events (or their laws and properties) as fixing certain probabilities about future behavior, such that facts about the whole have no causal influence in altering these probabilities. This account likewise bears the cost of excluding certain seemingly relevant facts about the whole from having causal significance in the performance of action. (Thanks to Rob Koons for pointing out this non-deterministic alternative.)

9 This proof against the mindedness of non-human animals was definitive for Descartes. He rejected any consideration by degrees, i.e. the thought that non-human

animals may be *less* minded than human beings and yet still possess some mental capacity, in his argument, that the inability to communicate by tongue proves 'not merely . . . the fact that animals have less reason than men but that they have none at all' (1637/1998: 33).

10 Davidson's distinctively Cartesian approach has little to do with Descartes's metaphysical doctrine of substance dualism. The dualism assumed by Davidsonian causalism is rather one concerned with sets or systems of concepts. One set of concepts is employed for describing matter, and another, very different set is used for describing mind; neither conceptual system makes reference to any of the concepts contained in the other. I have suggested that the problem arising from this view, the problem of causal exclusion, produces difficulties for the idea of a mental event or state as causing the event of a physical movement. That the latter idea encounters difficulties, in the case of contemporary causal theories, is not usually the fault of any *metaphysical* picture in which ghostly or incorporeal substances play a putative role in moving or pushing around physical entities (unless, of course, one really does subscribe to Descartes's metaphysical dualism). Instead, the difficulty of accepting physical causation by mental events is an *explanatory* difficulty, for the physical domain is believed to be wholly explicable in purely physical terms.

11 A complementary question might arise, although I do not address it here: why is it *not* a problem (for materialist proponents who accept explanatory exclusion of the mental) that a physical change or event might cause a mental event, e.g., a physiological change that causes an agent to become aware of that change?

12 Sometimes philosophers equate 'agency' with 'morally responsible agency'. Obviously, this assumption (usually) entails a denial of the claim that non-human animals have a membership in the class of true agents. I do not see why we should accept such an equivalency, at least not without argument.

13 One might include in considerations on this topic the scientific evidence suggesting human evolution from 'lower' animal forms. An important implication of such evidence is that, if the present evidence does confirm the theory of macro-evolution, naturally, we should see a certain observable continuity from humans to other animal species in aspects of behavior and learning. Such continuity can be granted without at all dispensing with the idea that human beings are themselves exceptional in the animal kingdom, in obvious (also observable) ways.

14 See research by Robert Pitman and John Durban, 'Cooperative hunting behavior, prey selectivity and prey handling by pack ice killer whales (*Orcinus orca*), type B, in Antarctic Peninsula waters', in *Marine Mammal Science* (Vol. 28, Issue 1, January 2012, pp. 16–36). Available at http://onlinelibrary.wiley.com/doi/10.1111/j.1748-7692.2010.00453.x/full.

15 Wave-washing is a well-documented hunting tactic among family pods. It has the goal of tipping ice floes, upon which the targeted prey has been spotted, by the group's swimming quickly in coordinated formation towards the floe, and then, in perfect unison, sharply diverging at just the right moment to create a flush of water that tips the prey into the water.

16 Christophe Guinet and Jérome Bouvier, 'Development of intentional stranding hunting techniques in killer whale (*Orcinus orca*) calves at Crozet Archipelago'. *Canadian Journal of Zoology*, 1995, Vol. 73, No. 1: pp. 27–33.

17 Studies of avian nest design represent another research area bearing rich insights into animal abilities. Natural scientists (arguably beginning with Aristotle) have long observed the highly sophisticated design and aims of nest building, and recognized the 'considerable cognitive abilities' required for accomplishing such ends (Mainwaring et al. 2014, in *Ecology and Evolution* 4(20): 3909–3928). Available at www.ncbi.nlm.nih.gov/pmc/articles/PMC4242575/. (See also Collias 1986; Muth and Healy 2011; Walsh et al. 2011).

18 Here, Steward explicitly notes some of the attributes of human action uniquely possessed by us: 'In order to exercise the forms of agency that we value so highly—moral choice, exercises of taste and skill, communication, self-disciplined attention to duties, personal development, creativity, etc.—we have to be able also to exercise forms that in themselves almost escape our notice—we have to be able to move our bodies in such a way as to make them carry out plans of our own devising, in the service of our ends' (2012: 5).

19 Although emergentism does not always entail exceptionalism, emergent theorists also frequently happen to be exceptionalists about emergent causation. They argue that human consciousness and free agency are uniquely emergent properties and *ipso facto* different in causal character from any other natural phenomena (Chalmers 2006; O'Connor 2000).

20 Steward's broader point is that the assumption of 'sufficient' as 'necessitating' is precisely one made by proponents of determinism. For my purposes, I presume a concept of 'sufficient condition' that does not entail 'necessitating'.

21 Steward's understanding of 'substance' is inclusive not only of animals and persons, but also 'stones and masses of air and water . . . as well as some of the smaller entities that go to make them up, like molecules and ions' (2012: 212). It may be that a different kind of argument is required for defending the independent causal efficacy of lower-level substances, as apart from the causation of events or facts. I will not address this issue here: my immediate concern is with substances that are animal agents.

22 Causation by agents involves a 'discretionary aspect' (2012: 246). Steward explains: 'It really is *up to the agent* which selections are made from the large repertoire of possible actions available: the agent counts as the settler of what is to happen with her body and thereby as the settler of what will happen to those parts of the world on which her body is able to impinge' (2012: 246).

23 It is not possible to review her objections here. Nor, arguably, is it necessary for my present purposes: the argument of this chapter is neutral with respect to the acceptance or denial of, e.g., the emergence defended by O'Connor and Wong (2005).

24 Kim's reply to the diachronic case, as opposed to the synchronic case, is too lengthy and tangential to be fully summarized here. As will shortly be seen, the synchronic case is accused of producing 'viciously circular' causation, given the simultaneity of downward and upward causation implicit in the emergentist's account. Because the diachronic case explains the higher-level causal exertion as occurring at a later time from the moment of upward determination, it escapes this criticism. However, since the diachronic case (like the synchronic version of emergentism) assumes the higher-level properties to be mental properties and the basal conditions to be physical properties, Kim accuses diachronic emergentism of violating the principle of causal exclusion (2000: 318). Because Steward's version of downward causation does not necessarily assume causation by mental versus physical properties, the objection does not bear much weight in the present discussion.

25 Steward deflects a potential charge that the difference between the causal types is merely explanatory; on her view, the differences between them are ontological as well as explanatory. She explains: 'Substances are simply causes of a different ontological type than either makers-happen or matterers, and it is entirely unsurprising that they do not play exactly the same role in explanation as causes of these other sorts' (2012: 220).

26 One could arguably add many other different causal types to the list, although it is unnecessary to plumb the possibilities here. For instance, Anscombe points out that the removal of a doorstop from underneath the door it is jamming may be one such addition, a *causa removens prohibens* (Anscombe 1983, 'The Causation of Action').

27 Aristotelians will recognize the origin of this idea to be that of Aristotle's material cause.

28 Robert Koons has offered a solution along these lines via his 'parts as sustaining instruments' (PASI) account, which purports to avoid the problem of circularity. On Koons' account, two kinds of dependency relations exist simultaneously between a whole and its constitutive parts, the synchronic and the diachronic: 'The synchronic dependency is top-down, with the powers of parts grounded in the powers of the whole, while the diachronic dependency is bottom-up, with the later existence of the whole dependent on the earlier activity of the parts. Hence, there is no circularity; instead, the dependency diagram is a zig-zag path, running down at each moment and up as time advances' (2014: 172). This appears to be an attractive solution, with one problem: as a self-identified Aristotelian account, PASI denies that there can be bottom-up synchronic causation, which one would assume is needed to retain the existence of an Aristotelian material cause. My solution to the problem of vicious causal circularity is that the material cause is synchronic with top-down substance causation, where these are distinct ontological types. Vicious circularity assumes identical causal types simply running in opposite directions, whereas here, the only relevant difference is the fact that material and formal causes are of absolutely diverse kinds.

29 Thanks to Rob Koons for this clarification.

30 It has not helped that some proponents of agent causation have characterized it in somewhat mystical terms. Farrer, for instance, describes the lower-level parts of an organism as 'the molecular constituents [that] are caught up and as it were *bewitched* by larger patterns of action, and cells in turn by the animal body' (1958: 57, italics mine). From Farrer's 1957 Gifford Lectures, published as *The Freedom of the Will* (1958).

31 This sort of example follows a liberal conception of events. Indeed, one may doubt whether such an example falls within a class of true events or merely facts masquerading as events.

32 A great deal could be said on the various linguistic ways to understand these statements. Steward argues that insofar as events behave as facts, neither are these the sorts of things that can cause actions (2012: 221).

33 Davidson's interpretation is that a relevant belief and desire pair (the event of an agent desiring an object *o* and believing relevant facts about *o*) are by themselves sufficient in causing physical movement, where 'causing' is in the sense of necessitating, and the physical movement being caused (provided that it is caused in the appropriate way) is also identified as an action. See his 1963 'Actions, Reasons and Causes'.

34 See especially Steward's Chapter 2, 'Up-to-Usness, Agency, and Determinism', in *A Metaphysics for Freedom* (2012).

35 To some extent, a Davidsonian causalist might recognize that he takes agential explanation for granted. (Davidson thinks it is simply obvious that mental events belong to agents; for this reason the criticism of disappearing agents does not impress him.) The problem for the causalist is that agential explanation does not reveal anything: it does not offer any kind of explanans, but is rather the explanandum, the thing to be explained. So the Davidsonian causalist needs to be persuaded that the concept of an agent as cause (i.e., that explanation in terms of agent cause) is intelligible and not merely a trivial re-statement of the problem. Some proponents of agent causation (e.g., Lowe) will face a very different problem: they will accept the explanation of agent causation as intelligible and informative, but reject the possibility that explanation in terms of causation by events or properties can be made available without damaging the view of agent as cause. It seems to me that a successful account of agent causation should

simultaneously address both of these objections, the Davidsonian causalist's and the agent causationist.

36 Aristotelians will also protest that substance causation, as a kind of formal or actualizing cause, is insufficiently described as the organizing principle of a supervenience base. As Koons notes, 'When Aristotle describes the soul as the form of the body (e.g., in *De Anima* II.1, 412a19-21), he clearly means more than just an arrangement or relationship among the parts of the body. A *form* (*morphe*) of a body is not analogous to the harmonious relations among a set of strings (*De Anima* I.4, 407b)' (2014: 152). Steward's puzzle of coincidence, on the other hand, seems unable to offer more than an explanation of the mere arrangement of the lower-level parts, in place of an account of why there is a substance with its own powers in existence at all.

37 Acknowledgements. I am grateful to the editors of this volume, especially Robert Koons, for helpful feedback and comments, and also to the Institute for Humane Studies at George Mason University for a research grant covering the completion of this article. Thanks is also owed to those who have reviewed drafts: Sarah Broadie, Helen Steward, Adrian Haddock, Raphael Mary Salzillo, OP, and Michael D. Breidenbach. All errors are solely the fault of the author.

References

Anscombe, E. (1983) 'The Causation of Action', in C. Ginet and S. Shoemaker (eds.) *Knowledge and Mind: Philosophical Essays* (New York: Oxford University Press).

Ayers, M. (1968) *The Refutation of Determinism* (London: Methuen).

Broadie, S. (2013) 'Agency and Determinism in a Metaphysics for Freedom', *Inquiry: An Interdisciplinary Journal of Philosophy* 56: 571–82.

Chalmers, D. (2006) 'Strong and Weak Emergence', in P. Clayton and P. Davies (eds.) *The Re-emergence of Emergence* (New York: Oxford University Press).

Chisholm, R. (1964) 'Human Freedom and the Self', The Lindley Lecture, published by the Department of Philosophy of the University of Kansas.

Collias, N. (1986) 'Engineering Aspects of Nest Building by Birds', *Endeavour* 10: 9–16.

Crane, T. (2010) 'What Is Distinctive About Human Thought?', Inaugural Lecture, Knightbridge Professorship at University of Cambridge.

Davidson, D. (1963) 'Actions, Reasons and Causes', *Journal of Philosophy* 60: 685–700.

———— (1982) 'Rational Animals', in E. Lepore and B. McLaughlin (eds.) *Actions and Events: Perspectives on the Philosophy of Donald Davidson* (New York: Basil Blackwell).

Descartes, R. (1649/1991) 'Letter to More', in John Cottingham and Robert Stoothoff (eds.), trans. by Dugald Murdoch, *The Philosophical Writings of Descartes* (New York: Cambridge University Press).

———— (1637/1998) *Discourse on Method*, trans. by Donald Cress (Indianapolis: Hackett).

———— (1641/1998) *Meditations on First Philosophy*, trans. by Donald Cress (Indianapolis: Hackett).

Farrer, A. (1958) *The Freedom of the Will* (London: A&C Black).

Guinet, C. and Bouvier, J. (1995) 'Development of Intentional Stranding Hunting Techniques in Killer Whale (*Orcinus orca*) Calves at Crozet Archipelago', *Canadian Journal of Zoology* 73: 27–33.

Humphreys, P. (1996) 'Aspects of Emergence', *Philosophical Topics* 24: 53–70.

———— (1997a) 'How Properties Emerge', *Philosophy of Science* 64: 1–17.

—— (1997b) 'Emergence, Not Supervenience', *Philosophy of Science* 64: 337–45.

Kim, J. (1999) 'Making Sense of Emergence', *Philosophical Studies* 95: 3–36.

—— (2000) *Mind in a Physical World: An Essay on the Mind-Body Problem and Mental Causation* (Cambridge, MA: MIT Press).

—— (2006) 'Being Realistic About Emergence', in P. Clayton and P. S. Davids (eds.) *The Re-Emergence of Emergence* (New York: Oxford University Press).

Koons, R. (2014) 'Staunch vs. Faint-hearted Hylomorphism: Toward an Aristotelian Account of Composition', *Res Philosophica* 91: 151–77.

Lowe, E. J. (2000) *An Introduction to the Philosophy of Mind* (Cambridge: Cambridge University Press).

—— (2008) *Personal Agency: The Metaphysics of Mind and Action* (Oxford: Oxford University Press).

Mainwaring, M., Hartley, I., Lambrechts, M., and Deeming, D. (2014) 'The Design and Function of Birds' Nests', *Ecology and Evolution* 20: 3909–3928.

McDowell, J. (1996) *Mind and World* (Cambridge, MA: Harvard University Press).

Muth, F. and Healy S. (2011) 'The Role of Adult Experience in Nest Building in the Zebra Finch, *Taeniopygia guttata*', *Animal Behaviour* 82: 185–189.

O'Connor, T. (2000) *Persons and Causes: The Metaphysics of Free Will* (New York: Oxford University Press).

O'Connor, T. and Wong, H. Y. (2005) 'The Metaphysics of Emergence', *Nous* 39: 658–78.

Pitman, R. and Durban, J. (2012) 'Cooperative Hunting Behavior, Prey Selectivity and Prey Handling by Pack Ice Killer Whales (*Orcinus orca*), Type B, in Antarctic Peninsula Waters', *Marine Mammal Science* 28: 16–36.

Steward, H. (2012) *A Metaphysics for Freedom* (New York: Oxford University Press).

Velleman, D. (1992) 'What Happens When Someone Acts?', *Mind* 403: 461–81.

Walsh, P., Hansell, M., Borello, W., and Healy, S. 'Individuality in Nest Building: Do Southern Masked Weaver (*Ploceus velatus*) Males Vary in Their Nest-Building Behaviour?', *Behavioural Processes* 88:1–6.

Wittgenstein, L. (2001/1953) *Philosophical Investigations*, trans. by G. E. M. Anscombe (Oxford: Basil Blackwell).

11 Psychology Without a Mental-Physical Dichotomy

William Jaworski

1. The Mental-Physical Dichotomy and Mind-Body Problems

Is there a mental-physical dichotomy? Many people seem to think so—both the many and the wise, as Aristotle would say. Ordinary folk frequently distinguish mental health and physical health or the mental aspects of athletic performance and the physical ones. Likewise, many standard definitions of psychology claim that it is the science of behavior and mental processes,[1] where behavior comprises observable bodily changes in humans and other animals, and mental processes supposedly comprise the unobserved inner causes of those changes such as thoughts, feelings, and imaginings.[2] Even if those inner causes turn out to be identical to physical occurrences, psychology does not take an interest in them insofar as they fall under physical concepts, but only insofar as they fall under mental ones. The concept of mentality is thus used to carve out the special subject matter of psychological science, and implicit in that use is a distinction between mental concepts and physical ones. Implicit, in other words, is a mental-physical dichotomy.

Despite its ubiquity, the mental-physical dichotomy is associated with mind-body problems: persistent philosophical problems understanding how mental phenomena are related to physical phenomena. The problem of psychophysical emergence is an example. The physical universe, it says, is a vast sea of matter and energy that can be exhaustively described and explained in principle by physics. We nevertheless have capacities—the capacities to think, feel, and perceive, for instance—that cannot be described and explained in any obvious way using only the conceptual resources of physics. It can thus be difficult to understand how thinking, feeling, perceiving, and other mental capacities can exist in the physical universe, as the following claims illustrate:

(1) We think, feel, and perceive.
(2) We are composed of physical particles.
(3) The properties of a composite whole are determined by the properties of the physical particles composing it.

(4) Physical particles do not think, feel, or perceive.
(5) No number of physical particles could combine to produce a composite whole that thinks, feels, or perceives.

Claims (1)–(5) are jointly inconsistent. Claim (1) implies that we can think, feel, and perceive, yet claims (2)–(5) imply that we cannot. The claims cannot all be true; at least one of them must be false, but it is not clear which is false, since there are good reasons to endorse each.[3]

The problem of emergence assumes that there is some type of categorical difference between thinking, feeling, and perceiving, on the one hand, and the kinds of facts that can be expressed using the concepts of physics, on the other. It and other mind-body problems assume that there is some more or less well understood distinction to be drawn between mental concepts, statements, facts, events, properties, or individuals, on the one hand, and physical ones, on the other.

Mind-body problems strike at the conceptual foundations of psychological science. They suggest that there is something conceptually problematic about the mental-physical dichotomy. If psychological science presupposes that dichotomy, then there is something conceptually problematic about psychological science. If the mental-physical dichotomy is somehow incoherent, or if it fails in some way to carve nature at its joints, then the questions psychologists ask, the theories they advance, and the research programs they pursue are all bound to be misguided in various respects—like the efforts of physicians whose approaches to health presupposed the four humors. Even if their efforts yield fruitful results, it will be difficult to understand exactly what those results mean: how they mesh with findings in correlative disciplines such as neuroscience and molecular biology, and what significance they have for a synoptic understanding of psychophysical subjects.

Not surprisingly, some philosophers have sought to reject the mental-physical dichotomy. According to John Dewey, for instance:

[T]he 'solution' of the problem of mind-body is to be found in a revision of the preliminary assumptions . . . which generate the problem.[4]

Consider likewise John Searle:

Both traditional dualism and materialism presuppose conceptual dualism . . . [Conceptual dualism] consists in taking the dualistic concepts very seriously . . . What I believe . . . is that the vocabulary [of dualism], and the accompanying categories, are the source of our deepest philosophical difficulties . . . [I]t would probably be better to abandon this vocabulary altogether.[5]

But Dewey's and Searle's proposals to reject the mental-physical dichotomy have not gained much acceptance. In fact, when it comes to stating his own

view, Searle himself makes use of the very dichotomy he takes to be problematic: mental states, he says, are both caused by and realized in physical
states of the brain.[6]

There might be many reasons why philosophers—even ones like Searle—
don't abandon the mental-physical dichotomy. An obvious one is that they
have no alternative: they don't have a metaphysical framework that enables
them to formulate their theories any other way. Dewey seems to be an exception. He suggests a framework that could provide an alternative:

> The difference between the animate plant and the inanimate iron mol-
> ecule is not that the former has something in addition to the physico-
> chemical energy; it lies in the *way* in which physico-chemical energies
> are interconnected and operate . . . Iron as a genuine constituent of an
> organized body acts so as to tend to maintain the type of activity of the
> organism to which it belongs. If we identify . . . the physical as such with
> the inanimate we need another word to denote the activity of organ-
> isms . . . Psycho-physical is an appropriate term . . . In the compound
> word, the prefix 'psycho' denotes that physical activity has acquired
> additional properties . . . Psycho-physical does not denote an abrogation
> of the physico-chemical; nor a peculiar mixture of something physical
> and something psychical . . . it denotes the possession of certain qualities
> and efficacies not displayed by the inanimate. Thus conceived there is
> no problem of the relation of physical *and* psychic. There are specifiable
> empirical events marked by distinctive qualities and efficacies. There is
> first of all, *organization* . . . Each 'part' of an organism is itself orga-
> nized, and so of the 'parts' of the part . . . '[M]ind' is an added property
> assumed by a feeling creature, when it reaches that organized interaction
> with other living creatures which is language, communication.[7]

According to Dewey, organization or structure is an irreducible ontological and explanatory principle, one that concerns both what things are and
also what they can do. He suggests, moreover, that what people think of as
mental phenomena (thought, feeling, and perception) can be understood
as species of structural phenomena. If he is right, then it's easy to see how
mind-body problems could be avoided: structural phenomena are uncontroversially part of the natural world; mental phenomena are just species
of structural phenomena, hence they must be uncontroversially part of the
natural world as well.

What's needed to cash in on Dewey's idea is a metaphysic of organization
or structure, one that squares the notion of structure with our best empirical methods, descriptions, and explanations, and that enables us to understand thinking, feeling, and perceiving as species of structural phenomena.
In what follows I outline a metaphysic along these lines and explain how
it enables us to make sense of psychology without requiring us to adopt a
mental-physical dichotomy. That metaphysic endorses hylomorphism.

2. Hylomorphism: A Metaphysics of Structure

Hylomorphism claims that structure (or organization, form, arrangement, order, or configuration) is a basic ontological and explanatory principle. Some individuals, paradigmatically living things, consist of materials that are structured or organized in various ways. You and I are not mere quantities of physical materials; we are individuals composed of physical materials with a certain organization or structure. That structure is responsible for us being and persisting as humans, and it is responsible for us having the developmental, metabolic, reproductive, perceptive, and cognitive capacities we have.

The hylomorphic notion of structure is not the same as others that have appeared in the literature. It is not the same, for instance, as the notion of structure that has been operative in discussions of grounding in metaphysics.[8] Nor is it the same as the notion that is operative in debates about scientific realism.[9] Nor is it the same as the notion David Chalmers sometimes employs when he speaks of structure and dynamics.[10]

To help illustrate the hylomorphic notion of structure I'll use a simple example; we can call it *the squashing example*. Suppose we put Gabriel in a strong bag—a very strong bag, since we want to ensure that nothing leaks out when we squash him with several tons of force. Before the squashing, the contents of the bag include one human being; after, they include none. In addition, before the squashing, the contents of the bag can think, feel, and perceive, but after the squashing, they can't. What explains these differences in the contents of the bag pre-squashing and post-squashing? The physical materials (whether particles or stuffs) remain the same—none of them leaked out. Intuitively, we want to say that what changed was the way those materials were structured or organized. That organization or structure was responsible for there being a human before the squashing, and for that human having the capacities it had. Once that structure was destroyed, there no longer was a human with those capacities. Structure is thus a basic ontological principle: it concerns what things there are. It is also a basic explanatory principle: it concerns what things can do.

When people think of structure they often think of something static such as the relatively unchanging spatial relations among atoms in a crystal.[11] But hylomorphists don't view structure so narrowly. Although we're free to call the sum of spatial relations among something's parts a 'structure' in some sense of the term, hylomorphic structures—the kind that, say, distinguish living things from nonliving ones—are not static spatial relations, but dynamic patterns of environmental interaction. They comprise programmatic sequences of changes over time, and often involve different kinds of changes under different kinds of conditions. The neurophysiologist Jonathan Miller brings out this idea of dynamic structure:

> [T]he physical universe tends towards a state of uniform disorder . . . In such a world the survival of form depends on . . . [either] the intrinsic

stability of the materials from which the object is made, or the energetic replenishment and reorganisation of the material which is constantly flowing through it . . . The configuration of a fountain . . . is intrinsically unstable, and it can retain its shape only by endlessly renewing the material which constitutes it; that is, by organising and imposing structure on the unremitting flow of its own substance . . . The persistence of a living organism is an achievement of the same order as that of a fountain . . . it can maintain its configuration only by . . . reorganising and renewing the configuration from one moment to the next. But the engine which keeps a fountain aloft exists independently of the watery form for which it is responsible, whereas the engine which supports and maintains the form of a living organism is an inherent part of its characteristic structure.[12]

It is because of their dynamic structures—their abilities to impose structures on incoming matter and energy—that composite individuals (paradigmatically living things) persist one and the same through the constant influx and efflux of matter and energy that characterize their interactions with the wider world.

The hylomorphic notion of structure is close to the notion of organization that many biologists and philosophers appeal to. Here is one example taken from a popular college-level biology textbook—note the references to organization, order, arrangement, and related things:

Life is highly organized into a hierarchy of structural levels . . . Biological order exists at all levels . . . [A]toms . . . are ordered into complex biological molecules . . . the molecules of life are arranged into minute structures called organelles, which are in turn the components of cells. Cells are [in turn] subunits of organisms . . . The organism we recognize as an animal or plant is not a random collection of individual cells, but a multicellular cooperative . . . Identifying biological organization at its many levels is fundamental to the study of life . . . With each step upward in the hierarchy of biological order, novel properties emerge that were not present at the simpler levels of organization . . . A molecule such as a protein has attributes not exhibited by any of its component atoms, and a cell is certainly much more than a bag of molecules. If the intricate organization of the human brain is disrupted by a head injury, that organ will cease to function properly . . . And an organism is a living whole greater than the sum of its parts . . . [W]e cannot fully explain a higher level of order by breaking it down into its parts.[13]

This passage suggests that the way things are structured, organized, or arranged plays an important role in them being the kinds of things they are and in explaining the kinds of things they can do.

Consider likewise the remarks of some philosophers about natural organization. David Armstrong, for instance, says that, "a man is a physical object distinguished from other physical objects only by the special complexity of his physical organization" (1968: 11). There is nevertheless a crucial difference between Armstrong's notion of organization and the hylomorphic one: Armstrong does not take the organization that characterizes living things to pose a challenge to physicalism, the claim that everything can be exhaustively described and explained by physics. Hylomorphists disagree for reasons I'll discuss in Section 5. For the moment, it's worth noting that many philosophers seem to concur with hylomorphists that biological organization poses a prima facie challenge to the explanatory completeness of physics. Among them is Philip Kitcher:

> [T]o the extent that we can make sense of the present explanatory structure within biology—that division of the field into subfields corresponding to levels of organization in nature—we can also understand the antireductionist . . . claim that . . . the current division of biology [is] not simply . . . a temporary feature of our science stemming from our cognitive imperfections but [is] the reflection of levels of organization in nature.[14]

John Heil sometimes employs a notion of organization similar to the hylomorphic one as well: "the world presents us with endless levels of complexity and organization" (2003: 245). Heil explicitly rejects the existence of so-called higher-order properties—logical constructions expressed by predicates whose definitions quantify over other properties. He insists for good reason that these predicates do not correspond to genuine properties—roughly, to the causal powers things have. But if that it is the case, then it seems that the levels of organization he mentions cannot be mere logical constructs, and if that is true, it becomes difficult to avoid the conclusion that he is implicitly committed to organization being a real ontological principle—something like hylomorphic structure. It's somewhat surprising, then, that he does not embrace the hylomorphic view.

The notion of organization—of, say, the biological organization that distinguishes living things from nonliving ones—does not come for free, at least not if we endorse *ontological naturalism*, the idea that when it comes to determining what exists, empirical investigation—paradigmatically science—is our best guide.[15]

Ontological naturalism can be understood as the conjunction of a broadly Quinean thesis about ontological commitment with a broad empiricism. The broadly Quinean thesis maintains that we are committed to all the entities postulated by our best descriptions and explanations of reality, and the broad empiricism maintains that our best descriptions and explanations of reality derive from empirical sources such as the natural and social sciences. Suppose we take the natural-language sentences in which our best descriptions and explanations are formulated and reformulate them in a quantifier-variable idiom the way Quine (1948) suggests. In that case, says the Quinean thesis, we would be committed to the existence of all the entities needed to make those descriptions and explanations true.

Ontological naturalism puts pressure on philosophers like Heil who invoke a notion of organization or structure. Ontological naturalism implies that if our best empirical descriptions and explanations posit various kinds of organization or structure, then we have good prima facie reason to think those structures exist. Those descriptions and explanations thus make of us a serious ontological demand. The most straightforward way of meeting this demand takes empirical claims about structure at face value. It says that structure really is an irreducible ontological and explanatory principle, and this is what makes descriptions and explanations that appeal to structure true. This straightforward realist approach to structure is the one favored by hylomorphists. From their perspective, philosophers like Heil and Armstrong try to use talk of organization or structure without paying the necessary ontological bill. But what exactly is involved in paying that bill? What exactly is hylomorphic structure?

The concept of structure is primitive or basic within a hylomorphic framework: it cannot be defined in terms of any categories that are more basic. The only way of defining a framework's basic concepts is, as Kit Fine says, "to specify the principles by which [they are] governed."[16] The remarks about structure surveyed earlier gesture toward some of those principles. Together, they specify some of the theoretical roles that structure is supposed to play. Structure can be defined as what plays the following roles:

> *Structure matters*: it operates as an irreducible ontological principle, one that accounts at least in part for what things essentially are.
> *Structure makes a difference*: it operates as an irreducible explanatory principle, one that accounts at least in part for what things can do, the powers they have.
> *Structure counts*: it explains the unity of composite things, including the persistence of one and the same living individual through the dynamic influx and efflux of matter and energy that characterize many of its interactions with the wider world.

In what follows, I'll outline a notion of hylomorphic structure that plays these roles.

3. Powers, Composition, and Emergence

The past decade has witnessed a resurgence of interest in hylomorphism. Kit Fine, Mark Johnston, David Oderberg, Kathryn Koslicki, Michael Rea, Anna Marmodoro, Robert Koons, Simon Evnine, and myself have all articulated hylomorphic views to be added to those attributed to thinkers of the past such as Aristotle, Aquinas, Leibniz, and Merleau-Ponty.[17] Among these hylomorphic theories, naturalistic ones, such as those defended by Mike Rea and myself, claim that hylomorphic structures are powers.

There are many competing theories of powers in the literature.[18] Elsewhere I've argued in favor of a version of the identity theory of powers—the kind of theory defended by C. B. Martin and John Heil.[19] The identity theory of powers claims that properties are essentially dispositional; each essentially empowers its individual possessor to interact with other individuals in various kinds of ways. A diamond's hardness empowers it to do a variety of things—to scratch glass, for instance. We describe this power-conferring role in many different ways. We say that the diamond is hard, that the diamond is able (or has the power or potential or capacity) to scratch glass, or that the diamond would scratch that mirror if raked across its surface. These different vocabularies create the impression that there are different kinds of properties: dispositional and categorical (or qualitative). According to the identity theory, however, these vocabularies describe the very same properties; they just represent different ways of conceptualizing those properties—ways that make explicit or leave implicit the various theoretical roles those properties play. Dispositional descriptions such as 'The diamond would scratch that mirror if raked across its surface' bring out the roles the diamond's hardness plays as a power. Nondispositional descriptions such as 'The diamond has a tetrahedral arrangement of carbon atoms' bring out the property's role as a stable manifestation of the power the carbon atoms have to be arranged tetrahedrally. The one property is thus simultaneously both a stable manifestation of a power and a power itself, both an actuality and a potentiality.

The identity theory of powers claims that powers are essentially directed toward their manifestations. This directedness has led some philosophers to draw analogies between dispositionality and intentionality.[20] Intentional mental states are said to be directed at things. My desire is essentially a desire *for* something, my fear is essentially a fear *of* something. Something analogous is true of powers; they are essentially powers *for* various manifestations. The property of fragility, for instance, is essentially directed toward breaking. Likewise, just as my desire can remain unfulfilled and my fear unrealized, so too a power can remain unmanifested. A quantity of table salt has the power to dissolve in water, but it might never actually be dissolved, and a fragile vase might never actually break.[21]

The identity theory also claims that powers are manifested only in specific circumstances and typically only in conjunction with individuals that have reciprocal powers—what Martin calls 'reciprocal disposition partners'. Powers can be manifested both actively and passively: both in the ways individuals affect things and in ways they are affected by them. In general, powers are manifested only when individuals with reciprocal powers are conjoined in the right circumstances. Water, for instance, can exercise its power to dissolve things only in conjunction with things that have the power to be dissolved by it.[22]

In addition, the same power can manifest itself differently in conjunction with different disposition partners. To use Heil's example: a ball will roll on a hard surface on account of its roundness, and it will make a concave depression in a soft surface on account of that same roundness. The same property, the ball's roundness, manifests itself in different ways in conjunction with different disposition partners. Likewise, the diamond's hardness empowers it to scratch glass and also to scratch jade, and the batter's power to hit a baseball 400 feet also empowers him to hit a bigger, heavier softball 300 feet.

Hylomorphic structures are powers to configure (or organize, order, or coordinate) things. What sets hylomorphic structures apart from other powers is that they cannot exist unmanifested. They are manifested essentially. Structured individuals are essentially and continuously engaged in configuring the materials that compose them. I configure the materials that compose me, and you configure the materials that compose you. Our continuous structuring activity explains our unity and persistence through the dynamic influx and efflux of matter and energy that characterizes our interactions with the surrounding world. This is what it means to say that *structure counts*: it explains the unity of composite things.

The hylomorphic view of composition is similar to Peter van Inwagen's.[23] Van Inwagen presents his view as an answer to the Special Composition Question: under what conditions do many things compose one thing? Van Inwagen's answer is that composition happens exactly if the activities of physical particles constitute a life. This implies that something qualifies as a part only if it is caught up in a life.[24] The expression "caught up in a life" is one that van Inwagen borrows from the biologist J. Z. Young.[25] Van Inwagen explains with an example:

> Alice drinks a cup of tea in which a lump of sugar has been dissolved. A certain carbon atom . . . is carried along with the rest of the sugar by Alice's digestive system to the intestine. It passes through the intestinal wall and into the bloodstream, whence it is carried to the biceps muscle of Alice's left arm. There it is oxidized in several indirect stages (yielding in the process energy . . . for muscular contraction) and is finally carried by Alice's circulatory system to her lungs and there breathed out as a part of a carbon dioxide molecule . . . Here we have a case in which a thing, the carbon atom, was . . . caught up in the life of an organism, Alice. It is . . . a case in which a thing became however briefly, a *part* of a larger thing when it was a part of nothing before or after . . . .[26]

What exactly is a life? Van Inwagen's descriptions of lives stay largely at the level of metaphor and analogy. The reason is that providing the literal details about what lives are and what characteristics they have is, he thinks, a job for biologists.[27] He does nevertheless offer some general characteristics. Lives, he says, are self-maintaining events like flames and waves except that unlike flames and waves they are *well-individuated* and *jealous*.

Flames are not as well-individuated as lives, van Inwagen argues: "If I light seven candles from one taper, has a spatially connected flame become a scattered flame, or have seven new flames come into existence? Presumably, there are no answers to these questions."[28] Waves are better individuated than flames, but waves for their part are not jealous:

> Consider two waves . . . which are moving in opposite directions and which pass through each other . . . I think we must say . . . that both the waves exist at the moment of superposition and that each is at that moment constituted by the activities of the same water molecules. We may describe . . . the possibility of two waves' being simultaneously constituted by the activities of the same objects . . . by saying that a wave is not a *jealous* event. Lives, however, are jealous. It cannot be that the activities of the *x*s constitute at one and the same time two lives . . . When two waves impinge upon the same water molecules, the activities that each demands of these molecules . . . sum neatly according to the rules of vector addition . . . A life, on the other hand, does not deposit and withdraw sequentially an invariant sum of energy . . . A life takes the energy it finds and turns it to its own purposes.[29]

Lives are thus a special kind of self-maintaining event on van Inwagen's view, and importantly, they play precisely the kinds of theoretical roles that hylomorphic structures are supposed to play. *Lives matter* on van Inwagen's view; they are ontological principles: whether the *x*s constitute a life makes a difference to whether a composite individual exists. Likewise, *lives make a difference*; they are explanatory principles: living beings can do things that cannot be exhaustively described and explained using the conceptual resources used to describe and explain the materials that compose them.[30] Finally, *lives count*; they operate as principles of unity[31] and persistence:[32] what binds the simples that compose me into a single being is that their activity constitutes a life, and what enables me to persist through changes in those simples is the persistence of that life. Because van Inwagen's lives play these roles, it is easy to use his view of composition as a basis for understanding the hylomorphic view.

Configuring materials and being composed of materials are co-foundational concepts on the hylomorphic view, just as having a life and being composed of simples are co-foundational concepts on van Inwagen's. Likewise, just as van Inwagen restricts composition to living things, hylomorphists restrict it to structured things in general. According to hylomorphists, composition occurs when and only when an individual configures materials.

Structured individuals are *emergent* individuals on the hylomorphic view: there are empirically describable conditions that are sufficient to bring into existence a new structured individual where previously no such individual existed. Suppose, for instance, that $b_1, b_2,..., b_n$ are physical particles or materials of some sort. On the hylomorphic view, there are changes the *b*s can undergo which will result in there being a new individual, *a*, which is composed of the *b*s. In the natural course of human events, for instance, changes of this sort regularly happen in utero: physical materials that didn't compose a human organism at time t_1 come by a series of changes to compose a human organism at time t_2. A new human individual comes to exist where previously no such individual did.

Once a structured individual comes into existence it is continuously engaged in configuring materials, and the materials it configures are precisely those that compose it. The individual *a* comes into existence exactly with the start of its configuring activity—exactly when that configuring activity begins. When it comes to characterizing that activity, hylomorphists can adopt most of what van Inwagen says about lives, at least when it comes to the configuring activities of living things, the paradigmatic structured individuals. My life is identical to my configuring various fundamental physical materials at various times—an event that has the characteristics van Inwagen attributes to lives and that has many other characteristics it is the business of the biological sciences to describe.

An individual living thing does not configure the same materials for very long; the materials composing it are in constant flux. If *a*'s existence commences with its configuring the *b*s, it will not take long for it to exchange some of the *b*s for other things. Yet despite this, *a* maintains itself one and the same through these changes on account of its ongoing configuring activity. That activity is what unifies various materials into a single individual, both synchronically and diachronically, just as lives do on van Inwagen's account.

Van Inwagen is well known for embracing the Denial, the claim that many objects in a commonsense ontology do not exist, including artifacts and natural bodies such as mountains and planets. According to van Inwagen, there is no table occupying the region of space before me—no single, unified individual. There are instead many physical particles spatially arranged tablewise. Since the hylomorphic account of composition is similar in its outlines to van Inwagen's, this raises an important question: how do we know which quantities of physical materials compose unified wholes on the hylomorphic account, and which are mere spatial arrangements of materials? How do we know, for instance, that the physical materials located in this region of space actually compose a human being, that they are not instead diverse materials that are merely spatially arranged human-wise and that do not compose a unified whole at all? What prevents us from concluding that, in fact, there are no human beings just as there are no tables and mountains?

In response, hylomorphists take a cue from van Inwagen: structured individuals have non-redundant causal powers that mere spatial arrangements of physical materials do not have.[33] Suppose, for instance, that it initially seems to us as if *a* is a structured whole composed of the *b*s. This initial impression could be accurate or not. Determining which is a matter of determining whether a has powers not had by the bs, and determining whether a has any such powers is a matter of determining whether the theories or conceptual frameworks that we use to describe and explain a's behavior are reducible to the ones that we use to describe and explain the behavior of the bs.

Reduction is a primarily relation between theories or conceptual frameworks.[34] It occurs when one theory or conceptual framework can take over the descriptive and explanatory roles of another. Suppose that TB is a conceptual framework whose predicates 'F_1', 'F_2',..., 'F_n' apply to the bs, and that TA is a conceptual framework whose predicates 'G_1', 'G_2',..., 'G_m' apply to a but not to the bs. Even though the predicates of TA do not apply to the bs it might still turn out that a's having this or that G-property is something that can be given an exhaustive account in terms of the bs. It might turn out that what makes it true that a is G_i is that various bs stand in the relation F_j. In that case, a's being G_i is what van Inwagen calls a "disguised cooperative activity" performed by some of the bs.[35] By analogy, we are not tempted to say that when Alice and Benny perform a tango they bring into existence a third entity of which they are proper parts. The reason is that any agency we might attribute to such an entity can be understood in terms of the cooperative activity of Alice and Benny alone: the power of each to modulate his or her own behavior in coordination with the behavior of the other.

Suppose now that all the properties we attribute to a are like G_i. In that case, there is nothing happening in the region we take to be occupied by a that cannot be exhaustively described and explained by appeal to the bs alone. The conceptual framework TA is reducible to TB. If there is nothing to a's being G_i other than some bs standing in F_j, then we can in principle dispense with or replace descriptions and explanations that attribute G_i to a with descriptions and explanations that attribute F_j to the bs. If the replacement of TA-descriptions and explanations by those of TB is possible across the board, then TB can take over in principle all the descriptive and explanatory roles that TA performs. Reduction is what we would expect, therefore, if a is not a unified individual in its own right.

Consider now the converse case. Suppose that a is a unified individual composed of the bs, and that the predicate 'G_i' expresses a power of a not had by the bs. There is, then, more to a's having G_i than simply some bs standing in an F-relation. As a result, we cannot dispense with or replace descriptions and explanations that attribute G_i to a with descriptions and explanations that appeal to the bs alone. TA resists reduction to TB; the latter is not able to take over, even in principle, all the descriptive and explanatory roles that TA performs.

Because a's ontological status is reflected in the conceptual situation in the ways I've described, the discovery that TA is reducible to TB gives us good reason to think that a is not a unified whole, and conversely, the discovery that TA is irreducible to TB gives us good reason to think that a is a unified whole with powers that the bs lack. According to hylomorphists, the conceptual frameworks we use to describe and explain the behavior of artifacts and natural bodies are reducible to the frameworks we use to describe and explain the behavior of physical materials alone, and this gives us good reason to think that strictly speaking there are no artifacts and natural bodies, but only physical materials spatially arranged in various ways. According to hylomorphists, however, the conceptual frameworks we use to describe and explain the behavior of living things like us resist this kind of reducibility, and this resistance gives us good reason to think that living things like us are unified wholes with powers distinct from those of the materials composing them.[36]

The hylomorphic view clearly implies a kind of property pluralism since structured individuals have properties of at least two sorts: properties due to their structures (or their integration into individuals with structures), and properties due to the materials composing them independent of the way those materials are structured. This is illustrated by the squashing example considered earlier. Gabriel's powers to think, feel, and perceive are clearly structure-dependent properties: destroying his structure destroys those properties. By contrast, the squashed contents of the bag have the same mass that Gabriel has despite losing Gabriel's human structure. Mass is thus a structure-independent property.

Similarly, subatomic particles, atoms, and molecules have properties such as mass irrespective of their surroundings, but when they are integrated into structured wholes, they become genes, growth factors, and metabolic and behavioral regulators. Each thus admits of two types of descriptions: a description in terms of the contribution it makes to the structured system, and also a description in terms of the properties it would possess independent of any such contribution. Descriptions of the former sort express structure-dependent properties, while descriptions of the latter sort express properties had independently of being integrated into a structured whole. A strand of DNA might always have various atomic or fundamental physical properties regardless of its environment, but it acquires new properties when it is integrated into a cell and begins making contributions to the cell's activities.

Some philosophers and biologists call the new properties of structured systems *emergent properties*. Emergent properties have three characteristics:

1. They are first-order properties, not higher-order ones; that is, they are not logical constructions with definitions that quantify over other properties; they are rather powers in their own right.
2. They are not epiphenomenal, but make distinctive causal or explanatory contributions to the behavior of the individuals having them.
3. They are possessed by an individual on account of its organization or structure.

Notice: it is not a characteristic of emergent properties (at least not on the hylomorphic view) that they are generated or produced by lower-level systems. As a result, hylomorphists do not need an account of how lower-level systems generate emergent properties. Emergent properties are due to something's structure, and structure is a basic principle on the hylomorphic view; it is not generated by something else.

4. Functional Analysis

Metaphysicians like van Inwagen are not the only ones attracted to a view of composition like the foregoing. Philosophers of biology and neuroscience have been attracted to a view like this as well because it is suggested by actual work in biology and neuroscience—both the methods of those sciences and the kinds of explanations they employ. Of central importance is a method of scientific investigation philosophers sometimes call *functional analysis*. Biologists, cognitive scientists, engineers, and others frequently employ this method to understand how complex systems operate. They analyze the activities of those systems into simpler subactivities performed by simpler subsystems.[37]

Consider a complex human activity such as running. Functional analysis reveals that running involves among other things a circulatory subsystem that is responsible for supplying oxygenated blood to the muscles. Analysis of that subsystem reveals that it has a component responsible for pumping the blood—a heart. Analysis of the heart's pumping activity shows that it is composed of muscle tissues that undergo frequent contraction and relaxation, and these activities can be analyzed into the subactivities of various cells. Analyses of these subactivities reveal the operation of various organelles that compose the cell and that are composed in turn of complex molecules. We can continue to iterate this analytic process until we reach a level at which no further functional analysis is possible. If, for instance, electrons contribute to the activities of things by virtue of having negative charges, and they have those charges not on account of the activities of some yet lower-level subsystems, but as an unanalyzable matter of fact, then no further functional analysis is possible. We reach a foundational level of functional parts.

Functional analysis provides a way of supplying empirical content to the idea that parts contribute to the activities of their respective wholes—that they are caught up into the lives of those wholes. If we want to know how a part contributes to the activity of a whole, hylomorphism leaves it to the relevant empirical disciplines to tell us. When we look at these disciplines, we find that they describe those contributions in terms of the operation of functional parts—the kinds of parts revealed through functional analysis. Even though it is possible to divide a human along, say, purely spatial lines into thirds, or fifths, or tiny metric cubes, empirical practitioners are typically more interested in dividing them functionally.[38] Given ontological naturalism, this provides a basis for understanding the kinds of parts that structured individuals have. Those parts are subsystems that contribute in empirically specifiable ways to the activities of the wholes to which they belong.

Two clarifications are in order about functional analysis. First, a remark about the name: 'functional analysis' is a name that has been used by philosophers, but biologists often call the method 'reduction.'[39] This notion of reduction is different from the notion typically discussed in connection with the philosophy of science and the philosophy of mind.[40] The latter notion concerns the ability of one conceptual framework to take over the descriptive and explanatory roles of another. To claim that, say, psychology is reducible to neuroscience implies that it is possible in principle for neuroscience to take over all the descriptive and explanatory roles psychology currently plays. By contrast, when biologists speak of reduction they are typically not speaking of the relation between conceptual frameworks I've just described, but of a *method* for studying complex systems—what I've been calling 'functional analysis.' A commitment to employing this method does not imply a commitment to reduction in the philosophical sense. It might be impossible for neuroscience to take over the descriptive and explanatory roles of psychological discourse even though it is possible and even necessary to use functional analysis to understand how humans can engage in psychological activities. In fact, this is precisely what hylomorphists claim. Explanations of living behavior are not reducible to descriptions of the lower-level mechanisms revealed by functional analysis because of the distinctive explanatory contributions a living thing's structure makes.

A second note about functional analysis: the notion of function that gives functional analysis its name is different from the notion of function discussed in connection with functionalism in philosophy of mind. According to classic functionalist theories of mind, mental states are postulates of abstract descriptions framed in terms analogous to those used in computer science—descriptions that ignore a system's physical details and focus simply on a specific range of inputs to it, outputs from it, and internal states that correlate the two.[41] When it comes to functional analysis, by contrast, the notion of a function is not abstract in this way, and it has a teleological dimension: subsystems contribute to the activities of the wholes to which they belong, and that contribution is their reason or purpose for belonging to the system: the purpose of the spark plug is to ignite the fuel; the purpose of the heart is to pump the blood, and so on.[42]

Teleological functionalism is a type of functionalist theory that appeals to a teleological notion of function along these lines. William Lycan's homunctionalism is an example.[43] Like functionalist theories of all sorts, however, teleological functionalism claims that higher-level discourse is abstract discourse: higher-level properties are higher-*order* properties— logical constructions that quantify over lower-order properties. Saying that something has a belief, for instance, amounts merely to saying that it has *some* internal state that correlates inputs with outputs in appropriate ways. Hylomorphists reject this understanding of higher-level properties;

they claim that higher-level properties are first-order properties in their own right. So although teleological functionalists and hylomorphists both claim that a system's components contribute teleologically to its overall operation, they disagree about how the notion of contribution is to be understood. Teleological functionalists say that descriptions of higher-level phenomena are simply abstract descriptions of lower-level occurrences. Hylomorphists deny this: higher-level descriptions correspond to distinctive natural structures, ones that factor into descriptions and explanations of living behavior in ways that cannot be eliminated, reduced to, or paraphrased in favor of lower-level descriptions and explanations.

5. Activity-Making Structures

So far I've focused on composite individuals and their structures—*individual-making structures*, the kinds of things Medieval hylomorphists called 'substantial forms.' But individuals are not the only composite entities on the hylomorphic view, nor are individual-making structures the only structures. There are composite events as well. The activities in which structured individuals engage have structures too: *activity-making structures*.

The activities of structured individuals involve coordinated manifestations of the powers of their parts. When we walk, talk, sing, dance, run, jump, and engage in the various other activities we do, we impose an order on the ways our parts manifest their powers. My parts needn't manifest their powers in an ordered way. It is possible for my neurons to fire or my muscles to contract in ways that do not compose an activity of, say, throwing a baseball or playing an instrument. Fatigue, injury, insufficient training, and many other factors can result in uncoordinated manifestations of the powers of my parts. But when I succeed in throwing or playing, I succeed in imposing a structure on the way my parts (and in many cases surrounding things) manifest their powers: I structure their manifestations throwing- or playing-wise. In some cases, the structuring is conscious and intentional, as in throwing a baseball or producing the precise limb movements in a dance. But in many cases, the structuring is neither conscious nor intentional, as in digesting food or increasing blood flow to the legs in response to something fearful. In whatever way it occurs, whether consciously and intentionally or not, the result of this structuring is not a new individual but rather an activity, another manifestation of the power structured individuals have of imposing order on things.

Activity-making structures unify diverse events in something analogous to the way individual-making structures unify physical materials in composite individuals. The very same muscle fibers that contract in my shoulder when I throw a baseball might also contract when I experience an uncontrolled muscle spasm. What unifies or coordinates the contractions of the muscle fibers along with changes in surrounding things, such as the baseball, is what I do when I try to make an out, or try to knock down cans at the county fair, or try to accomplish whatever I try to do when I throw a baseball. In undertaking these activities I impose a unified order on the way my parts and surrounding things manifest their powers.

On the hylomorphic view, structured activities include thinking, feeling, and perceiving. When, for instance, I experience an emotion, I am engaging in an activity in which various parts of my nervous system and various objects in the environment manifest their powers in a coordinated way that unifies them into a single event. It is possible to describe the unifying role of activity-making structures in terms of a notion of activity composition analogous to the notion of composition for individuals. Just as physical materials compose an individual exactly if they have the right kind of individual-making structure, various events compose an activity exactly if they have the right kind of activity-making structure. An individual *a* engages in the activity of *F*-ing exactly if *a*'s parts and surrounding things manifest their powers *F*-wise. I throw a baseball exactly if my parts and surrounding materials manifest their powers throwing-a-baseball-wise. Similarly, I experience anger or enjoyment exactly if my parts and surrounding things manifest their powers anger- or enjoyment-wise.

Given reasonable assumptions, activity composition implies that the behaviors of structured individuals never violate the laws governing their fundamental physical components. According to hylomorphism, the activities of structured wholes are composed of the structured manifestations of the powers of their lower-level components and surrounding things. If those components or things were to lose their powers, or were to become incapable of manifesting them, they would become incapable of composing the activities of structured wholes. Those activities depend on lower-level items retaining and manifesting the powers they have. By analogy, it is only because bricks and timbers retain their shapes under compression that they can be recruited as components of buildings. Similarly, it is only because lower-level materials retain their distinctive powers that structured individuals can recruit them as components for their own activities. This is one thing that sets the hylomorphic view apart from those classic emergentist theories such as Roger Sperry's, which claim that higher-level powers trump or nullify the powers of lower-level things.[44]

Activity composition also makes it clear in what sense a structured individual has the power to engage in various activities *because of* its parts. Those parts form a subset of the individuals with powers whose coordinated manifestations compose its activities. We can express this idea by saying that a structured individual's parts *embody* its powers. My visual system embodies my power to see; your circulatory system embodies your power to bring oxygenated blood to various parts of yourself; Gabriel's limbic system embodies his power to experience emotions, and so on.

According to the hylomorphic theory I've been describing, all the powers of structured individuals are essentially embodied in their parts; the activities in which they engage are essentially composed of the coordinated manifestations of the powers of their parts and surrounding things. It is impossible, not just nomologically but metaphysically, for me to engage in the activity of throwing a baseball unless my parts manifest their powers in the right coordinated way. Likewise, it is impossible, not just nomologically but metaphysically, for Gabriel to experience anger or enjoyment unless his parts manifest their powers in the right coordinated way.[45]

On the hylomorphic view, then, thought, feeling, and perception are essentially embodied in the physiological mechanisms that compose us, yet it is not possible to reduce explanations of them to explanations of physiological mechanisms.[46] The reason is that there is more to these activities on the hylomorphic view than the operations of physiological mechanisms: there is also the way those operations are coordinated or structured, and structure in general is something different from things that get structured. It is possible for parts of our nervous systems to be activated in the ways they are when we are experiencing an emotion, for instance, even though we are not experiencing the emotion in fact. Patients with pseudobulbar affect suddenly and unpredictably cry or laugh in ways that are indistinguishable from the ways they would if they were experiencing sadness or mirth, and yet they do not feel sad or amused (Parvisi et al. 2006). Parts of their nervous systems are activated in the ways they would be during a real emotional episode, and yet their activation fails to be coordinated in the way necessary to compose an emotion. The hylomorphic view is thus robustly antireductive despite its commitment to essential physical embodiment.

I've already explained how hylomorphism differs from Lycan's homunctionalism and other nonreductive physicalist views that endorse functionalism. It will perhaps be helpful to say more about how it differs from physicalist theories in general. Physicalism is the claim that everything can be exhaustively described and explained by the most empirically adequate theories in current or future physics. Philosophers sometimes use the term 'physicalism' to refer to much weaker claims, such as the claim that everything has physical properties, that everything is composed of physical parts, or that everything is necessitated by or supervenes upon physical events or facts. Elsewhere I've argued that these definitions are inadequate because each fails to imply the core physicalist thesis that everything is physical.[47] Each is compatible with the existence of nonphysical properties, and because of that each is compatible with dual-attribute theories such as emergentism or epiphenomenalism.[48] This is true especially of many varieties of nonreductive physicalism. As John Bickle observes, "Much current 'nonreductive physicalism' is not physicalism at all. It is instead . . . a dualism not of substances but of their properties."[49]

Once physicalism is properly defined in the way I've suggested it should be evident why hylomorphism is incompatible with it. According to hylomorphists, there are structures which can't be described using the conceptual resources of physics alone—structures of the sort described by special sciences such as biology and psychology. The claim that there are such structures is, we've seen, largely an empirical one on the hylomorphic view I've outlined. That view, like physicalism, operates as a "high-level empirical hypothesis," to use Hartry Field's expression.[50] It is thus open to empirical falsification. Perhaps biological or psychological structures can ultimately be identified with complex relations that are exhaustively describable by physics. In that case, the kind of hylomorphic theory I've outlined will be false. At present, however, the question is not whether hylomorphism is true, but whether it is compatible with physicalism. It should be evident why it isn't: if hylomorphism is true, then there are individual- and activity-making structures which cannot be identified with any structures that might be described and explained exhaustively by physics. If there are such structures, then physics cannot exhaustively describe and explain everything, and if that is the case, then physicalism must be false. Hylomorphism is thus incompatible with physicalism.

Hylomorphism is nevertheless compatible with many weaker claims that have sometimes been labeled 'physicalism.' We've seen, for instance, that hylomorphism implies that structured individuals and their activities are exhaustively decomposable into the activities and subactivities of their parts and surrounding materials, and that these parts and materials are in turn exhaustively decomposable into fundamental physical materials. But this kind of exhaustive physical decomposition does not imply physicalism in the strong sense that everything can be exhaustively described and explained by physics. It is possible for individuals that are exhaustively decomposable into fundamental physical materials to have first-order nonphysical properties and for their behavior to be governed by emergent laws in addition to those governing their fundamental physical constituents. Claims like these are hallmarks of dual-attribute theories such as emergentism and epiphenomenalism.

Exhaustive physical decomposition would imply that all properties are physical if it were combined with a thesis like the following:

> *Property exhaustion thesis:* Necessarily, for any x, if x is exhaustively decomposable into y_1, y_2,..., y_n, and the activities of x are exhaustively decomposable into the manifestations of the powers of the ys, then x has no properties other than those of the ys.

We have seen, however, that hylomorphists reject any such thesis. They claim that some of a composite individual's properties depend on its structure. Even if x is exhaustively decomposable into the ys, x will still have some properties that the ys lack, namely the properties due to its structure. The same is true mutatis mutandis of x's activities on the hylomorphic view. If my throwing a baseball is exhaustively decomposable into the manifestations of the powers of my parts, there is still the way those manifestations are structured or coordinated. That coordination is not a property of the parts and materials taken on their own; it is rather a property of the individual as a whole—a structure that I impose on them. Hylomorphism thus implies that anything like the property exhaustion thesis is false.

6. Hylomorphism and the Mental-Physical Dichotomy

The hylomorphic view I've outlined takes thoughts, feelings, perceptions, and other prototypical mental or psychological phenomena to be coordinated manifestations of the powers of our parts and surrounding things. When, for instance, Gabriel sees something—a ripe tomato, say—he and the tomato both manifest powers they possess. He manifests the power to see the tomato, and it manifests the power to be seen by him. Gabriel and the tomato are reciprocal disposition partners. The powers of both are mutually manifested in each other's presence when the surrounding conditions are right, just as water and salt mutually manifest their powers to dissolve and be dissolved when conditions are right.

Moreover, Gabriel has the power to see by virtue of having the parts he has. The coordinated manifestations of the powers of some of his parts (intuitively those composing his visual system) contribute to his seeing, and those parts form a subset of the individuals whose powers, when manifested in the right way, compose his seeing. The same is true mutatis mutandis of the tomato. Intuitively, the parts of the tomato by virtue of which it has the power to be seen are those composing its surface, the ones which reflect light to Gabriel's eyes. In addition, there are other environmental factors involved in Gabriel's seeing the tomato such as the direction and intensity of the light, the condition of the air through which he sees it, and so on. Gabriel's seeing the tomato is thus a complex structured activity composed of the coordinated manifestation of the powers of his parts and those of surrounding things.

Let us now return to our original desiderata. We wanted a metaphysical framework that did three things: first, it would enable us to make sense of an empirically-informed notion of organization or structure like Dewey's; second, it would enable us to understand thoughts, feelings, and perceptions as species of structured phenomena, and third, it would not require us to adopt a mental-physical dichotomy. It should be evident how the hylomorphic theory I've outlined satisfies the first two desiderata. Hylomorphic structure carves out composite individuals (paradigmatically living things) from the otherwise undifferentiated sea of matter and energy that is or will be described by our best physics, and it confers on those individuals powers that distinguish what they can do from what unstructured materials can do. The activities of those individuals—including the thinking, feeling, and perceiving in which humans engage—are essentially embodied in the powers of their parts. Details about what parts those are, what powers they have, and how the manifestations of those powers must be structured to compose our activities, are all to be supplied through empirical methods like functional analysis.

What of the third desideratum? Based on what's been said, we can begin to appreciate how hylomorphism satisfies it too, for within the hylomorphic framework, whether we decide to call some of the powers and activities of structured individuals 'mental' or 'nonmental,' or 'physical' or 'nonphysical' is orthogonal to the project of understanding what they are, and why and how they operate as they do. Consider some ways of drawing a mental-physical distinction:

- Something is mental if and only if it displays intentionality.
- Something is mental if and only if it is subjective or has a subjective point of view.
- Something is mental if and only if it can be described or explained using a psychological vocabulary.
- Something is mental if and only if we have privileged access to it.
- Something is physical if and only if it can be exhaustively described by physics.
- Something is physical if and only if it is composed of materials that can be exhaustively described by physics.
- Something is physical if and only if it belongs to the causal order of the world.
- Something is physical if and only if we do not have privileged access to it.

There is nothing to stop hylomorphists from adopting one of these definitions or another, but nothing about their framework of powers and manifestations forces them to do so. The hylomorphic framework does not imply a commitment to categorizing powers or manifestations or individuals or their parts, as mental or nonmental, physical or nonphysical. According to hylomorphists, we can get on with the empirical investigation of powers, parts, and manifestations without ever employing these categories. Whether we decide to draw these distinctions or not is purely a function of our descriptive and explanatory interests. There is nothing built into the nature of things on the hylomorphic view that forces us to draw the distinction one way or another, or that forces us to draw any such distinction at all.

If we want to understand Gabriel's running, we might look to locate his running within the broader rational structure of his intentional activities: he is running because he wants to stay fit, and he wants to stay fit because he wants to continue playing baseball, and he wants to continue playing baseball because, etc. Or, we might look to understand how he can engage in a complex activity like running by analyzing that activity into simpler sub-activities whose coordinated occurrences compose his running. Whichever way we take our inquiry—whether we look to describe the reasons for Gabriel's running or the physiological subsystems that enable him to run—the important thing is that we do not at any point need to introduce a distinction between mental factors and nonmental ones or between physical factors and nonphysical ones. Once we accept the hylomorphic framework I've described, nothing forces us to accept a mental-physical dichotomy.

Nothing forces us to reject a mental-physical dichotomy either. The choice to accept or reject such a dichotomy is contingent upon whatever theoretical or practical ends we are looking to achieve. Within a hylomorphic framework the mental-physical dichotomy is an artifact of our descriptive and explanatory interests.

Because there is nothing canonical about the mental-physical dichotomy on the hylomorphic view, the latter provides an alternative way of understanding the nature of psychological science. Within a hylomorphic framework, sciences can be defined by the kinds of structures their methods enable us to investigate. Psychological science can be defined, then, simply by appeal to the actual methods that working psychologists employ. Those include methods that yield descriptions of higher-level structures in animal behavior such as the structures that make thinking, feeling, and perceiving what they are, as well as methods like functional analysis that yield descriptions of various lower-level subsystems whose coordinated operations compose higher-level behavior. Because this way of defining psychological science makes no appeal to a mental-physical dichotomy, it gingerly sidesteps the mind-body problems which the dichotomy generates. It thus has the potential to provide a stable conceptual foundation for understanding how psychological science meshes with biology and neuroscience. An example involves the science of perception.

7. Hylomorphism and the Science of Perception

On the hylomorphic view I've been developing, the manifestation of perceptual powers—of, say, Gabriel's power to see and the tomato's power to be seen—is a temporally extended process. To understand this idea, it is helpful to contrast it with the approach to perception that has tended to dominate the empirical literature. For decades, that approach has been inspired by the functionalist thinking described in Section 4. David Marr's theory of vision is an example.[51]

According to Marr, a perceiver receives sensory stimulation from the environment. That stimulation is nevertheless insufficient to tell the perceiver exactly what objects in the environment have produced it since the same retinal image could be produced by an infinite number of distinct shapes at various distances (Figure 11.1). The perceiver must therefore supplement the meager sensory input with internally stored assumptions about its likely environmental causes. Perception is thus a process of constructing an internal model of the external world based on a combination of sensory stimuli and internally stored assumptions about the environment. That process, moreover, does not depend in any essential way on the perceiver's movements through the environment, nor does it depend on the perceiver having a specific bodily organization since the same manipulations

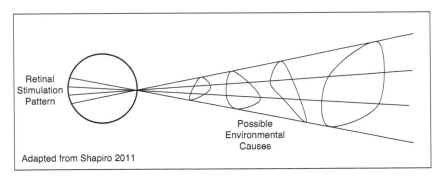

Figure 11.1 The Retinal Stimulation Pattern Underdetermines Its Environmental
Cause

of internal information might be carried out in physical systems of very
different kinds.

On the hylomorphic view I've been developing, by contrast, we come to
know the perceptible properties of things through a temporally extended
process of sensorimotor interaction of the sort described by J. J. Gibson
and, more recently, Alva Noë.[52] Objects in the environment reflect light
to a perceiver in different ways. The differences in reflected light provide
information about the objects reflecting it—information picked up by perceivers at various points of observation. Those points of observation can be
understood as points at which pyramids of reflected light converge. The sum
of converging pyramids of light at a point is what Gibson calls the *ambient
optic array* (Figure 11.2). That array changes as a function of a perceiver's
movements—what Gibson calls the *optic flow*.

Some appearances nevertheless remain invariant across the optic flow,
and these invariants provide information about objects in the environment.
Invariant proportions among the angles and lines on surfaces, for instance,
provide information about the relative sizes of objects, while horizon cuts
and occluding edges provide information about the relative positions of
objects.[53] Because information about objects in the environment is conveyed
by the reflected light itself, there is no need for sensory information to be
supplemented with internal representations; all that's required is that the
perceiver be equipped with a way of picking up invariants in the optic flow.

Pyramids of reflected light converge at different points. As perceivers
move through the environment, they pick up information about the surrounding layout from features that remain invariant across changes in the
optic flow.

As Gabriel moves relative to the tomato, for instance, he gains an implicit
understanding of how its various facets come into and go out of view as a
function of his movements: he knows implicitly that moving this way or

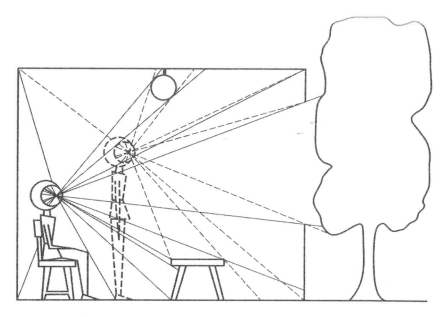

Figure 11.2 The Ambient Optic Array (Gibson, 1979, p. 72)

that will bring into view these or those facets which he expects to look these or those ways under the present conditions. In this way, Gabriel comes to know the tomato's perceptible properties—its uniform redness, for instance. That redness is revealed through a series of appearances none of which is uniformly red. The young child who depicts a tomato by applying a single shade of red paint across the canvas fails to capture how it really looks. The skilled painter, by contrast, uses a variety of colors to depict the tomato: a bit a red here, a bit of gray there, white toward the top, and so on. The result is a more accurate depiction of how the tomato really looks from a particular vantage point. If Gabriel's vision were limited to the way the tomato looks from that point, he might never know its uniform color; there would be no explanation for what psychologists call *color constancy*, the ability of perceivers to discern that an object has a uniform color despite changes in its nonuniform appearance. But Gabriel's vision is not limited in this way. By moving in relation to the tomato he grasps its uniform redness through the shifting appearances.

Perception is not a passive process, therefore, of receiving sensory stimuli and constructing internal representations of external objects, but an active, temporally extended process of coming to know the perceptible properties of objects through a series of appearances that vary as a function of the perceiver's movements through the environment. Movement is thus essential to perception. So too is a perceiver's specific bodily organization since the way appearances vary as a function of movement depends on that organization. The changes in visual experience that vary as a function of turning one's head side to side, for instance, depend on having a head that can turn in that way and having eyes situated in particular locations on that head. On the hylomorphic account I've described, then, perception is both enactive and embodied.

Some empirical work supports an account of perception along these lines. One prediction the account makes is that disrupting the interplay of sensation and movement will disrupt perception. This is precisely what experiments with inverted goggles appear to demonstrate. Inverted goggles are equipped with lenses that alter the trajectory of light to the eye: light reflected from an object on the left, for instance, is made to enter the eye from the right. If perceiving were simply a matter of passively receiving sensory stimuli, we would expect that subjects wearing inverted goggles would simply perceive objects on the left as if they were on the right. In fact, their perceptual abilities are disrupted in a much more radical way.[54] Subjects wearing inverted goggles initially fail to grasp what is in the environment and where it is. They undergo what Noë calls *experiential blindness*: although their retinas receive sensory stimulation, that stimulation doesn't enable them to gain knowledge of their environment.[55] An enactive account of perception explains why: perception depends not just on sensory stimulation, but on the regular interplay of sensory stimulation and movement. Inverted goggles disrupt that interplay: head movements that used to alter retinal stimulation in one way no longer do. As a result, subjects are unable to perceive things until they master the new patterns of sensorimotor correlation.

In addition, Ballard and his colleagues have shown that an embodied account of perception yields models of cognitive behavior that are more elegant than those based on functionalist assumptions.[56] Their experiments involve tracking test subjects' eye movements. Subjects are presented with an arrangement of colored blocks on a computer screen and asked to copy the arrangement in a workspace using a computer mouse by taking blocks from a supply area. Marr's functionalist account suggests that subjects construct an internal representation of the external environment and guide their hand movements by reference to that internal representation. Ballard's findings challenge this view. Subjects' eye movements suggest that they employ a deictic strategy in performing the task;[57] that is, they guide their actions not by reference to an internal representation of the external environment, but by reference to a fixation point in the environment itself. The environment-centered deictic model not only accounts for the experimental data, it also simplifies the amount of computational overhead needed to accomplish the perception-guided task. An account like Marr's, which posits an internal representation of the external environment, involves a great deal of algorithmic complexity which the environment-centered alternative sidesteps entirely.

The upshot of empirical work like the foregoing is reminiscent of the criticism Aristotle advanced against a philosopher he calls 'Socrates the Younger,' who appears to have claimed that human activities could be defined abstractly the way we define geometrical objects. The definitions of those objects make no essential reference to any realizing materials. Defining a circle as an infinite number of points equidistant from a single point does not specify that the points be in wood, plastic, metal, air, or any other material. Socrates the Younger appears to have claimed that definitions of human activities and capacities were like the definition of a circle: they made no essential reference to particular bodily parts. Aristotle disagreed:

> . . . Socrates the Younger was wrong in always comparing an animal with the circle and bronze . . . [I]t supposes that a man can exist without his parts, as a circle can exist without the bronze. But in fact, the two cases are not similar, for an animal . . . cannot be defined without reference to parts in the right condition.
>
> (*Metaphysics* 1037a22–31;
> cf. *On the Soul* 403a3-b15 and *Physics* 194a1–27)

Empirical work of the foregoing sort seems to support Aristotle's contention. It suggests that the human cognitive capacities are essentially embodied in a specific kind of body plan, that the best models of human cognitive performance presuppose that body plan.

A Neo-Aristotelian framework of powers and structured activities rejects the canonical status of the mental-physical dichotomy. It leaves it unmysterious how thought, feeling, and perception can exist in the natural world, and how psychology and biological subdisciplines such as neuroscience contribute to a synoptic vision of what kinds of things we are and what kinds of capacities we have.

Notes

1 Coon and Mitterer (2012: 14), Kalat (2013: 3), Griggs (2014: 1).
2 Plotnik and Kouyoumdijian (2013: 3).
3 Claims (1) and (2) seem to be well-supported empirically, and many examples seem to support claim (3): I have the mass I have because I am composed of physical particles with smaller masses that collective add up my bigger mass. Likewise, I have the position and velocity I do because the particles composing me are located in such-and-such places and are moving with such-and-such velocities. Change their positions or velocities, and you succeed in changing mine. Given the range of properties that are like this, it's not implausible to suppose that all the properties of composite wholes are determined by the properties of the particles composing them. It seems, moreover, that the behavior of those particles can be described and explained exhaustively by physics. We don't need to use a psychological or even a biological vocabulary to describe and explain what they are and what they can do. This lends some support to claim (4). There are also, it seems, good reasons to endorse claim (5). One particle by itself does

not have the power to think, feel, or perceive. If it did, then thought, feeling, and perception would have emerged much earlier in the universe's history than we think they did, and they would also be more widespread—even rocks and tables would be thinkers, feelers, and perceivers. But if one particle by itself does not have the power to think, feel, or perceive, then it is difficult to see how any number of these particles could combine to form a whole that has these powers. Suppose that some number of particles, N, do not compose a whole with power to think, feel, or perceive. If one particle cannot make a difference to whether or not something has these powers, then clearly, N+1 particles cannot compose a whole that has them. Since N can be any number one likes, it seems to follow that no number of physical particles has the power to compose a whole that thinks, feels, or perceives. Each claim, (1)–(5), is therefore plausible.

4 Dewey (1958: 263).
5 Searle (1992: 26, 54–5).
6 Searle (2004: 79).
7 Dewey (1958: 253–8).
8 Schaffer (2009), Sider (2012).
9 Worrall (1989), Ladyman and Ross (2007).
10 Chalmers (2002: 258).
11 In fact, even some contemporary hylomorphists use the term 'structure' this way. Oderberg (2014: 177) is an example.
12 Miller (1978: 140–1).
13 Campbell (1996: 2–4).
14 Kitcher (1984: 369, 373).
15 John Dupré has endorsed a similar thesis: "I place myself firmly in the philosophical tradition that sees empirical, often scientific, inquiry as providing the most credible source of knowledge of how things are" (1993: 1). Replacing the phrase 'how things are' in Dupré's statement with 'what there is' yields what I am calling 'ontological naturalism.'
16 Fine (2008: 112).
17 Fine (1999), Johnston (2006), Oderberg (2007), Koslicki (2008), Rea (2011), Marmodoro (2013), Koons (2014), Evnine (2016), Jaworski (2014, 2016).
18 The hylomorphic theory I develop here assumes a substance-attribute ontology which takes substances or *individuals*, as I'll typically call them, and attributes or *properties*, to be fundamental entities. Individuals act on other individuals and are acted on by them on account of their properties. Properties are sparse, not abundant, in Lewis's (1983) sense (Jaworski 2014, 2016). The only properties that exist are ones that empower individuals to enter into causal relations. Properties are particulars, not universals. They are tropes—also called 'unit properties,' 'property instances,' 'individual accidents,' and 'modes,' among other things. I've defended this metaphysical framework in detail elsewhere (Jaworski 2016).
19 Martin (1996a, b, 1997, 2007), Heil (2003, 2005), Martin and Heil (1998, 1999), Jaworski (2014, 2016).
20 Martin and Pfeifer (1986), Place (1996a, b), Molnar (2003).
21 Martin (1996a) defends this idea with an example: there might be fundamental physical particles in the universe that have the power to interact in various ways with particles around here, and yet that are so far away that they reside outside the light cones of the particles around here. The two groups of particles never actually interact, yet it seems obvious that the distant particles still have the power to interact with the local ones.
22 Harré and Madden's (1975) examples of radioactive decay and ammonium triiodide seem initially to provide counterexamples to the general rule that powers are manifested or exercised in pairs, or triples, or *n*-tuples. But even here it

might be possible to understand the cases in a way that conforms to the general reciprocity model. At the very least, the environment surrounding the radioactive nuclei or the ammonium tri-iodide cannot include any agents that inhibit the exercise of their powers to decay or explode, respectively. Environments that are free of inhibitory factors might then be viewed as reciprocal disposition partners for the decaying nuclei and the explosive compound.

23 Van Inwagen (1990).
24 Ibid.: 94.
25 Young (1971).
26 Van Inwagen (1990: 94–5).
27 Ibid.: 84.
28 Ibid: 88.
29 Ibid.: 88–9.
30 Ibid.: 122, 180.
31 Ibid.: 121.
32 Ibid.: 145, 148.
33 This response implies a certain meta-mereology which Kathrin Koslicki describes as follows: "I take the mereologist's job to be to devise an appropriate conception of parthood and composition which accurately reflects the conditions of existence, spatio-temporal location and part/whole structure of those objects to which we take ourselves to be already committed as part of the presupposed scientifically informed, commonsense ontology. The question of which kinds [of objects] there are I take to be . . . answered [not] by the mereologist proper, but by the ontologist at large, in conjunction with . . . science and common sense, which . . . have something to contribute to the question, 'What is there?'" (2008: 171).
34 Churchland (1986), Bickle (1998), Jaworski (2011, 2016).
35 Van Inwagen (1990: 122).
36 See also van Inwagen (1990: 118, 122).
37 Fodor (1968), Cummins (1975), Dennett (1978), Lycan (1987), Craver (2007), Bechtel (2007, 2008).
38 Bechtel (2007, 2008), Craver (2007: ch. 5) calls purely spatial parts 'pieces' and parts in the functional sense 'components.' John Heil (2003: 100) also suggests something like the distinction between merely spatial parts and parts of other sorts, which he calls 'substantial parts.'
39 Campbell et al. (1999: 4).
40 Jaworski (2011: 277).
41 Putnam (1967).
42 Lycan (1987), Sober (1985).
43 Ibid.
44 Sperry (1984).
45 Many hylomorphists of the past have denied that all our powers are essentially embodied in the powers of our parts. Aristotle himself appears to deny it in *De Anima*, book 3, chapter 4, where he apparently argues that understanding or *nous*, the power to grasp the essences of things, has no organ and is in general unmixed (*amigēs*) with a body (429a10-27). A commitment to the essential embodiment of our capacities is nevertheless the default position for hylomorphists. In fact, Aristotle treats embodiment as the default position as well (403a16–19, 24–27; 403b17–18), and he claims that the emotions are essentially embodied (403a16–19, 24–27; 403b17–18; cf. 413a4–6). Elsewhere, I've argued in line with other commentators that the argument of *De Anima* is flawed in multiple ways (Jaworski 2016: 162–70).
46 The term 'reduction' is used in a variety of ways in philosophy and the sciences. The notion of reduction that interests us here is intertheoretic reduction

(Churchland 1986: 278–9). Intertheoretic reduction is a synchronic relation between theories or conceptual frameworks in which one of them, the reducing theory or framework, is able to take over the descriptive and explanatory roles played by the other, the reduced theory or framework.
47 Jaworski (2016: 221–49).
48 Jaworski (2011: 202–45).
49 Bickle (1998: 8).
50 Field (1972: 357).
51 Marr (1982).
52 Gibson (1979), Noë (2004).
53 Gibson (1979).
54 Taylor (1962), Kohler (1964).
55 Noë (2004).
56 Ballard et al. (1992), Ballard (1996).
57 Agre and Chapman (1987).

References

Agre, P. E. and Chapman, D. (1987) 'Pengi: An Implementation of a Theory of Activity', *Proceedings of the American Association for Artificial Intelligence* 87: 268–72.
Armstrong, D. M. (1968) *A Materialist Theory of the Mind* (London: Routledge & Kegan Paul).
Ballard, D. (1996) 'On the Function of Visual Representation', in Akins (ed.) *Perception* (New York: Oxford University Press): 111–31.
Ballard, D. et al. (1992) 'Hand-eye Coordination During Sequential Tasks', *Philosophical Transactions of the Royal Society: Biological Sciences* 337: 331–9.
Bechtel, W. (2007) 'Reducing Psychology While Maintaining Its Autonomy via Mechanistic Explanations', in M. Schouten and H. L. de Jong (eds.) *The Matter of the Mind* (Oxford: Blackwell Publishing): 172–98.
——— (2008) *Mental Mechanisms: Philosophical Perspectives on Cognitive Neuroscience* (New York: Routledge).
Bickle, J. (1998) *Psychoneural Reduction: The New Wave* (Cambridge, MA: MIT Press).
Campbell, N. A. (1996) *Biology*, 4th edition (Menlo Park, CA: The Benjamin/Cummings Publishing Company, Inc.).
Campbell, N. A., Reece, J. B. and Mitchell, L. G. (1999) *Biology*, 5th edition (Menlo Park, CA: The Benjamin/Cummings Publishing Company, Inc.).
Chalmers, D. (2002) 'Consciousness and Its Place in Nature', in Chalmers, D. (ed.) *Philosophy of Mind* (Oxford: Oxford University Press): 247–72.
Churchland, P. S. (1986) *Neurophilosophy* (Cambridge, MA: MIT Press).
Coon, D. and Mitterer, J. O. (2012) *Introduction to Psychology: Gateways to Mind and Behavior*, 13th edition (Wadsworth Cengage Learning).
Craver, C. F. (2007) *Explaining the Brain: Mechanisms and the Mosaic Unity of Neuroscience* (New York: Oxford University Press).
Cummins, R. (1975) 'Functional Analysis', *Journal of Philosophy* 72: 741–64.
Dennett, D. C. (1978) *Brainstorms* (Cambridge, MA: Bradford Books).
Dewey, J. (1958) *Experience and Nature* (New York, NY: Dover Publications).
Dupré, J. (1993) *The Disorder of Things: Metaphysical Foundations of the Disunity of Science* (Cambridge, MA: Harvard University Press).
Evnine, S. J. (2016) *Making Objects and Events: A Hylomorphic Theory of Artifacts, Actions, and Organisms* (Oxford: Oxford University Press).
Field, H. (1972) 'Tarski's Theory of Truth', *Journal of Philosophy* 69: 347–75.

Fine, K. (1999) 'Things and Their Parts', *Midwest Studies in Philosophy* 23: 61–74.

Fine, K. (2008) 'Form and Coincidence', Proceedings of the Aristotelian Society Supplementary Volume 82: 101–118.

Fodor, J. (1968) 'The Appeal to Tacit Knowledge in Psychological Explanation', *Journal of Philosophy* 65: 627–40.

Gibson, J. J. (1979) *The Ecological Approach to Visual Perception* (Prospect Heights: Waveland Press, Inc.).

Griggs, R. A. (2014) *Psychology: A Concise Introduction*, 4th edition (New York, NY: Worth Publishers).

Harré, R. and Madden, E. H. (1975) *Causal Powers: A Theory of Natural Necessity* (Oxford: Basil Blackwell).

Heil, J. (2003) *From an Ontological Point of View* (Oxford: Clarendon).

——— (2005) 'Dispositions', *Synthese* 144: 343–56.

Jaworski, W. (2011) *Philosophy of Mind: A Comprehensive Introduction* (Malden, MA: Wiley-Blackwell).

——— (2014) 'Hylomorphism and the Metaphysics of Structure', *Res Philosophica* 91: 179–201.

——— (2016) *Structure and the Metaphysics of Mind: How Hylomorphism Solves the Mind-Body Problem* (Oxford University Press).

Johnston, M. (2006) 'Hylomorphism', *Journal of Philosophy* 103: 652–98.

Kalat, J. W. (2013) *Introduction to Psychology*, 10th edition (Belmont, CA: Cengage Learning).

Kitcher, P. (1984) '1953 and All That: A Tale of Two Sciences', *Philosophical Review* 93: 335–73.

Kohler, I. (1964) *Formation and Transformation of the Perceptual World* (New York: International Universities Press).

Koons, R. (2014) 'Staunch vs. Faint-hearted Hylomorphism: Toward an Aristotelian Account of Composition', *Res Philosophica* 91: 151–78.

Koslicki, K. (2008) *The Structure of Objects* (Oxford: Oxford University Press).

Ladyman, J. and Ross, D. (2007) *Every Thing Must Go: Metaphysics Naturalized* (Oxford University Press).

Lewis, D. K. (1983) 'New Work for a Theory of Universals', *Australasian Journal of Philosophy* 61: 343–77.

Lycan, W. G. (1987) *Consciousness* (Cambridge, MA: MIT Press).

Marmodoro, A. (2013) 'Aristotelian Hylomorphism Without Reconditioning', *Philosophical Inquiry* 36: 5–22.

Marr, D. C. (1982) Vision: A Computational Investigation into the Human Representation and Processing of Visual Information (San Francisco, CA: W. H. Freeman and Co.).

Martin, C. B. (1996a) 'Properties and Dispositions', in Crane (ed.) *Dispositions: A Debate* (London: Routledge): 71–87.

——— (1996b) 'Replies to Armstrong and Place', in Crane (ed.) *Dispositions: A Debate* (London: Routledge): 126–46.

——— (1997) 'On the Need for Properties: The Road to Pythagoreanism and Back', *Synthese* 112: 193–231.

——— (2007) *The Mind in Nature* (Oxford: Oxford University Press).

Martin, C. B. and Heil, J. (1998) 'Rules and Powers', *Philosophical Perspectives* 12: 238–312.

——— (1999) 'The Ontological Turn', *Midwest Studies in Philosophy* 23: 34–60.

Martin, C. B. and Pfeifer, K. (1986) 'Intentionality and the Non-Psychological', *Philosophy and Phenomenological Research* 46: 531–54.

Miller, J. (1978) *The Body in Question* (New York, NY: Random House).

Molnar, G. (2003) *Powers: A Study in Metaphysics*, ed. by S. Mumford (Oxford University Press).
Noë, A. (2004) *Action in Perception* (Cambridge, MA: The MIT Press).
Oderberg, D. S. (2007) *Real Essentialism*. New York: Routledge.
—— (2014) 'Is Form Structure?', in D. Novotný and L. Novak (eds.) *Neo-Aristotelian Perspectives in Metaphysics* (New York: Routledge): 164–80.
Parvisi, J. et al. (2006) 'Diagnosis and Management of Pathological Laughter and Crying', *Mayo Clinic Proceedings* 81: 1482–6.
Place, U. T. (1996a) 'Intentionality as the Mark of the Dispositional', *Dialectica* 50: 91–120.
—— (1996b) 'Dispositions as Intentional States', in Crane (ed.) *Dispositions: A Debate* (London: Routledge): 19–32.
Plotnik, R. and Kouyoumdijian, H. (2013) *Introduction to Psychology*, 10th edition. (Belmont, CA: Cengage Learning).
Putnam, H. (1967) 'Psychological Predicates', in W. H. Capitan and D. D. Merrill (eds.) *Art, Mind, and Religion* (University of Pittsburgh Press): 37–48.
Quine, W. V. (1948) 'On What There Is', in Quine, W. V. (1953) *From a Logical Point of View* (Cambridge, MA: Harvard University Press): 1–19.
Rea, M. C. (2011) 'Hylomorphism Reconditioned', *Philosophical Perspectives* 25: 341–58.
Schaffer, J. (2009) 'On What Grounds What', in D. Chalmers, D. Manley and R. Wasserman (eds.) *Metametaphysics: New Essays on the Foundations of Ontology* (Oxford: Oxford University Press): 347–83.
Searle, John R. (1992) *The Rediscovery of the Mind* (Cambridge, MA: MIT Press).
—— (2004) *Mind: A Brief Introduction* (Oxford: Oxford University Press).
Shapiro, L. (2011) *Embodied Cognition* (New York: Routledge).
Sider, T. (2012) *Writing the Book of the World* (Oxford: Oxford University Press).
Sober, E. (1985) 'Panglossian Functionalism and the Philosophy of Mind', *Synthese* 64: 165–93.
Sperry, R. (1984) 'Emergence', in Weintraub, P. (ed.) *The Omni Interviews* (New York: Ticknor & Fields): 187–207.
Taylor, J. G. (1962) *The Behavioral Basis of Perception* (New Haven, CT: Yale University Press).
Van Inwagen, P. (1990) *Material Beings* (Ithaca, NY: Cornell University Press).
Worrall, J. (1989) 'Structural Realism: The Best of Both Worlds?', *Dialectica* 43: 99–124.
Young, J. Z. (1971) *An Introduction to the Study of Man* (Oxford: Clarendon Press).

12 Hylomorphism and the New Mechanist Philosophy in Biology, Neuroscience, and Psychology

Daniel D. De Haan

> Scholasticism did not know how to draw from its principles the physics which could and should flow from them. So our first duty today is to be more faithful to the demands of realism than the Middle-Ages were, and giving each order of reality its due. In each order, the reality of the form should be preserved, since without it one cannot account for structures, and it remains the principle of reality's intelligibility.
>
> Etienne Gilson, *Methodical Realism*, 103

Many philosophers and scientists believe that the turn to mechanistic explanations in the seventeenth century dealt the final death blow to Aristotelian hylomorphism.[1] While this might be the correct interpretation of the historical shift from hylomorphic to mechanistic explanations, in this essay I argue that contemporary versions of Aristotelian hylomorphism and the "new mechanist philosophy" in biology, neuroscience, and psychology share significant commitments about the reality of the organized causal components of mechanisms. My aim is to challenge the well-known narrative that hylomorphic and mechanistic ontologies are fundamentally incompatible by establishing that the new mechanist philosophy and Neo-Aristotelian hylomorphism are not only complementary, but are defending many of the same ontological claims.

I begin with a brief sketch of the fundamental claims of hylomorphism (§1). I then situate the new mechanist philosophy (*NMP*) within recent developments in philosophy of science (§2.1), before introducing the basic framework of *NMP* (§2.2). In the last two sections of the paper I argue for the compatibility of hylomorphism and *NMP*. I start with the major points of agreement between hylomorphism and *NMP* (§3). Significantly, I establish that *NMP* is committed to organization or structure realism (a touchstone of hylomorphism), and Neo-Aristotelian hylomorphism is committed to the reality of mechanisms or causal powers that produce, underlie, or maintain the behavior or capacity of (i) *phenomena* that are constituted through the (ii) spatial, temporal, and active *organization* of their (iii) *component entities* and (iv) *component activities* (the four hallmarks of *NMP*). In the last section

(§4) I introduce some possible points of disagreement between these two positions pertaining to hylomorphism's substance-attribute ontology, emergence, downward causation, and teleology. I show that the disagreements about these topics do not distinguish hylomorphists from new mechanists, but represent disagreements among hylomorphists and among new mechanists. I conclude that Neo-Aristotelian hylomorphism should not been seen as fundamentally opposed to mechanisms, but that it can and should embrace the many complementary features of the new mechanist philosophy in biology, neuroscience and psychology. If correct, this is a significant advance over and against an influential narrative that should be rejected.

I begin with a few preliminary points of clarification for this comparative study. My principal aim is to encourage a constructive conversation between hylomorphists and new mechanists. I therefore intentionally avoid trying to settle significant points of disagreement among hylomorphists or new mechanists that would prematurely alienate some hylomorphists from some new mechanists (and vice versa) on issues that I believe are not pertinent to the compatibility of hylomorphism and the new mechanist philosophy. My presentation of both hylomorphism and *NMP* focuses more on the fundamental principles that unite rather than divide hylomorphists under a common banner, and that bring together the diverse views of exponents of *NMP*. Finally, in order to achieve a synoptic comparison of these two complex and nuanced philosophical frameworks, my exposition forgoes many of the arguments for and against particular positions maintained by hylomorphists and the new mechanists.

1. Neo-Aristotelian Hylomorphism

Hylomorphism originates with Aristotle's philosophy of nature and takes its name from his account of matter (*hyle*) and form (*morphe*), that is, the two integrated and intrinsic principles or grounds for all physical substances. Aristotle's doctrine of hylomorphism is worked out alongside his account of physical substances, attributes, forms of natural change and composition, act and potency, the four causes (material, formal, agential, and final), the distinction between inanimate and animate beings, and his application of these positions throughout his philosophical biology and zoology. Contemporary exponents of hylomorphism often introduce it as a "third way" alternative to substance dualism and various forms of eliminative, reductive, and non-reductive physicalism.

1.1 Hylomorphism's Organization Realism

Contemporary hylomorphism is often characterized by its realist account of higher-levels of what is variously called "form," "organization," or "structure" and causal powers. William Jaworski, a contemporary exponent of hylomorphism, argues that the fundamental distinguishing feature of hylomorphism is its *structure realism*—I call this *form* or *organization* realism. Following Jaworski, we can sum up this insight about formal organization in the following slogans:

> *Formal organization matters*: it operates as an irreducible ontological principle, one that accounts at least in part for what things essentially are.
>
> *Formal organization makes a difference*: it operates as an irreducible explanatory principle, one that accounts at least in part for what things can do, the powers they have.
>
> *Formal organization counts*: it explains the unity of composite things, including the persistence of one and the same living individual through the dynamic influx and efflux of matter and energy that characterize many of its interactions with the wider world.[2]

The formal organization realism defended by hylomorphists like Jaworski, Robert Koons, David Oderberg, Anna Marmodoro, Bernard Lonergan, and others distinguish form from the mere static spatial relation of parts to each other.[3] For hylomorphism, form, organization, or structure is a *dynamic intrinsic ordering principle* of the organized materials, especially in the case of the organizational form of living beings. As Oderberg points out:

> [M]ere structure in the sense of configuration of parts is far too *static* a concept to tell you all there is about the form of an animal: There are its characteristic functions and behaviour, its dispositions, instincts, tendencies, actions and reactions, and all the rest of which ethology is made. These *dynamic* notions have to be added to the relatively static structural notions to get us to something like an account of the form of a living thing.[4]

Aristotelian hylomorphism maintains that organization is not an additional part standing alongside material components that provides a relation of unity to material parts; furthermore, form is not a mere extrinsic organization imposed on material parts from an extrinsic (efficient) cause; rather, formal organization intrinsically *transforms* the matter, thereby constituting a new unity, identity, whole.[5]

1.2 Hylomorphism's Substance–Attribute Ontology

This brings us to hylomorphism's connection to Aristotle's substance-attribute ontology. Hylomorphists distinguish between two kinds of organization: what many Neo-Aristotelians call *substantial* and *accidental forms*, what Jaworski calls *individual-making* and *activity-making structures*, and what Bernard Lonergan identifies as central and conjugate forms.[6] I employ Lonergan's terminology of central and conjugate formal organization to articulate the main contentions of hylomorphism.[7] A central form is a fundamental intrinsic principle that actually organizes and transforms the materials that compose a fundamental physical entity, that is, a substance. Hylomorphists maintain that the central forms of physical substances are essentially and continuously organizing and transforming the material components that constitute the substance, and it is this organizing activity of central forms that unifies and enables the same physical substances to persist through changes in their materials.[8] The same is true, *mutatis mutandis*, of conjugate forms, which organize the constituent materials of a substance's powers and activities. Following Jaworski, we can also frame three slogans that capture what is specific to conjugate forms.

> *Conjugate formal organization matters*: it is an irreducible ontological principle that organized activities possess essentially and which, in part, accounts for the very nature of what the activities are that they organize.
>
> *Conjugate formal organization makes a difference*: because it is the form of a power which provides an irreducible explanatory principle; and powers enable individuals to engage in organized activities they could not perform without such powers.
>
> *Conjugate formal organization counts*: it confers unity on diverse events in a way that is similar to how substantial forms confer unity on physical materials that compose organized individual substances.[9]

In addition, conjugate formal organization also *minds*: sensation, perception, affectivity, understanding, rational reflection, and intentional action are species of organized activities of psychological powers. The psychological powers and activities of humans and other animals are composed from the formal organization of their material parts and the manifestation of any requisite powers of the organized materials in the surrounding environment. Animals have the psychosomatic powers and operations they do in virtue of the organization of specific zones of their material parts by their conjugate forms.[10]

The central and conjugate forms of a substance are bound up with its fundamental central matter and organized conjugate potencies. The former is what is actualized and organized by a substance's central form, the latter pertains to zones of organized material components that are constituted by the conjugate forms of an individual substance. Accordingly, there are formal and material properties that pertain both to substances and to their conjugate attributes.

1.3 Staunch Hylomorphism: Fundamental Entities, Properties, and Powers

Hylomorphism's account of central and conjugate forms is connected to Rob Koons's division of contemporary proponents of hylomorphism into faint-hearted and staunch hylomorphists.[11] According to Koons, staunch hylomorphists accept the following Aristotelian claims, whereas faint-hearted hylomorphists deny one or more.[12] I am only concerned with staunch hylomorphism.

(1) *A sparse theory of fundamental entities.* A soul is a substantial form, and only substances have substantial forms. Socrates is a substance, but sitting Socrates is only an accidental unity, and so there is no substantial form corresponding to Socrates's sitting as there is to Socrates's living.

(2) *A sparse theory of fundamental properties.* Only substances have essences or natures in the strictest sense. An essence or nature is a fundamental property, which accounts for both the possibility and actuality of all other properties, acting as a 'principle' (*arche*) of motion (change) and rest.

(3) *A powers ontology.* The natures of substances confer fundamental causal powers on those substances, and those powers (both active and passive) are the ultimate grounds for explaining all change and activity.[13]

Claims (1) and (2) pertain to the nature of substances and their substantial or central form and matter; claim (3) concerns the powers that are conjugate attributes of a substance.

Along with an armament of classical and contemporary arguments for its organization realism, hylomorphism defends an anti-reductionist naturalist ontology that draws upon empirical science, especially the life sciences, to fortify and support its position. Indeed, some hylomorphists, like Jaworski, even cite proponents of *NMP*, like Carl Craver and William Bechtel, to provide empirical support for hylomorphism's ontological realism.

> Hylomorphism is committed to ontological naturalism, the claim that when it comes to determining what exists, empirical investigation is our best guide. Since many of our best empirical descriptions, explanations, and methods appear to posit various kinds of organization or structure, those descriptions, explanations, and methods give us prima facie reason to think that organization or structure exists. Structure realism takes empirical appeals to structure at face value. Structure, it says, is a basic ontological and explanatory principle: descriptions, explanations, and methods that posit structure cannot in general be reduced to, or paraphrased, or eliminated in favor of nonstructural descriptions and explanations. This straightforward approach to structure is the one favored by hylomorphists.[14]

2. New Mechanist Philosophy

2.1 *The Rise of the New Mechanists in the Philosophy of Science*

The *new mechanist philosophy* represents a more recent movement in philosophy of science; it is tied to a number of positions in philosophy of science that began to coalesce around the end of the twentieth century as a reaction to logical empiricism. In contrast to logical empiricism's focus on logical and mathematical idealizations of the theories of physics, *NMP* attends more to actual scientific practice, especially among the special sciences of biology, neuroscience, and psychology. "Many new mechanists developed their framework explicitly as a successor to logical empiricist treatments of causation, levels, explanation, laws of nature, reduction, and discovery."[15] While the discovery of mechanisms has always been important to scientists, mechanisms were neglected in twentieth-century philosophy of science due to the overwhelming influence of logical empiricism's covering-law account of explanations, like the deductive-nomological model. Against the framework of logical empiricism, more recent philosophers of science have argued that scientific explanations of a phenomenon, especially outside of physics, do not consist in showing that a phenomenon was predictable on the basis of laws of nature or other generalizations. Carl Craver argues that neuroscientists are instead interested in "showing how a phenomenon is produced by its causes. To explain neurotransmitter release, one shows that the depolarization *opens* the Ca^{2+} channels, that opening the Ca^{2+} channels *allows* Ca^{2+} to *diffuse* into the cell, that vesicles *dock* to the membrane by forming SNARE complexes and that the *influx* of Ca^{2+} *triggers* the formation of a fusion pore."[16] Robert Cummins argues that psychology is not concerned with discovering explanatory laws. Following his detailed survey of the major explanatory frameworks employed in empirical psychology (belief-desire-intention folk psychology, computationalism, connectionism, neuroscience, and evolutionary psychology), he concludes:

> Explanation in psychology, like scientific explanation generally, is not subsumption under law. Such laws as there are in psychology are specifications of effects. As such, they do not explain anything but themselves require explanation. Moreover, though important, the phenomena we typically call effects are incidental to the primary *explananda* of psychology, viz., capacities. Capacities, unlike their associated incidental effects, seldom require discovery, though their precise specification can be nontrivial. The search for laws in psychology is therefore the search for *explananda*, for it is either the search for an adequate specification of a capacity or for some capacity's associated incidental effects. Laws tell us what the mind does, not how it does it. We want to know how the mind works, not just what it does.[17]

The consolidation of these ideas and others culminated in a new orientation in philosophy of science that focused on the role of multi-level mechanisms in scientific explanation. In their seminal paper "Thinking about Mechanisms," Peter Machamer, Lindley Darden, and Carl Craver (MDC), begin with the following claim:

> In many fields of science what is taken to be a satisfactory explanation requires providing a description of a mechanism. So it is not surprising

that much of the practice of science can be understood in terms of the discovery and description of mechanisms.[18]

For *NMP*, "this new sense of mechanism is deeply anti-reductionist; science may uncover explanatory or ontological *connections* between higher and lower levels, but does not thereby either eliminate or *reduce* the higher levels thus connected."[19] Craver contends that the new mechanist approach to neuroscience supports what he calls the mosaic unity of neuroscience.

> This mosaic view of the unity of neuroscience is broader in scope than reduction because it covers both the integration of fields in research at a given level and in research that crosses levels. This mosaic view also provides a more accurate and elaborate view of interlevel interfield integration. Where reductionists understand the unity of science in terms of stepwise reduction to lowest levels, the mosaic view treats the unity of science as the collaborative accumulation of constraints at multiple levels. Whereas reduction focuses on relations of identity, supervenience, and ontological reductive links, the mechanistic mosaic view emphasizes the importance of explanatory relevance as the bridge between levels. Finally, whereas reduction models emphasize the importance of explanatory reduction to fundamental levels, the mosaic view can be pluralistic about levels, recognizing the genuine importance of higher-level causes and explanations. The mosaic unity of science is constructed during the process of collaboration by different fields in the search for multilevel mechanisms. One task for the philosophy of neuroscience is to show how that research ought to proceed.[20]

Many proponents of *NMP*, influenced by Wesley Salmon's account of causal explanations, are also committed to an *ontic conception of scientific explanation*, where mechanistic explanations pick out realties that are causal structures in the world.[21] Craver distinguishes four distinct but related senses of *explanation* employed by English speakers. It can be used "(1) to refer to a communicative act, (2) to refer to a cause or a factor that is otherwise responsible for a phenomenon (the ontic reading), (3) to refer to a text that communicates explanatory information, and (4) to refer to a cognitive act of bringing a representation to bear upon some mysterious phenomenon."[22] Craver recognizes that scientists use *explanation* in all these senses; his contention is that, contrary to certain anti-realist views in philosophy of science, it is not only perfectly legitimate, but even necessary for scientists to make use of the second sense of explanations that depend upon and are evaluated in light of the ontic structure of the world.[23] Indeed, "the norms of scientific explanation fall out of a prior commitment on the part of scientific investigators to describe the relevant ontic structures in the world."[24] If Craver is correct, then we have good reasons for taking seriously the underlying realism of certain ontic explanations in biology and neuroscience that provide accounts (even if inadequate) of the real or ontic structure of the world—especially those that appeal to mechanisms constituted by organized causal components. I shall not enter into the debates over Craver's complex and contentious defense of ontic explanations.[25] Nonetheless, it is worth pointing out that his defense of scientific realism about ontic explanations and structures found in nature provides one more point of agreement between hylomorphism and some proponents of *NMP*.

2.2 New Mechanists on Mechanisms

There have been many conceptions of mechanisms throughout the history of philosophy and science, but *NMP* places some distance between its understanding of mechanisms from the conceptions of mechanisms of Democritus, Descartes, Boyle, and others. *NMP* does not endorse the metaphysically austere accounts of the natural world defended by Cartesianism or more recent ontologies that endeavor to reduce all causal activities to a few fundamental forces. Unlike many accounts of mechanisms, *NMP* resists the idea that mechanisms are machines or are straightforwardly analogous to them.[26] Furthermore, according to Craver and other proponents of *NMP*, mechanisms are *not* essentially: deterministic (many are stochastic), reductionistic (most require multilevel irreducible ontic structures), *localizable* (many brain mechanisms are highly distributed), and are not mere fictions, metaphors, or explanatory heuristics (but describe and explain ontic structures in the world). So what are mechanisms, according to *NMP*?

NMP prefers qualitative accounts of the explanatory mechanisms actually employed by scientists over retailing their necessary and sufficient conditions.[27] A number of accounts of mechanisms are on offer from different proponents of *NMP*.

> Mechanisms are entities and activities organized such that they are productive of regular changes from start or set-up to finish or termination conditions.[28]
>
> A mechanism is a structure performing a function in virtue of its component parts, component operations, and their organization. The orchestrated functioning of the mechanism is responsible for one or more phenomena.[29]
>
> Mechanisms are entities and activities organized such that they exhibit the *explanandum phenomenon*.[30]
>
> *Minimal mechanism*: A mechanism for a phenomenon consists of entities (or parts) whose activities and interactions are organized so as to be responsible for the phenomenon.[31]

In what follows, I draw on the work of the philosopher of neuroscience, Carl Craver, and other exponents of *NMP* to summarize the basic claims of *NMP*. Craver presents a detailed account of mechanistic explanations used in neuroscience in his 2007 monograph, *Explaining the Brain: Mechanisms and the Mosaic Unity of Neuroscience* and in his 2013 *In Search of Mechanisms: Discoveries Across the Life Sciences* co-authored with Lindley Darden. Craver contends that "Explanations in neuroscience describe mechanisms, span multiple levels, and integrate multiple fields."[32] In other words, neuroscientific explanations appeal to mechanisms that encompass a complex hierarchy of organized component entities and activities that require the integration of such scientific fields as biochemistry, genetics, microbiology, electrophysiology, neurobiology, neuroscience, neuropsychology, and many others.[33] Craver's account of a mechanism draws attention to four fundamental elements: (1) a phenomenon, (2) its component entities, (3) component activities, and (4) the organization of these components.

2.2.1 Phenomena of Mechanisms

A *phenomenon* is variously described as the behavior or the manifestations of a capacity or power of the mechanism taken as a whole. The proper demarcation of a phenomenon depends in part on the level of investigation relevant to one's descriptive and explanatory goals: what counts as a phenomenon for the biochemist might be a component in a mechanism for the microbiologist, what is a phenomenon to be explained by a mechanism for the microbiologist, might be a component within a mechanism for the neuroscientist, and so on.[34] In every case, the phenomenon is the behavior of the mechanism as a whole: the mechanism for protein synthesis synthesizes proteins; the mechanism for an action potential generates an action potential, and a mechanism for opening ion channels causes ion channels to open. For *NMP*, phenomena are not adequately described as mere input–output relations, even though mechanisms often do involve many inputs and outputs. Some mechanisms operate in a linear causal process from one stage to the next, but many more operate in cycles (like the Krebs cycle) and feedback loops. Phenomena can range from being products of productive mechanisms, to being systems dependent on underling mechanisms, or systems that are sustained by maintenance mechanisms. Roughly speaking, mechanisms explain *how* organized entities and activities (i) *produce*, (ii) *underlie*, or (iii) *maintain* some phenomenon.[35]

(i) *Productive* mechanisms typically consist in a causal sequence and often explain some terminating end-product. The phenomenon explained by a mechanism in the case of protein synthesis is an end-state: the production of a protein.[36] Other products of mechanisms include activities and events, like digestion. (ii) Some mechanisms do not produce but *underlie* phenomena, such as the capacity or behavior of a whole system that is educed from the organized interactions of its implementing components. For instance, the underlying mechanism of an action potential does not produce it. The mechanism that underlies a neuron's action potential—the rapid depolarization and repolarization of the neuronal membrane that implements the electrical potential charge and communicates it down the neuron's axon—involves the whole neuron and the activities of its component parts such as membranes, neurotransmitters, ion channels, and ions. "This mechanism involves the choreographed opening and closing of sodium- and potassium-channel proteins and the diffusion of ions across the nerve cell's membrane."[37] (iii) Mechanisms also *maintain* phenomena, like regulatory or homeostatic mechanisms, which preserve a stable equilibrium among certain properties or relationships within a system in response to the alteration of internal or external conditions. For instance, "cells have mechanisms to maintain concentrations of metabolites, cardiovascular systems have mechanisms to maintain stable blood pressure, or warm-blooded animals have mechanisms to maintain constant body temperature."[38]

Another feature of mechanisms flagged by *NMP* is *regularity*. A mechanism is regular if it operates "always or for the most part in the same ways under the same conditions."[39] This classic Aristotelian adage does not exclude some mechanisms from being a one-off, such as mechanisms that initiate an epidemic or speciation. *NMP* also distinguishes the regularity of mechanisms from determinism. "Determinism is only the limit of regularity in most biological mechanisms." Many mechanisms, especially biological ones, are not deterministic but stochastic where indistinguishable conditions, at least for us, can have very different results. "For example, when the action potential arrives at the axon terminal, one, two, or more quanta of neurotransmitters may be released, or none at all. One can generate frequencies at which each of the outcomes will occur, but one cannot predict with certainty on this basis just which outcome will be in a particular case."[40]

2.2.2 Components of Mechanisms: Entities and Activities

NMP distinguishes the *entities*, *activities*, and *organizational features* of mechanisms. Entities and their activities or operations constitute the two kinds of component parts of a mechanism. Proponents of *NMP* disagree among themselves about which mereological account of parthood best captures the wide range of organized components that constitute a mechanism, especially for the diversity of parts present in biological mechanisms. Similarly, there are various views about how to characterize the causation of causal mechanisms. According to Craver, the ongoing *NMP* debate over causal mechanisms has focused on four different accounts of causation: conserved quantity accounts, mechanistic accounts, activity accounts, and counterfactual accounts. It remains an open question which, if any, of these accounts, will persevere among proponents of *NMP*. In the absence of space, I shall not enter into this debate.[41] Despite such disagreements over these details, Craver identifies a few points of agreement among proponents of *NMP* on causality.

New mechanists have in general been at pains both (1) to liberate the relevant causal notion from any overly austere view that restricts causation to only a small class of phenomena (such as collisions, attraction/repulsion, or energy conservation), and (2) to distance themselves from the Humean, regularist conception of causation common among logical empiricists.[42]

For *NMP*, entities are component parts of mechanisms that interact with other entities through their reciprocal activities. A mechanistic explanation in neuroscience, for instance:

> includes various entities (N-type $Ca2+$ channels, $Ca2+$ ions, active zones, a host of intracellular molecules such as Rab3A, Rab3C, VAMP/synaptobrevin, SNAP-25, and syntaxin, vesicles containing neurotransmitters, fusion pores, and neural membranes) and their various activities (opening, clamping, diffusing, docking, fusing, incorporating, phosphorylating, and priming).[43]

Activities induce causal changes in entities with the properties exigent for various actions or passions. "Activities are the producers of change. They are constitutive of the transformations that yield new states of affairs or new products. Reference to activities is motivated by ontic, descriptive, and epistemological concerns."[44] For example, a neurotransmitter and its post-synaptic receptor are two entities that are able to engage in a reciprocal activity of *binding* by virtue of such properties as their structure and charge distributions. *Entities* have properties like location, size, structure, duration, and orientation; they typically have masses, carry charges, and transmit momentum.[45] Enzymes, neurotransmitters, neurons, organs, and organisms are all entities that engage in activities—the causal components of mechanisms—including:

> productive behaviors (such as opening), causal interactions (such as attracting), omissions (as occurs in cases of inhibition), preventions (such as blocking), and so on. In saying that activities are *productive*, I mean that they are not mere correlations, that they are not mere temporal sequences, and, most fundamentally, that they can potentially be exploited for the purposes of manipulation and control . . . There are many kinds of activity, and it is the task of science rather than philosophy to sort them out. The mechanism of neurotransmitter release includes different forms of chemical bonding, conformation changes, diffusion, attraction and repulsion.[46]

Activities are individuated by their spatiotemporal location, rate, duration, the kinds of reciprocal entities and properties required for such activities, their setup, start and terminating conditions, modes of signaling (via autocrine, juxtacrine, paracrine, endocrine signaling), range, and energy requirements.[47]

2.2.3 *Organization of Mechanisms*

NMP unequivocally draws attention to the *ontic organization* of component parts—entities and activities—of mechanisms. The mechanisms that underlie an action potential are, for instance:

> organized together spatially, temporally, causally, and hierarchically such that transmitters are released when the axon terminal depolarizes. The voltage-sensitive ion channels are located in the terminal, they span the membrane, and they open to expose a channel. Biochemical cascades in the cytoplasm have sequences or cycles of interactions, they are organized in series and in parallel, and their steps have different orders, rates and durations. The components in the mechanism often stand in mechanism/component relations, a species of part–whole relation. As a result the mechanism is hierarchically organized. The behavior of the mechanism as a whole requires the organization of its components.[48]

The component parts of a mechanism are *organized* spatially, temporally, and actively, and this organization makes a difference to the mechanism as a whole. *Spatial* organization consists in locations, sizes, shapes, positions and orientations of component entities; *temporal* organization pertains to the orders, rates, and durations of entities and their activities. *Active* organization comprises the diverse ways the properties of components—entities and activities—enable components to make a difference to other components.

> Active organization distinguishes mechanisms from mere aggregates (or heaps) of matter, such as piles of sand. The parts act and interact with one another in such a way that the whole is literally not a mere sum of its parts. Mechanisms are in this sense nonaggregative: the parts of the mechanism are organized in ways that go beyond, e.g., the contribution made by the mass of a grain of sand to the mass of the pile. Mechanisms are not mere sums of properties of their component parts . . .[49]

The *active* organization of a mechanism's components is essential to the way *NMP* distinguishes organized mechanisms from aggregates. Aggregate properties are simply a sum of the properties of their parts, which change through the addition and subtraction of parts; the parts of aggregates can be rearranged or intersubstituted and the whole can be decomposed and recomposed without altering the properties or behavior of the whole. "These features of aggregates hold because organization is irrelevant to the property of the whole."[50] In contrast to aggregates, mechanisms are non-aggregative, they are "not mere static or spatial patterns of relations, but rather patterns of allowance, generation, prevention, production, and stimulation. There are no mechanisms without active organization, and no mechanistic explanation is complete or correct if it does not capture correctly the mechanism's active organization."[51] *NMP* contends that nature is replete with a spectrum of forms of organization that extends from truly aggregative properties, which are relatively rare, through a whole range of organized parts and causes to very complex organized mechanisms, sometimes characterized as forms of mechanistic emergence by *NMP*.[52] "Organization is the interlevel relation between a mechanism as a whole and its components. Lower-level components are made up into higher-level components by organizing them spatially, temporally, and actively into something greater than a mere sum of the parts."[53] Craver argues at length that mechanisms, "by virtue of their organization, are able to do things that their parts cannot do individually. They can respond to inputs that the parts alone cannot detect. They can produce behaviors that their parts alone cannot produce. *There are generalizations about causal relevance that are true of mechanisms and false of their parts.*"[54]

2.2.4 Levels of Mechanisms

The commitment to ontic organization of mechanistic components by *NMP* is connected to its antireductionist account of the hierarchy of integrated *levels* in the mechanisms of biology and neuroscience. Talk of levels, however, is ambiguous; in order to distinguish levels of mechanisms from other types of levels, Craver provides an extensive field guide to levels, the rich details of which exceed the scope of this article.[55] In brief, Craver contrasts levels of causality, which are relations between distinct entities, from levels of mechanisms, which are relations between a whole and its parts. The distinctive feature of levels of mechanisms is *mechanistic composition*, which Craver contrasts with levels of size, formal mereology, aggregativity, and spatial containment. *Mechanistic composition* takes the behaving components of a mechanism to be salient for the individuation of levels over any distinguishable boundaries between levels of objects or their sizes. Consequently, two or more items belong to the same level of a mechanism if they belong to the same mechanism and neither item is a component of the other. Working components of a mechanism, however, are at a lower mechanistic level than the mechanism taken as a whole. Craver defines levels of mechanisms in terms of the relationship between system S's ψ-ing behavior, taken as a whole, and some φ-ing activity of a component entity X of the system S, where the relata are the *behaving mechanism* at the higher-level and its component *acting-entities* at lower-levels. Accordingly, X's φ-ing is at a lower-level of mechanistic organization than S's ψ-ing if and only if X is a component entity of S and X's φ-ing is a component activity in S's ψ-ing.[56]

2.2.5 Causal and Constitutive Relations

Craver's account of levels of mechanisms is connected to his sharp distinction between *intralevel causal relations* and *interlevel constitutive relations*.

Levels of mechanisms are a kind of part–whole constitutive relation that is *not* causal. There is no "interlevel causation" or "causal interaction" between items at different levels for Craver. If we accept "the common assumptions that causal relationships are contingent and that cause and effect must be wholly distinct" then there can be no conflation of the levels of mechanisms with causal relationships.[57] The component parts of the mechanism are not *causes* that are contingently related to distinct *effects* identified with the behavior of the mechanism as a whole; rather, the higher-level behavior of the whole mechanism is itself *constituted* or *composed* from the organization of its lower-level components, which are themselves constituted from the organization of their own lower-level components. *NMP*'s denial of "interlevel causation" between the behavior of a mechanism as a whole and its organized component parts does not entail the rejection of interlevel dependency; this is captured by constitutive relationships.

> The behavior of the whole is dependent on the behavior of the components in such a way that interventions to change the components can change the behavior of the whole and vice versa. While there are not interlevel causal relations in [levels of mechanisms], there are many interlevel relations of dependency, and thereby interlevel relations of regularity and predictability. One can disrupt spatial memory by ablating the hippocampus or knocking out NMDA receptors.[58]

In short, causal relations concern the intralevel causes of entities and their activities at the same level, constitutive relations pertain to the interlevel forms of dependency among the behaviors of whole mechanisms and their organized component parts at different levels.

2.2.6 *"Top-Down Causation Without Top-Down Causes"*

This brings us to the revisionary account of top-down causation and emergentism of Craver and Bechtel. They accept that the language of top-down and bottom-up interlevel "causation" employed by scientists often does identify "perfectly coherent and familiar relationships" between the activities of wholes and the activities of their component parts. However, for Craver and Bechtel, it is a category mistake to conceptualize the interlevel whole–part dependency between levels of mechanisms as a form of top-down or bottom-up causes, given standard assumptions about (efficient) causation. They argue that any unobjectionable cases of "interlevel causation" referenced in the scientific literature can be reinterpreted, without remainder, as "appeals to mechanistically mediated effects."

> Mechanistically mediated effects are hybrids of constitutive and causal relations in a mechanism, where the constitutive relations are interlevel,

and the causal relations are exclusively intralevel. Appeal to top-down [or bottom-up] causation seems spooky or incoherent when it cannot be explicated in terms of mechanistically mediated effects.[59]

The account of mechanistically mediated effects of Craver and Bechtel not only provides an alternative interpretation of reputed cases of interlevel bottom-up or top-down causation, but also establishes concrete grounds for distinguishing *mechanistic emergence* from *spooky emergence*, via the presence or absence, respectively, of mechanistic hybrids of *constitutive interlevel* and *causal intralevel* relations. "Mechanistic (or organizational) emergence thus understood is ubiquitous and banal but extremely important for understanding how scientists explain things."[60] According to *NMP*, *mechanistic* or *organizational emergence* is common to the entire biological world, for organizational emergence pervades the hierarchy of multilevel constitutive and causal relationships described and explained by the mechanisms discovered by scientists. Unlike mechanistic organizational emergence, *spooky* or *strong emergence* lacks mechanistic explanations for it fails to provide the explanatory hybrid of constitutive and causal relations that are essential to mechanistic emergence. Consequently, the sense of "level" connected with strong emergence must be distinguished from *NMP's* account of levels of mechanisms.[61] "Levels of mechanisms are constitutive levels; levels of strong emergence are not. For this reason, the notion of strong emergence can borrow no legitimacy from its loose association with the levels of mechanisms so ubiquitous in biology and elsewhere."[62] Craver and Bechtel conclude:

> This hybrid framework provides a way to understand most, if not all, the cases for which appeal to top-down causes seems compelling. There may be cases that cannot be handled by this account, but if there are, those who invoke the notion of top-down causation for them owe us an account of just what is involved . . . Although our explication of interlevel causation in terms of mechanistically mediated effects renders reference to top-down causation unproblematic, it does not show that the phenomenon is unimportant. The biological world, and much of the world besides, is populated by multilevel mechanisms. Talk of interlevel causation is merely a misleading way to talk about an explanatory interlevel relationship that, upon close inspection, does not involve interlevel causes.[63]

In sum, the *NMP* in biology, neuroscience, and psychology defends an account of mechanistic ontic explanations. This account consists in explanations of how a multilevel hierarchy of organized intralevel causal interactions among components entities can constitute the interlevel organizational emergence of mechanisms and their manifestation of various phenomena.

3. Points of Agreement Between Hylomorphism and New Mechanisms

Those familiar with the basic principles of hylomorphism will recognize in the foregoing summation of the new mechanist philosophy many points of agreement—and even some identical doctrines expressed in a different nomenclature—with Neo-Aristotelian hylomorphism. I shall take note of some of the more fundamental points of agreement, beginning with hylomorphism's organization realism. As was noted before, exponents of hylomorphism are minimally committed to some account of organization or structure realism. Hylomorphism's organization realism touches upon a variety of interconnected philosophical topics, such as Aristotle's substance-attribute ontology, property, causal and explanatory pluralism, theory of composition and parthood, and Aristotle's metaphysics of act and potency—which overlaps with significant features of contemporary views in the metaphysics of causal powers. Let us begin with *NMP's* and hylomorphism's shared commitment to the reality of organization.

3.1 Organization Realism

Hylomorphists understand the *organization* of material things to be a real and fundamental ontological and explanatory principle that accounts both for the very nature of what a distinct physical substance is and what its powers enable it to do.

> Hylomorphism claims that structure (or organization, form, arrangement, order, or configuration) is a basic ontological and explanatory principle. Some individuals, paradigmatically living things, consist of materials that are structured or organized in various ways. You and I are not mere quantities of physical materials; we are quantities of physical materials with a certain organization or structure. That structure is responsible for us being and persisting as humans (as opposed to, say, dogs or rocks), and it is responsible for us having the particular developmental, metabolic, reproductive, perceptive, and cognitive capacities we have.[64]

The *organization realism* of hylomorphism corresponds to *NMP's* account of the ontic structure of spatial, temporal, and active *organization* in biological organisms as a significant factor—along with the organized component entities and activities—that constitute any mechanism. One way *NMP* distinguishes itself from classical mechanisms is by emphasizing that organization is an ineliminable and irreducible feature of mechanistic explanations. For *NMP*, spatial, temporal, active organization is a fundamental explanatory factor that accounts for the way the causal component parts *constitute* a mechanism as a whole. While the interaction of component entities and their activities explain the intralevel causal relations in mechanisms, it is the active organization of these components that provides the most salient explanatory factor of the interlevel constitutive relation between the parts of a mechanism and the mechanism as a whole. It is this constitutive relation that enables and so explains, in part, what the mechanism as a whole is, what it is able to do, and accounts for the persistent unity and organization of the component parts of some mechanistic functional whole. In short, for *NMP*, as with hylomorphism, *organization matters, makes a difference,* and it *counts.*

3.2 Property, Causal, and Explanatory Pluralism

Hylomorphism's commitment to property, causal, and explanatory pluralism resonates with numerous features of *NMP*. For hylomorphism, every organized physical entity and its activities possess two kinds of properties:

> *Organizational or Formal Properties*: Properties due to organization or being integrated within an organized entity
> *Material Properties*: Properties due to matter irrespective of its being integrated within an organized entity[65]

Because hylomorphism maintains that all physical entities are constituted by the organization of their material components, a complete description and explanation of any physical, chemical, biological, neurophysiological, psychological, ethological, or anthropological phenomenon must detail both the formal properties due to *organization* and the material properties due to the *materials* that are organized. Jaworski provides a number of illustrations of these two kinds of properties. Subatomic particles, atoms, and molecules have *material properties* like mass regardless of whether they are taken up into the life of an organism and integrated into its formal organization. But whenever such entities do become organized constitutive parts of a living organism, they contribute formal properties to the animated activities of these living things; for instance, nucleic acids, hormones, and neural transmitters have such formal properties as being genes, growth factors, and metabolic and behavioral regulators. The atomic or fundamental material properties of a strand of DNA might not be modified by its surrounding environment, but it exhibits new formal properties when it is integrated into the organization of a cell and is enlisted in the cell's activities.[66]

Hylomorphism is also committed to causal and explanatory pluralism. There are many different kinds of causes and causal relations; causes are explanatory factors and explanatory relations target causal relations. In the case of rational animals, like human persons, Aristotelian hylomorphists distinguish between personal and sub-personal level descriptions and explanations.[67] Personal level explanations address rational patterns of organized human activities that are best explained by reasons for action, but personal level explanations also appeal to perceptions, desires, emotions, and other forms of psychosomatic cognitive and motivational factors that humans share with other animals. Subpersonal level descriptions and explanations provide the *material causes* of these personal level descriptions and explanations, and this includes the kinds of mechanisms treated by *NMP*.

While most proponents of *NMP* do not speak in such terms, their commitment to multilevel ontic explanations of mechanisms constituted by the organization of a plurality of component entities with different intralevel causal activities, can be explicated in terms that are consistent with hylomorphism's commitment to property, causal, and explanatory pluralism. For as we have seen, the entities at distinct levels of mechanisms have different properties that enable them to engage in different causal interactions with other intralevel entities. The properties and causal activities of the components of a mechanism are typically very different from the behavior and causal capacities of the mechanism as whole, which is constituted from the organization of such components. The coordinated opening and closing of transmembrane ion channels (along with the interaction of many other component entities) that constitute the underlying mechanism of a neuron's action potential, are very different from the phenomenon of an action potential itself, such as the rapid and transient changes in the electrical potential difference across a neuron's membrane and the unidirectional propagation of the action potential from the cell body to the axon terminal.

For *NMP*, mechanistic explanations enlist a hybrid of constitutive and causal relations that are consonant with hylomorphism's distinction between material and formal properties. While there are important differences among the organized components that underline the mechanism for an action potential as well as variations among action potentials, nonetheless, the action potentials of neurons do possess a range of similar *material properties, causes,* and *explanations* that constitute organized components for a myriad of higher-level neural assembles and neural systems (e.g., peripheral, central, motor, sensory, and a host of more fine-grained systems for recognition, memory, language, attention, etc.). And these latter wholes or systems have components with distinct *formal properties, causes,* and *explanations* in virtue of the way neurons and patterns of action potentials are organized constituents of these higher-level systems.

Explanations for *NMP* and hylomorphism aim to account for the way reality is, to capture its ontic structure. For *NMP*, "Mechanistic decomposition cuts mechanisms at their joints."[68] This does not mean higher-level phenomena are explanatorily reduced to the mechanisms of a lower-level; rather, such joint-carving requires integrative explanations of the hybrid of constitutive and causal relations that comprise mechanisms. In short, hylomorphism and *NMP* agree that descriptions and explanations must take into account the plurality of causal and organizational factors that constitute living organisms.

3.3 Composition, Wholes, and Parts

As with *NMP*, there are a range of disagreements among hylomorphists about the best accounts of causality, the metaphysics of powers, and theories of composition and parthood. Because such internal disagreements do not distinguish hylomorphism from other views, they do not provide grounds for showing the incompatibility of hylomorphism and *NMP*. One area of agreement pertains to some general claims about composition and parthood that follows from the organizational realism and plurality of properties, causes, and explanations maintained by *NMP* and hylomorphism. Hylomorphism is committed to a number of positions pertaining to the nature of composition and mereology, that is, a theory about parts and wholes.[69] Let us take note of widespread agreement between hylomorphism and *NMP* on such issues. First, for hylomorphism, organized entities and activities are more than the sum of their parts; an entity includes both its *material parts* and the *organization* of these parts. New mechanists also maintain "Every complex is a mereological sum, but mechanisms are always literally more than the sum of their parts. Any account of the composition relation in [levels of mechanisms] must accommodate this fact."[70]

Second, hylomorphism holds that the *organization* of an entity and its activities can persist and remain the same notwithstanding the continuous ebb and flow of its material parts through its interactions with the environment. The persistence conditions of the entity are contingent on the continuity of its *organization* of some material parts, not on the continuity of *these* material parts. As Jaworski points out, when an animal breathes, the oxygen atoms it inhales become organized in a particular way by becoming integrated into the organized material parts of the animal and its metabolic activities. Conversely, when the animal exhales, carbon atoms cease to be organized material parts of the overall organization of the animal and its metabolic activities.[71]

Third, the nature of parthood requires being integrated into the organization of the whole and contributing to the overall functioning of the whole and its activities, and parts are distinguished according to the different ways in which they contribute to the different activities of the whole. "Genes and messenger molecules are both parts of cells; they both contribute to the cell's activities, and what distinguish them from each other are the different roles they play in protein synthesis: their different jobs within the cell qualify them as different parts of it."[72] Many hylomorphists accept the existence of a range of proper parts, especially biofunctional parts in living things (from organelles and cells to distinguishable organs and their organic parts), and recognize the organic parts identified by our best empirical descriptions and explanations.

Fourth, even though hylomorphists are divided over the details of the best theory of parthood, most hylomorphists would grant the following general account of parthood from Jaworski: "roughly, *x* is a part of *y* if *x* contributes to the activities of *y*. An electron is a part of me, for instance, if it contributes to my activities—if, say, it depolarizes one of my cellular membranes. This notion of composition dovetails with work in biology, philosophy of biology, and philosophy of neuroscience."[73] Robert Koons's version of staunch hylomorphism defends an account of *Parts as Sustaining Instruments* (*PASI*). According to *PASI*, "the persistence of the whole is grounded in the ongoing cooperation of the parts, and the active and passive powers of the parts are grounded in corresponding primary powers of the whole. In addition, the whole acts through the parts, as teleologically subordinate instruments." Furthermore, *PASI* "ties the whole and parts together in such a way that the whole is neither existentially separate from its parts nor able to act in a way that is separate from the actions of its parts. The staunch hylomorphist must accomplish two things: (1) ensure that the persistence through time of the composite whole is grounded in the cooperation of its parts, and (2) ensure that the whole cannot act or be acted upon except, at least in part, doing so 'through' the powers of its parts." Accordingly, Koons offers the following "Definition of 'proper part': *x* is a proper part of *y* at *t* iff *x* is a sustaining instrument of *y* at *t*."[74]

Fifth, hylomorphism is committed to an account of fundamental entities constituted by a complex hierarchy of parts and causal powers comprised of organized systems and sub-systems with corresponding activities and subactivities. This is paradigmatically the case with living organisms.[75]

As we have seen, *NMP* is committed to a similar dynamic account of the unity of the organization that constitutes the distinct levels of mechanism, and an account of parthood and levels of mechanisms whereby the organized activities of component entities contribute to the activities of the whole. Indeed, the general hylomorphic account of parthood is nearly identical to the aforementioned account of the individuation of component parts within the levels of a mechanism from Craver.[76]

3.4 Levels of Organization and Emergence

The causal powers of a higher-levels of organization emerge or are educed from the potentialities of the materials they organize and integrate. Because higher-level systems of, say, living organisms are composed from the organized materials of lower-level systems, the activities of an organism's higher-level powers cannot violate the fundamental activities that belong to the lower-level organic systems and chemical and physical subsystems. Indeed, the novel activities of higher-level powers depend upon the activities of lower-level activities that compose them. "It is because fundamental physical entities behave in stable, characteristic ways that they can be recruited to play in organisms the higher-level roles they do. It is because electrons have a characteristic mass and charge, for instance, that they are able to operate as membrane depolarizers within certain structures."[77] Hylomorphism's understanding of the eduction or emergence of higher-level properties is therefore radically different from more well-known accounts of strong emergence.

First, hylomorphism rejects standard accounts of strong emergentism that maintain higher-level emergent properties are products generated or caused by the organization of lower-level properties. For hylomorphism, educed or emergent properties are not distinct from the organization of lower-level organization; rather, the hylomorphic emergent properties are constituted from the organization of their components. Second, it is the central form of a physical substance that grounds all of its conjugate properties, whether they be higher-level psychosomatic powers or the lower-level biochemical powers that are components within psychosomatic powers. Finally, hylomorphic emergence is not unique to psychological powers or phenomenal consciousness, but is ubiquitous to the entire biological world.

The hylomorphic account of hierarchies of organized levels that constitute an organism—where a higher-level phenomenon is educed or emerges from the organization of lower-level components—is nearly identical to Craver's presentation of mechanistic emergence of higher-levels of mechanisms from the organization of lower-level mechanistic components.

> When one says that atoms compose molecules, which are organized into cells, which are linked into networks from which mental properties spookily emerge, the first three steps are upward steps in a hierarchy of levels of mechanisms, but the last is not. The ability of organization to elicit novel causal powers (that is, nonaggregative behaviors and properties) is unmysterious both in scientific common sense and common sense proper . . . Appeal to strong or spooky emergence, on the other hand, justifiably arouses suspicion.[78]

Both *NMP* and hylomorphism reject standard accounts of emergence in favor of complementary, if not equivalent, alternative conceptions of organizational emergence.

In sum, hylomorphism is committed to a range of issues concerning (1) organization realism, (2) property, causal, and explanatory pluralism, (3) the composition of wholes and parts, and (4) levels of organization that are compatible, if not in complete agreement, with stands that *NMP* takes on these same topics. In short, these fundamental features of hylomorphism set in relief a significant range of points of agreement, and the possibility for fruitful collaboration, between proponents of hylomorphism and *NMP*.

4. Points of Disagreement Between Hylomorphism and New Mechanisms

The synoptic character of this comparative essay prevents me from digging any deeper into the details and more substantive arguments for and against the connections I have been making between hylomorphism and *NMP*. My

presentation of hylomorphism will certainly leave the metaphysician unsat-
isfied, and my digest of *NMP* will have the philosophers of science yearning
for a more critical examination of empirical literature. I realize more needs
to be said to sift out the exact points of substantive agreement from other
mere naïve correlations between superficially similar positions. I conclude
by taking up some apparent objections and potential points of disagreement
between hylomorphism and *NMP*.

4.1 Substance-Attribute Ontology: Central and Conjugate Forms

The first concerns the substance-attribute ontology of hylomorphism, especially its distinction between the central and conjugate formal organization
of a substance. Are these features of staunch hylomorphism compatible with
NMP?

It is important to point out a difference of scope between the principal concerns of hylomorphism and *NMP*. *NMP* is a philosophy of science;
hylomorphism is an ontology. Hylomorphism aims to provide a general
ontology of the physical world. *NMP* is not specifically concerned with
addressing more general metaphysical questions about the nature of organisms, substances, fundamentality, and so forth; rather, it is focused on the
concrete macro and micro ontic structures of organisms that constitute
the mechanisms discovered in biology, neuroscience, and psychology. Few
of these forms of ontic organization will prove to be ontologically fundamental; most will comprise forms of conjugate attributes. Consequently,
the distinction between substantial organization and conjugate organization does not arise for the more specialized studies of *NMP*. Nevertheless,
given the basic contentions of *NMP*, there seems to be no principled reason
that militates against an exponent of *NMP* defending a substance–attribute
ontology that squares with the multileveled view of the new mechanists (as
Glennan, Maley, and Piccinini do),[79] or even from adopting hylomorphism's
full-fledged distinction between the substantial organization of substances
and the conjugate organization of its components. Likewise, there is nothing
to prevent hylomorphists from looking to *NMP* (as Jaworski does) to provide a more concrete empirical account of the organization of biofunctional
components of the mechanisms or capacities that comprise the conjugate
attributes of a substance's proper parts.

The general strategy of *NMP's* defense of hybrid causal and constitutive explanations is to acknowledge mechanistic emergence wherever
there are higher-level wholes constituted from the organization of causal
components. Insofar as the staunch hylomorphist employs a similar strategy to make the case that substances constitute distinct organized wholes
over and above their organized components, then the staunch hylomorphism stands on solid ground—by *NMP's* own lights—for distinguishing between substantial and conjugate forms of organization. Seen in this
light, a new mechanist-*cum*-staunch hylomorphism could hold that the

substantial form and matter of a substance, say, of an animal, explains a thing's fundamental nature and grounds the overall multilevel organization of its matter and a wide range of physical, chemical, biological, and psychological powers. A substance's higher-level psychological powers, say, memory, are form-matter conjugate properties that admit of mechanistic explanation via *NMP's* hybrid constitutive and causal relations. The causal relation of mechanistic explanations concerns the way the causal powers of these organized material component entities enable them to engage in intralevel interactions. The constitution relation of mechanistic explanation pertains to the way the organization of these material component parts compose form-matter conjugate properties. The conjugate form of a psychological power explains the active, spatial, and temporal *formal organization* of the zones of organized material components of some power, say, neural distributed within the medial temporal lobe (and perhaps elsewhere), which in turn are mechanistically explained by the formal organization of these component's organized material subcomponents, say, the patterned firing rate of individual pyramidal neurons in the hippocampus and the diffusion of glutamate into the synaptic cleft. Similarly, these zones of organized *material components* have various material and formal properties that explain the range of causal interactions these component parts are capable of engaging in, as well as the way their enabling, precipitating, inhibiting, and modulating conditions influence how these mechanistic components constitute (or fail to constitute) the manifestation of a higher-level mechanism or capacity, like the animal actually remembering.[80] In short, far from being incompatible, hylomorphism and *NMP* provide a wealth of insights for thinking about diverse forms of organization found among substances and their attributes.

4.2 Emergentism and Bottom-Up and Top-Down Causation

These points bring us to another apparent incompatibility between *NMP* and hylomorphism. Most defenders of *NMP* adopt Craver's and Bechtel's rejection of top-down and bottom-up interlevel causation in favor of their distinction between intralevel causation and interlevel constitutive relations. However, many hylomorphists defend accounts of top-down or downward causation. This conflict, I suggest, is less substantive and more terminological. The causal pluralism of hylomorphism opens it up to more forms of dependence than those captured by standard contemporary views on efficient causation. Classically, Aristotelians distinguish four kinds of dependence, and so four distinct kinds of *causes* and explanations: material, formal, efficient or agential, and final causes. The material stuff or components of some entity make a difference to what that entity is and can do; material causes explain this kind of dependency. As we have seen, the *organization* of these material components also *matters, makes a difference*, and *counts*; formal causes explain the dependency of a whole on the *organization* of its material components. Efficient causes explain the range of dependencies between events, properties, and substances that consist of moving, exerting force, attracting, repelling, producing, generating, corrupting, transforming, and so on. Final causes explain the range of functional dependencies, the way the formally organized material components are ordered towards constituting and enabling powers of efficient causation to perform or produce stereotyped, organized operations, products, properties, substances, and so forth in organized material objects with reciprocal powers.

NMP rightly distinguishes its constitutive relation between the component parts and the mechanism as a whole from standard accounts of (efficient) causation, which maintain that causes and effects are distinct. However, Aristotelian hylomorphism defends two forms of causal dependency that are equivalent to *NMP's* constitution relation, namely, its understanding of the way a whole is constituted or composed from the union of its formal and material causes, that is, the way the *organization* of *material components* cause the whole. This constituted whole is not a separate effect or entity produced and distinct from the formal and material causes, the whole as effect is united with and dependent on its intrinsic formal and material causes. In short, the disagreement here is terminological. The causal pluralism of *NMP* and hylomorphism permits both views to stipulate accounts of "causality" that *either* pragmatically appease widespread assumptions about causality among their interlocutors and introduce alternative terms to capture other forms of dependency (such as *constitution* or *grounding* or *realization*) *or* that maintain some version of Aristotelian causal pluralism, albeit recognizing that this approach hazards being misunderstood by contemporary philosophers.

A similar conclusion can be drawn with respect to downward causation. Neither *NMP* nor hylomorphism need to be committed to a strong emergentist view of downward causation, where the manifestation of higher-level causal powers efficiently cause lower-level effects. So, for some strong emergentists, psychological powers not only manifest psychological efficient causes like an intention to walk home, but also efficiently cause the motor cortex in the brain to generate the appropriate bodily movements of an intentional action. Instead, *NMP* and hylomorphism can maintain that intentional actions are comprised of a plurality of causes or a hybrid of constitutive and causal relations, where higher-level factors like practical reasoning and other psychological motivations can provide personal level explanations for an intentional action. Again, given causal and explanatory pluralism, personal level explanations do not compete with, but make a distinctive complementary difference to mechanistic explanations of an embodied intentional action. An embodied intentional action is itself a whole composed of a complex hierarchy of levels of mechanisms, all of which are integrated and engaged by the higher-level psychological powers for intentional action. On this score, intentional actions (like walking home) should not be conceived as mere efficient causes acting on sites of the nervous system. Rather, since higher-level psychological phenomena (like intentional actions) are enabled by and constituted from the organization of lower-level mechanisms (such as the nervous system), intentional actions consist of rationally coordinated patterns of activation of the organized causal components that underlie and constitute them. In this way, mechanistic explanation

> offers significantly more insight into what interlevel integration is, into the evidential constraints by which interlevel bridges are evaluated, and into the forces driving the co-evolution of work at different levels. Constraints on the parts, their causal interactions, and their spatial, temporal, and hierarchical organization all help to flesh out an inter-level integration. Finally, mechanists repeatedly recognize the need to not only look down to the constitutive mechanisms responsible for a given phenomenon (emphasized by classical reduction models), but also to look up and around to the context within which the phenomenon is embedded: interlevel integration is an effort to see how phenomena at many different levels are related to one another . . .[81]

A great deal more needs to be said about the way hybrid explanations from staunch hylomorphism and *NMP* would address issues of psychological operations and causal powers. I hope this sketch is at least suggestive of the possibility of fruitful exchange of perspectives from these two complementary approaches to multilevel organization realism and causality. *NMP* provides a rich way to think more concretely and empirically about the kinds of embodied causal powers of organized components which hylomorphism addresses in its more abstract ontological framework.

4.3 Teleology and Mechanisms

I conclude with what might seem to be the most controversial point of disagreement between staunch hylomorphism and the new mechanist philosophy, namely, the compatibility of mechanistic explanations with teleology and final causality. Aristotelian hylomorphism seems to be wed to some form of teleological explanation; after all, final causality is taken to be the cause of causes for Aristotelians. The difficulty is that a widely accepted script tells us that mechanistic science caused the final end to teleological explanations.[82] Consequently, if mechanist explanations are against teleological explanations and Aristotelian hylomorphism requires teleology, then the two positions seem to be fundamentally at odds. As with many disputes, the real point of disagreement is more complicated.

First, while it is the case that Aristotelian hylomorphists are committed to some robust realist doctrine of teleology, there are some hylomorphists that are neutral about the nature of teleological explanation. Surprisingly, a similar range of views about teleology has emerged among new mechanists, but the battle line that divides the different views among *NMPs* is not drawn where one might expect. The contention at issue is not about teleology versus no teleology; rather, the recent dispute among *NMPs* is between ineliminable perspectivalism about the function of teleological explanations in mechanistic explanations and realism about hybrid teleological-*cum*-mechanistic ontic explanations. Craver defends a version of perspectivalism on teleological functions. He holds that our explanatory interests render teleology ineliminable from our efforts to make the world intelligible. Craver distinguishes his mechanist design stance from the "associations with adaptationism and optimality" characteristic of Daniel Dennett's well known account of the intentional, design, and physical stances. Craver's conclusion is worth quoting at length, as it provides one among many stark examples of how misleading the received wisdom is on the incompatibility of mechanisms and teleology.

> In the contemporary mechanical philosophy, functional and mechanistic descriptions work in tandem to bring intelligible order to complex systems. By identifying functions within such systems, one approaches the system with some set of interests and perspectives in mind. One might be interested in understanding how parts of organisms work, how they break or become diseased, or how they might be commandeered for our own purposes. Regardless of which perspective one takes, the identification of functions is a crucial step in the discovery of mechanisms. We no longer speak of mechanisms simpliciter, but rather as mechanisms *for* some behavior. Mechanistic descriptions thus come loaded with teleological content concerning the role, goal, purpose, or preferred behavior of the mechanism. This teleological loading cannot be reduced to features of the causal structure of the word, but it is ineliminable from our physiological, and particularly neural, sciences, precisely because their central goal is to make the busy and buzzing confusion of complex systems intelligible and, in some cases, usable.[83]

Craver concludes that adopting his mechanist design stance makes intelligible the way a mechanism as a whole exhibits some behavior which is constituted from the organization and interaction of the components of the mechanism. It remains an open question for Craver whether or not these teleological aspects of the new mechanist worldview are reducible, without any remainder, to other features of the causal structure of the world; "here we have a, perhaps *the*, central puzzle that any properly mechanical understanding of mind must someday face."[84]

For many, Craver's rather rich perspectivalism about teleology will be real enough.[85] Functions are ineliminable features within mechanistic explanations, even if they are not real. Craver's account is honest about the exigency of employing teleological explanations in mechanistic explanations. More recently, Corey Maley and Gualtiero Piccinini have argued that Craver's perspectivalism is inadequate and that new mechanists should embrace a realist account of teleological explanations in psychology and neuroscience.[86] Perspectivalism understands all ascribed teleological functions to be "observer-dependent and hence subjective, and in no way objective."[87] For perspectivalism, the range of explanatory interests accounts for the proliferation of otherwise unchecked teleological attributions. If we adopt the perspective of an organism's survival, then the function of the heart is to pump blood, but seen from the perspective of diagnosing heart conditions, the heart's function is to make thumping noises, and so forth.[88] NMP proponents like Maley and Piccinini are not impressed with perspectivalism's line of argumentation against realism about functions. They argue, contrary to perspectivalism, that the function of the heart is to pump blood, and assigning the heart this function is independent from any observer's perspective or explanatory interests. Of course, some traits have multiple functions, but again, this is not due to multiple perspectives. According to Maley and Piccinini:

> Perspectivalism does not do justice to the perspectives we actually take in the biological sciences. If we could identify non-teleological truthmakers for teleological claims, we would avoid perspectivalism and deem functions real without deeming them mysterious. That is our project.[89]

Indeed, if one is willing to accept Craver's argumentative strategy in favor of taking mechanisms and organization to provide *ontic* explanations, it becomes difficult to resist the similar argumentative strategy in favor of taking teleology to provide ontic explanations as well. I leave the debate here.

My aim, once again, is to show that proponents of the *NMP*—of both perspectivalist and realist stripes—take teleology seriously. Within this debate between realists and perspectivalists the function of teleological descriptions and explanations remains controversial for new mechanists and hylomorphists. In short, these more restricted disagreements about teleology do not divide *NMP* and hylomorphism. *NMP* perspectivalism about function is compatible with a variety of non-Aristotelian versions of hylomorphism that are neutral or even skeptical about the reality of teleology. Similarly, a *NMP* defender of teleological realism, like Maley and Piccinini, can adopt an Aristotelian hylomorphism that endorses finality realism. This conclusion is significant, for this compatibility between hylomorphism and the new mechanists reveals that the long-standing script that teleology and mechanisms are fundamentally opposed, is mistaken. For the *NMP*'s recognition of organizational realism and the significance of teleology in biology, neuroscience, and psychology provide an amiable philosophy of science ally to Aristotelian hylomorphism.

Conclusion

Neo-Aristotelians and hylomorphists often participate in debates in philosophy of mind and metaphysics, but they rarely address topics in philosophy of biology, neuroscience, and psychology. My aim has been to introduce Aristotelian hylomorphists to the new mechanist philosophy in biology, neuroscience, and psychology, and, I hope, to convince hylomorphists that there is a real potential here for fruitful constructive engagement with scientific research, philosophy of science, and especially with the new mechanist philosophy. In this essay, I have argued that Neo-Aristotelian hylomorphism is perfectly compatible with the account of mechanisms presented by *NMP*. I have established that there are substantial points of agreement between the two views on fundamental issues that typically distinguish hylomorphism and *NMP* from other views. In light of such common points of agreement, I believe their integration would be mutually beneficial. Hylomorphism provides a robust abstract ontological framework for *NMP*, and *NMP* can enrich hylomorphism with its detailed concrete exposition of the complex hierarchy of organized mechanisms discovered by biologists, neuroscientists, and psychologists. Finally, I have shown that despite any historical controversies concerning the compatibility of teleology and mechanisms, there are no principled reasons for taking Neo-Aristotelian teleology and the mechanisms of the new mechanistic philosophy to be incompatible.[90]

Notes

1 Pasnau (2011, 2004), Cartwright (1999), Gilson (1995), Lonergan (1992).
2 Jaworski (2016: 3, 159). I have modified Jaworski's slogans for terminological consistency.
3 In order to elucidate the fundamental claims of hylomorphism, I draw on William Jaworski's recent presentation of hylomorphism because it is systematic, rigorous, and it engages a wide range of familiar issues in contemporary metaphysics, philosophy of mind, and philosophy of science (including proponents of *NMP*). I focus on the features of Jaworski's account that are shared, with some qualifications, by most contemporary hylomorphists and I avoid getting into more controversial topics and internal turf wars within hylomorphic circles. See also Brower (2014), Koons (2014), Lonergan (1992), Marmodoro (2013), Oderberg (2007), Ross (2008).
4 Koons (Forthcoming), Oderberg (2014: 177).
5 Hylomorphists disagree among themselves about the way formal organization *transforms* the material components of a whole. I say more about this disagreement in note 73.
6 Jaworski (2016), Lonergan (1992: 456–511).
7 Jaworski's terminology of "structure" aims to provide contemporary nomenclature for discussing hylomorphism; however, due to the many contemporary accounts of "structure," Jaworski spends as much time explaining his own technical philosophical vocabulary as Aristotelian hylomorphists do. I shall deploy a version of Lonergan's more modest revisions to the traditional Aristotelian terminology. Lonergan speaks of *central* matter, form, act of existence and *conjugate* potencies, forms, and operations, over substantial form, prime matter, act

and accidental form, secondary matter, and operations. Lonergan's transposed terminology liberates hylomorphism from unnecessary equivocations and infelicities. An illustration of each will be instructive. First, there is the equivocal and overlapping use of the term "accident" from the ten categories and the five predicables. For example, the color green and the intellect are both *accidents* in the *category* of quality, but from the perspective of the *predicables* the intellect is a *property* and green is an *accident* of a human. Second, there are many bizarre consequences that result from an excessive conservativism with respect to Aristotelian taxonomies. For example, in the case of a human, like Socrates, his separable attributes like having a particular hand (or even *these* skin cells) is a *substantial part*, and by resolution falls into the category of substance, but the intellect is an accident; it is a power belonging to the third species of the category of quality. A human can lose its substantial parts, like hands and skin cells, but it cannot lose its intellect, a kind of accident that is a necessary and inseparable concomitant of the specific difference of the human substance as a rational animal. I think such consequences are manifestly problematic and give us good reasons, among the many other philosophical and empirical reasons given by Lonergan, to prefer Lonergan's schemata where hands, skin cells, and the intellect are all distinct kinds of conjugate attributes and only the rational soul is a central form. Finally, Lonergan provides a rich account of the way his hylomorphic schemata explains how an individual thing or substance, constituted by central matter, form, and act, also consists of a complex and integrated hierarchy of explanatory conjugates from physics, chemistry, biology, psychology, and noetics and thereby anticipates the many insights about levels of organization detailed in Jaworski's hylomorphism and the new mechanist approach to genetics, molecular biology, cell biology, neuroscience, and psychology. See Lonergan (1992: ch. 15). "Elements of Metaphysics."

8 Jaworski (2016), Koons (2014), Lonergan (1992), Marmodoro (2013), Marmodoro and Page (2016), Oderberg (2007).

9 Jaworski (2016: 159–62). I have modified Jaworski's slogans for terminological consistency.

10 Jaworski (2016: 177), Lonergan (1992), Ross (2008).

11 "Many contemporary would-be defenders of hylomorphism fail to distinguish their position from contemporary materialism. I will label the resulting theories "faint-hearted hylomorphism." "A *staunch* hylomorphism involves a commitment to a sparse theory of universals and a sparse theory of composite material objects, as well as to an ontology of fundamental causal powers. *Faint-hearted* hylomorphism, in contrast, lacks one or more of these elements. On the staunch version of HM, a substantial form is not merely some structural property of a set of elements—it is rather a power conferred on those elements by that structure, a power that is the cause of the generation (by fusion) and persistence of a composite whole through time" Koons (2014: 151).

12 According to David Oderberg, hylomorphism, or his preferred *hylemorphism*, requires a commitment to much more. "The doctrine of substantial forms is but one part of a system of logically related concepts, principles, and distinctions; to lose one is to lose them all. Form and matter; species and genus; dichotomous classification; act and potency; properties and accidents; even the dreaded doctrine of prime matter—these are all part and parcel of the hylemorphic system, and together they provide a coherent and eminently plausible framework for understanding the essences of things" Oderberg (2007: xi). "Since things are constituted by their essences, those essences themselves must in some way be mixtures of actuality and potentiality. Hylemorphism says that they are—their actuality is form and their potentiality is matter" Oderberg (2007: 65).

13 Koons (2014: 152–3).

14 Jaworski (2016: 336).
15 Craver and Tabery (2016: § 1). It is also worth pointing out the influence Aristotelian philosophers of science, like Nancy Cartwright, have had on *NMP*. More recently, Cartwright and Pemberton have pointed out similarities with their work on powers and activities with *NMP*. "Our account of change-processes develops previous ideas of Pemberton (2011) that mesh Cartwright's account of nomological machines, involving powers and their contributions, with features taken from the closely related account of mechanisms by Machamer et al. (2000), most notably their account of *activities*. Activities are important for understanding what the nomological machine does in changing arrangements from one stage to another" Cartwright and Pemberton (2013: 96, n. 8).
16 Craver (2007: 8).
17 Cummins (2000: 140).
18 Machamer et al. (2000: 1). For brief histories and bibliographies concerning the development of *NMP*, see Bechtel (2007, 2009: 548–53), Craver (2007), Craver and Darden (2013: 26–9), Glennan (2016: § 2).
19 Andersen (2014: 276). See Darden (2016), Glennan (2010).
20 Craver (2007: 271).
21 Craver (2014), Craver (2007), Salmon (1984).
22 Craver (2014: 35).
23 ". . . the ontic mode of thinking about explanation, does not depend on the existence of intentional agents . . . A given ontic structure might cause, produce, or otherwise be responsible for a phenomenon even if no intentional agent ever discovers as much" Craver (2014: 36).
24 Craver (2014: 51).
25 Some proponents of *NMP*, like William Bechtel, are less sanguine about *ontic* explanations; see Bechtel (2007), Illari (2013), Wright (2012).
26 Craver (2007), Craver and Darden (2013: 48–50), Craver and Tabery (2016).
27 Craver and Tabery (2016).
28 Craver and Darden (2013: 15), Machamer et al. (2000: 3).
29 Bechtel and Abrahamsen (2005: 423), Bechtel and Richardson (2010).
30 Craver (2007: 6).
31 Glennan (2016: § 3).
32 Craver (2007: 1).
33 Craver (2007: ch. 7), Craver and Darden (2013: 167–72, 193–4).
34 Craver and Darden (2013: 21–2). In connection to what they call topping-off and bottoming-out, "In the practice of scientific explanation, decomposition in a given inquiry bottoms out when the investigation reaches entities and activities that are viewed as unproblematic or for which investigators lack tools for further decomposition. Few decompositions extend more than a small number of levels" Craver and Bechtel (2007: 549, n. 5).
35 Craver and Darden (2013).
36 Craver and Darden (2013), Darden (2006), Tabery et al. (2017).
37 Craver and Darden (2013: 19).
38 Glennan (2016).
39 Machamer et al. (2000: 3).
40 Craver and Darden (2013: 20).
41 Bogen (2005, 2008), Craver (2007: ch. 3), Glennan (2016). Aristotelians will be interested in Bogen's efforts to wed Anscombe on causality and productive activities to *NMP*.
42 Craver and Tabery (2016: § 2.3).
43 Craver (2007: 5).
44 Glennan (2016: § 5), Machamer et al. (2000: 4).
45 Craver (2007: 5–6).

46 Craver (2007: 6).
47 Craver and Darden (2013: 16–20).
48 Craver (2007: 6).
49 Craver and Darden (2013: 20).
50 Craver and Tabery (2016: § 2.4.1), Wimsatt (1997).
51 Craver (2007: 136). "Action potentials cannot be explained by mere temporal sequences of events utterly irrelevant to the phenomenon, but one can derive a description of the action potential from descriptions of such irrelevant phenomena. Action potentials cannot be explained by mere patterns of correlation that are not indicative of an underlying causal relation. Irrelevant byproducts of a mechanism might be correlated with the behavior of the mechanism, even perfectly correlated such that one could form bridge laws between levels, but would not thereby explain the relationship. Merely finding a neural correlate of consciousness, for example, would not, and is not taken by anyone to, constitute an explanation of consciousness" Craver and Tabery (2016: § 3.1).
52 Craver and Tabery (2016: § 2.4.1).
53 Craver (2007: 189).
54 Craver (2007: 227). My emphasis.
55 See Craver (2007: ch. 5) "A Field Guide to Levels."
56 Craver (2007: ch. 5).
57 Craver (2007: 179).
58 Craver (2007: 183), Glennan (2010).
59 Craver and Bechtel (2007: 547).
60 Craver and Tabery (2016: § 4.2).
61 Craver (2007: ch 5).
62 Craver and Bechtel (2007: 551).
63 Craver and Bechtel (2007: 562).
64 Jaworski (2016: 8).
65 Jaworski (2016: ch. 14), Stump (2012: 60).
66 Jaworski (2016: 106).
67 I employ this distinction akin to the way it is used by Hornsby, McDowell, and Hacker's related account of the mereological fallacy. A similar distinction between the animal level and sub-animal level descriptions and explanations can be employed for non-rational animals. See Hacker and Bennett (2003), Hornsby (2000), McDowell (1994).
68 Craver (2007: 188).
69 Jaworski notes: "Fine and Johnston conceive of hylomorphic structures as relations among something's parts. I conceive of them rather as relations between wholes and their parts: a whole configures or structures its parts. This has important implications for how my hylomorphic theory avoids some versions of Williams' worry" Jaworski (2016: 96, n. 2).
70 Craver (2007: 186).
71 Jaworski (2011: 210).
72 Jaworski (2011: 275).
73 Jaworski (2011: 269). Roughly, staunch hylomorphists can be divided up into *minimalists* and *substantivists* about the way formal organization *transforms* the ontological identity of material components. *Minimalists*, like Jaworski (and perhaps pluriformists like Avicenna and Duns Scotus), hold that the formal identity of the materials remain fundamentally the same whether they exist in the wild or as integrated material parts that depend upon the organization of the whole. *Radical minimalists* about formal transformation of materials go so far as to say that sometimes these material parts, say fundamental physical particles, are individual substances integrated within more complex substances. Minimalists argue substantivists cannot account for the continuity of the substrate that

is required for change. *Substantivists* deny that material components—even the ground-floor physical stuff—can actually remain *what they are* when they are incorporated into the organization of an inclusive whole: electrons are transformed into animated material components when integrated into the life of an animal. Substantivists contend the minimalist view undermines the integrity, unity, and identity of the organized whole, since parts of a whole cannot possess their own central or substantial formal identity. *Radical Substantivists* about the formal transformation of materials hold that there cannot be any *actual parts* in hylomorphic individuals, only potential parts. I leave the debate here.

74 Koons (2014: 171). For a detailed account of various hylomorphic accounts of composition and a defense of *PASI*, see Jaworski (2016: chs. 6–7, 10, 14), Koons (2014).

75 Jaworski (2016), Lonergan (1992: chs. 6, 8, 15), Ross (2008).

76 Craver (2007: ch. 5).

77 Jaworski (2011: 274–5).

78 Craver (2007: 217).

79 Glennan (2010), Maley and Piccinini (2017).

80 Craver (2007: 122–8).

81 Craver and Tabery (2016: § 5.2).

82 Gilson (2009).

83 Craver (2013: 155).

84 Craver (2013: 156).

85 Darden and Craver use William Harvey's (1578–1657) argument for his revolutionary account of the circulation of blood as an illustration of the various constraints (locations, structures/entities, abilities, activities, timing, roles/functions, production, global organization) on mechanism schemata that guide empirical discoveries and the verification of mechanisms. In their section on the importance of roles or functions, they note, "The teleological form of thought embodied in reasoning about an item's role, a form of thought that asks what the different components and activities are for, or what they contribute to a mechanism, is not merely a *façon de parler* maintained as yet another vestige of Aristotelianism in Harvey's thinking. Rather, it is crucial to understanding how biological mechanisms work: to seeing how a part fits into the organization of the mechanism as a whole" Craver and Darden (2013: 115).

86 In their excellent account of teleological functions as stable causal powers that contribute towards the (objective and subjective) goals of organisms, Maley and Piccinini provide an ontologically serious account of functional or teleological mechanisms that "grounds a system's functions in objective properties of the system or the population to which it belongs, as opposed to features of the epistemic or explanatory context of function attribution." On their account, "functions are an aspect of what a system *is*, rather than an aspect of what we may or may not say about that system" Maley and Piccinini (2017).

87 Maley and Piccinini (2017).

88 Maley and Piccinini (2017).

89 Maley and Piccinini (2017).

90 I would like to thank the editors of this volume for their helpful comments on previous drafts of this article.

Bibliography

Andersen, H. (2014) 'A Field Guide to Mechanisms: Part I', *Philosophy Compass* 9(4): 274–83.

Bechtel, W. (2007) *Mental Mechanisms: Philosophical Perspectives on Cognitive Neuroscience* (New York: Psychology Press).
—— (2009) 'Constructing a Philosophy of Science of Cognitive Science', *Trends in Cognitive Sciences* 1(3): 548–69.
Bechtel, W. and Abrahamsen, A. (2005) 'Explanation: A Mechanist Alternative', *Studies in History and Philosophy of Science Part C: Studies in History and Philosophy of Biological and Biomedical Sciences* 36(2): 421–41.
Bechtel, W. and Richardson, R. C. (2010) *Discovering Complexity: Decomposition and Localization as Strategies in Scientific Research* (Cambridge, MA: MIT Press).
Bogen, J. (2005) 'Regularities and Causality; Generalizations and Causal Explanations', *Studies in History and Philosophy of Science Part C: Studies in History and Philosophy of Biological and Biomedical Sciences* 36(2): 397–420.
—— (2008) 'Causally Productive Activities', *Studies in History and Philosophy of Science Part A* 39(1): 112–23.
Brower, J. (2014) *Aquinas's Ontology of the Material World: Change, Hylomorphism, and Material Objects* (Oxford: Oxford University Press).
Cartwright, N. (1999) *The Dappled World: A Study of the Boundaries of Science* (Cambridge: Cambridge University Press).
Cartwright, N. and Pemberton, J. (2013) 'Aristotelian Powers: Without them, What Would Modern Science Do?', in J. Greco and R. Groff (eds.) *Powers and Capacities in Philosophy: The New Aristotelianism* (London: Routledge): 93–112.
Craver, C. (2007) *Explaining the Brain: Mechanisms and the Mosaic Unity of Neuroscience* (New York: Oxford University Press).
—— (2013) 'Functions and Mechanisms: A Perspectivalist View', in P. Huneman (ed.) *Functions: Selection and Mechanisms* (Dordrecht: Springer): 133–58.
—— (2014) 'The Ontic Account of Scientific Explanation', in M. I. Kaiser, O. R. Scholz, D. Plenge and A. Hüttemann (eds.) *Explanation in the Special Sciences: The Case of Biology and History* (Berlin: Springer): 27–52.
Craver, C. and Bechtel, W. (2007) 'Top-down Causation Without Top-down Causes', *Biology and Philosophy* 22(4): 547–63.
Craver, C. and Darden, L. (2013) *In Search of Mechanisms: Discoveries Across the Life Sciences* (Chicago: University of Chicago Press).
Craver, C. and Tabery, J. (2016) 'Mechanisms in Science', in E. N. Zalta (ed.), *The Stanford Encyclopedia of Philosophy* (Winter 2016), at http://plato.stanford.edu/archives/win2016/entries/science-mechanisms/ [last accessed 10.4.17].
Cummins, R. (2000) '"How Does It Work" Versus "What Are the Laws?": Two Conceptions of Psychological Explanation', in F. Keil and R. A. Wilson (eds.) *Explanation and Cognition* (Cambridge, MA: MIT Press): 117–45.
Darden, L. (2006) *Reasoning in Biological Discoveries: Essays on Mechanisms, Interfield Relations, and Anomaly Resolution* (Cambridge: Cambridge University Press).
—— (2016) 'Reductionism in Biology', in Angus Clarke, (ed.) *eLS* (Chichester: John Wiley and Sons): 1–7.
Gilson, E. (1995) *Methodical Realism: A Handbook for Beginning Realists* (San Francisco: Ignatius Press).
—— (2009) *From Aristotle to Darwin and Back Again: A Journey in Final Causality, Species, and Evolution* (San Francisco: Ignatius Press).
Glennan, S. (2010) 'Mechanisms, Causes, and the Layered Model of the World', *Philosophy and Phenomenological Research* 81(2): 362–81.
—— (2016) 'Mechanisms and Mechanical Philosophy', in P. Humphreys (ed.) *The Oxford Handbook of Philosophy of Science* (Oxford: Oxford University Press): 796–816.

Hacker, P. M. S. and Bennett, M. R. (2003) *Philosophical Foundations of Neuroscience* (Malden, MA: Blackwell Publishing).

Hornsby, J. (2000) 'Personal and Sub-personal: A Defence of Dennett's Early Distinction', *Philosophical Explorations* 3(1): 6–24.

Illari, P. (2013) 'Mechanistic Explanation: Integrating the Ontic and Epistemic', *Erkenntnis* 78(2): 237–55.

Jaworski, W. (2011) *Philosophy of Mind: A Comprehensive Introduction* (Malden, MA: Wiley-Blackwell).

—— (2016) *Structure and the Metaphysics of Mind: How Hylomorphism Solves the Mind-Body Problem* (Oxford: Oxford University Press).

Koons, R. (2014) 'Staunch vs. Faint-hearted Hylomorphism', *Res Philosophica* 91(2): 151–77.

—— (Forthcoming) 'Forms Are Not Structures But the Grounds of Structure: How Grounding Theory Illuminates Hylomorphism (and Vice Versa)'.

Lonergan, B. (1992) *Insight: A Study of Human Understanding* (Toronto: University of Toronto Press).

Machamer, P., Darden, L. and Craver, C. (2000) 'Thinking About Mechanisms', *Philosophy of Science* 67(1): 1–25.

Maley, C. and Piccinini, G. (2017) 'A Unified Mechanistic Account of Teleological Functions for Psychology and Neuroscience', in D. Kaplan (ed.) *Explanation and Integration in Mind and Brain Science* (Oxford: Oxford University Press): ch.11.

Marmodoro, A. (2013) 'Aristotle's Hylomorphism Without Reconditioning', *Philosophical Inquiry* 37(1/2): 5–22.

Marmodoro, A. and Page, B. (2016) 'Aquinas on Forms, Substances and Artifacts', *Vivarium* 54(1): 1–21.

McDowell, J. (1994) 'The Content of Perceptual Experience', *The Philosophical Quarterly* 44(175): 190–205.

Oderberg, D. (2007) *Real Essentialism* (London: Routledge).

—— (2014) 'Is Form Structure?', in D. Novotny and L. Novak (eds.) *Neo-Aristotelian Perspectives in Metaphysics* (London: Routledge): 164–80.

Pasnau, R. (2004) 'Form, Substance, and Mechanism', *Philosophical Review* 113(1): 31–88.

—— (2011) *Metaphysical Themes 1274–1671* (Oxford: Oxford University Press).

Pemberton, J.M. (2011). 'Integrating mechanist and nomological machine ontologies to make sense of what-how-that evidence'. http://personal.lse.ac.uk/pemberto

Ross, J. (2008) *Thought and World: The Hidden Necessities* (University of Notre Dame Press).

Salmon, W. (1984) *Scientific Explanation and the Causal Structure of the World* (Princeton: Princeton University Press).

Stump, E. (2012) 'Emergence, Causal Powers, and Aristotelianism in Metaphysics', in R. Groff and J. Greco (eds.) *Powers and Capacities in Philosophy: The New Aristotelianism* (London: Routledge): 48–68.

Tabery, J., Piotrowska, M. and Darden, L. (2017) 'Molecular Biology', in E. N. Zalta (ed.) *The Stanford Encyclopedia of Philosophy* (Spring 2017), at https://plato.stanford.edu/archives/spr2017/entries/molecular-biology/ [last accessed 10.4.17].

Wimsatt, W. C. (1997) 'Aggregativity: Reductive Heuristics for Finding Emergence', *Philosophy of Science* 64(4): 372–84.

Wright, C. D. (2012) 'Mechanistic Explanation Without the Ontic Conception', *European Journal of Philosophy of Science* 2(3): 375–94.

About the Contributors

Christopher J. Austin is a Research Fellow in Metaphysics at the University of Oxford. He received his MA (BPhil) from the University of Oxford, and his PhD from the University of Nottingham. His research focuses on the metaphysics of science with a particular emphasis on evolutionary developmental biology, and he has recently published on issues relating to ontology, causation, essentialism, and natural kinds.

Janice Chik Breidenbach is a Visiting Research Scholar at the University of Oxford, Blackfriars Hall, and Assistant Professor of Philosophy at Ave Maria University. Her specialty is in action theory, philosophy of biology, and Aristotelian metaphysics, with interests in the applications of the metaphysics of agency to areas in political theory, philosophy of law, and ethics. Selected publications include articles and reviews published in the Review of Metaphysics and the Cambridge Companion to First Amendment and Religious Liberty (forthcoming). She holds degrees from Princeton University, the University of Texas at Austin, and the University of St Andrews, UK.

Edward Feser is an Associate Professor of Philosophy at Pasadena City College in Pasadena, California. He is the author of Aquinas, Scholastic Metaphysics: A Contemporary Introduction, and many other books and academic articles, and the editor of Aristotle on Method and Metaphysics.

Daniel D. De Haan is a postdoctoral research associate at the University of Cambridge. He is working in the neuroscience strand of the TWCF Theology, Philosophy of Religion, and the Sciences Project in the Faculty of Divinity and the Translational Cognitive Neuroscience Laboratory, Department of Psychology, University of Cambridge. His research focuses on philosophical anthropology, the philosophy of psychology and neuroscience, the philosophy of religion, and medieval philosophy.

William Jaworski is an Associate Professor of Philosophy at Fordham University in New York City. He is the author of Structure and the Metaphysics of Mind: How Hylomorphism Solves the Mind-Body Problem (Oxford University Press 2016) and Philosophy of Mind: A Comprehensive Introduction (Wiley-Blackwell 2011). He works on topics in meta-physics, philosophy of mind, and the philosophy of religion.

Robert C. Koons is a Professor of Philosophy at the University of Texas at Austin, and has a PhD from UCLA and an MA from Oxford (First Class Honours). He is the co-author of four books: *Paradoxes of Belief and Strategic Rationality* (Cambridge University Press, 1993), *Realism Regained* (Oxford University Press, 2000), *Metaphysics: The Fundamentals*, with Timothy H. Pickavance (Wiley-Blackwell, 2014), and *The Atlas of Reality: A Comprehensive Guide to Metaphysics*, with Timothy H. Pickavance (Wiley-Blackwell, 2017). He is the co-editor (with George Bealer) of *The Waning of Materialism* (Oxford University Press, 2010) and the author of over 50 articles in academic journals. He specialises in metaphysics, philosophical logic, philosophy of mind, and the philosophy of religion.

Xavi Lanao is a Postdoctoral Fellow in the Philosophy Department at the University of Notre Dame, specializing in philosophy of science and meta-physics. He received his PhD from the University of Notre Dame in 2017. His research explores the ontological consequences of scientific theories and models. More specifically, Xavi is interested in probing what scientific theories and models tell us, not only about what there is, but also about what is possible and necessary. He also works on the metaphysics of causal powers and dispositions and their relationship with causality, laws of nature, and actualist accounts of modality.

Anna Marmodoro is an Official Fellow of Corpus Christi College at Oxford and will later this year take up the Chair of Metaphysics at the University of Durham. She has published monographs, edited books, and journal articles in metaphysics, ancient, late antiquity, and medieval philosophy, the philosophy of mind, and the philosophy of religion. She currently directs a large-scale multidisciplinary research group at Oxford funded by the European Research Council and the Templeton World Charity Foundation.

David S. Oderberg is a Professor of Philosophy at the University of Reading, England. He is the author of many articles in metaphysics, ethics, philosophy of religion, and other subjects. He is the author of The Metaphysics of Identity over Time (Palgrave 1993), Moral Theory: A Non-Consequentialist Approach (Blackwell, 2000), Applied Ethics: A Non-Consequentialist Approach (Blackwell, 2000), and Real Essentialism (Routledge, 2007). He is the editor of Form and Matter: Themes in Contemporary Metaphysics (Blackwell, 1999), The Old New Logic (MIT Press, 2005), and Classifying Reality (Blackwell, 2011). He is a co-editor of Human Lives: Critical Essays on Consequentialist Bioethics (Macmillan/St. Martin's Press, 1997) and of Human Values: New Essays on Ethics and Natural Law (Palgrave, 2004). In 2013, he delivered the

Hourani Lectures in Ethics at SUNY Buffalo, and in 2003 was a Visiting Scholar at the Social Philosophy and Policy Center, Bowling Green, Ohio.

Alexander Pruss is a Professor of Philosophy at Baylor University and has a PhD in each of mathematics and philosophy. He works in metaphysics, philosophy of religion, philosophy of mathematics, formal epistemology, and applied ethics. His first book is *The Principle of Sufficient Reason: A Reassessment* (Cambridge, 2006); his latest is *Necessary Existence*, co-authored with Joshua Rasmussen (Oxford, forthcoming). He is currently working on a manuscript on causal paradoxes of infinity.

William M. R. Simpson is a doctoral student in the faculty of philosophy at the University of Cambridge and a Research Associate at the University of St Andrews. He was a former postdoctoral research fellow in theoretical physics at the Weizmann Institute of Science, having completed a PhD in Theoretical Physics at the University of St Andrews in 2014, for which he was awarded the Springer Thesis Prize. He obtained a Master's Degree in the History and Philosophy of Science at Cambridge in 2016, where is he currently pursuing a second doctorate in philosophy. He is the author of Surprises in Theoretical Casimir Physics (Springer, 2014) and contributing co-editor of Forces of the Quantum Vacuum (World Scientific, 2015).

Tuomas E. Tahko is an Associate Professor in Theoretical Philosophy and Academy of Finland Research Fellow at the University of Helsinki. He has worked in Helsinki since 2011. He has also held visiting appointments at the University of Sydney, University of Toronto, University of Reading, New York University, University of North Carolina at Chapel Hill, and the Eidos Centre in Metaphysics of the University of Geneva. He earned an MA in philosophy from Helsinki in 2005 and a PhD in philosophy from Durham in 2008 (under E. J. Lowe's supervision). Tahko is the author of *An Introduction to Metametaphysics* (Cambridge, 2015) and the editor of *Contemporary Aristotelian Metaphysics* (Cambridge, 2012). He has published over 30 articles in metaphysics, philosophy of science, philosophical logic, epistemology, and related areas.

Nicholas J. Teh is a philosopher of physics at the University of Notre Dame. He received his PhD from the University of Cambridge in 2012 and has held postdoctoral fellowships at the University of Oxford, the University of Cambridge, and Princeton University. He works on the philosophy of spacetime, the philosophy of field theory, and the philosophy of theoretical equivalence and dualities.

Index

abstraction: degrees of 45; process of 188, 199, 205; as theoretical tool 30–1

accident 56n6, 118–19, 230n7; form of 116, 214, 220–1, 295, 320n7; as trope 287n18; *see also* essence; substance

act and potency 35, 294, 308, 321n12

action: chances of 69, 111; chemical potential 301–2, 304, 310, 232; at a distance 162; for an end or purpose 212, 257, 296; *versus* movement 239, 241, 257; rational 69, 237–8, 296, 309

agency: of animals 235–7, 239, 248, 251; collective 213; disappearing agent 252, 258; of humans 235–6, 259; and moral choice 256–7

animals: as agents 236, 238–9; behaviour and learning 241; as organised structures 265, 282, 286, 295, 315, 319, 324; as 'settlers of matters' 253; as substances 242, 257

Anscombe, Elizabeth 7, 230, 245, 257, 322

anti-fundamentalism 16, 18–19, 21, 32–3

anti-realism 50–1, 54

Aquinas, Thomas 267, 325–7

Aristotle: act and potency 35, 308; biology 189; coincidence 245; continuity 154–5; embodied cognition 286, 288; homonymy principle 6; hylomorphism 294–5; monsters 180n30; soul as principle of unity 211, 259; substance 148, 178

artifact 113, 169, 178n7, 250, 271, 273

A-theory of time 50–1, 64, 68, 74

atomism 138–9, 148

attribute 212, 287n18, 294–6, 308, 314–15, 320n7; *see also* substance

autopoietic 192

Bacciagaluppi, Guido 5, 117, 122, 143, 152, 162–4

Bechtel, William 205, 297, 306–7

behaviorism 70, 79

belief: and language 236–7, 255; principle of charity 76, 80; in a true model 75–8, 80, 84, 88–9, 255

Bell, John Stewart 114, 117, 122, 131, 141–3

Berto, Francesco 157

bioattractors 207

biofilm 224–6

biology: autonomy of 18, 92; of cells 201; developmental 169, 174–5, 179–80, 185, 188, 190, 202–3, 205; and mechanisms 293, 305, 307, 314; natural unities 172, 211, 216; and substantial form 115, 185, 188

blindness, experiential 285

body 20, 43, 74, 111; living 190, 218–19, 221, 238–43, 252–3, 257n22, 258n30, 259n36, 263, 286, 288n45; *see also* mind/body dichotomy

Bohm, David 152

Bohmian mechanics 60, 93, 114, 116, 138–9, 142–4; Bohm-Hiley 158; de Broglie-Bohm theory 138, 152

Boltzmann brain 108, 115

Born rule 4, 68, 107–10, 114, 121, 131

Bourne, Craig 44, 56, 58

branching structure 96–7, 101, 108–9, 112, 121n4

B-theory of time 47, 54

Cartwright, Nancy: anti-
fundamentalism 16–17, 19, 21, 31–3;
capacities 134, 322n15; doppled
world 3; models 24–30; natural laws
91, 106, 142n24
categories: Aristotle's 154, 321; causal
248, 254; metaphysical 147, 161,
216, 267, 281
causal capacity: to dissolve 123–4; in
isolation 134–5; to persist 169,
177–8; to re-organise or develop 169,
174, 177–8, 192–3, 198, 201; for
speech 238, 256; for thought
236, 239
causalism 237, 241, 251–2, 256
causal pluralism 239, 242, 307–8
causal power: cognitive 256, 287n3,
282–6, 296, 315–16; conferred
by processes 132, 135; context-
dependent 251; downward directed
188, 246, 258n28, 316; embodiment
277–8; emergent 249, 272–3, 286–7,
312–13; fundamentality of 2, 15;
as grounding laws 298; and human
agency 246; Humean 5, 123–8, 134–
40; individuation of 129, 131, 133,
137; localized 9, 103; manifestation
of mechanism 278, 301; multifaceted
177; non-redundant 271, 273; novel
202; of organisms, animals 6–7,
246–7, 296, 315, 324n86; of organs
219–21; potentiality 136; in quantum
mechanics 5, 95, 97–8, 101–2,
119–20, 129; structural 169–78;
structured manifestation of 276–7,
280; unifying (*see* unifying power);
see also causal capacity
causal process 95, 124, 126–9, 187,
190–3, 221, 243, 301; fusion of
131–7, 140–1, 143n55
causation: agent 111, 236–7, 252–4,
255n1, 258n30, 258n35; *versus*
counterfactual dependence 74, 85;
downward 7, 111, 243, 245–6,
248–9, 257n24, 294, 306–7, 315–
16; emergent 257n19; final cause
211, 213, 294, 307, 316 (*see also*
teleology); fundamental/irreducible
1–2; immanent 211–13, 224;
macroscopic 97; mechanisms 302;
mental 110, 256n10; in modern
physics 49, 85, 153; pluralism 8
(*see also* pluralism, causal); reality
of 46; substance 227, 234–5,

238–40, 242–9, 250–3, 258n28,
259n36
chance 68, 102, 107, 108–9, 111,
121n4; *see also* probability
change: enacted by agents 248;
entangled 124; induced by activities
303, 322n15; in materials 270–1,
296, 323n73; morphological 174–5,
192; produced by causal powers
123–4, 215, 297; reality of 44–6, 49,
56n5; in relativity 36, 41, 45, 50,
55; substantial 116, 271; temporal
passage 35, 37–40, 43
circularity, vicious 249–51, 258
collection of substances 215–19, 222–3,
226, 229, 231n60
collective agency 213
collectively stable collection 191
colony 213, 217–18, 223–5
component: of activities 277, 293, 300–1,
303–6; material 295–6, 306–18,
320n5, 323n73; of organ 274, 276;
of organism 8, 171, 265, 293, 302; of
quantum system 149–51, 162n21; of
wavefunction 119
compositional hierarchy 98, 178, 265,
301, 305, 307, 312–13
concretism 16, 21, 31
conditional, subjunctive 71–4, 76
configuration space 130–1, 143, 149,
152, 156
contrariety 154–6
Craig, William Lane 44, 51–2, 57,
59–60
Craver, Carl 205n58, 297–302, 305–7,
312–13, 318–19, 322–4
cyclopoietic activity 171–2, 174, 177–8

dappled world 18, 25–6, 33, 142n33
Davidson, Donald 7, 75, 236–9, 255,
258
degeneracy 172, 179n17
Dennett, Daniel 76, 93, 95, 318
Descartes, René 46, 215, 238, 240,
256n10, 300
determinable 117–19, 121
determinism 19, 116, 119, 255, 257–8,
302
Deutsch, David 62, 64, 66, 68–9, 102,
107, 121n3
developmental constraint 172, 186, 198
developmental module 175, 184–6,
190, 291–3
Dewey, John 262–3, 280

diachronic unity 162–3, 168–70
disposition 70–1, 169, 171, 185, 295;
   global 134; Humean analysis of 2, 124;
   identity of properties 268; localized 5;
   partners 269, 280, 287n22
dispositionalism 123–4, 126, 129, 131,
   134–8, 140–1
domain of a theory 19–23, 25–33, 62,
   76–7, 87, 96–7, 256n10
domain of formal model 80–1, 83
dualism 15, 97, 112–13, 237–8,
   256n10, 262, 278, 294
Dupré, John 216–17, 287n15
dynamic systems theory 180n34, 194

efficient cause 101, 215, 295, 315–17
Einstein, Albert 56n22, 57n36, 58n50,
   141n11, 144, 162n21; special
   relativity 35–6, 45–6, 51
Eleatic Principle 202
emergence 3, 11, 70, 96, 150–3, 161,
   189, 225, 261–2; activity 187; causal
   15, 257n19; of macroscopic world
   62–3, 73, 76, 101, 157, 162n28;
   ontological 95, 158, 162n26;
   property 202, 243, 248–9, 265;
   theory 72, 77–8, 80–93
enactive perception 285
epistemology 46–7, 69; of models
   16–17, 21–2, 24–32, 33n7
Ereshefsky, Marc 225
Esfeld, Michael 138–6, 142
essence 2, 4, 6, 54, 101, 230n7,
   288n45, 297, 321n12; complex
   212–13; and existence 35, 55n1,
   56n6, 63, 77, 93–8; nominal 203n2
essentialism 127, 230, 327
event 50–4, 64–5, 68–9, 245, 247;
   causation by 248, 251–4, 316 (*see
   also* causation); composite 276–7,
   296; mental 74, 237–41, 256n10,
   323n51; ontology of 124, 134, 139;
   quantum 86, 107; self-maintaining
   269–71
Everett, Hugh 4, 61–3, 106, 121
Everettian interpretation 62–3, 66, 80,
   84, 89–90, 95, 98, 101–3, 158
Evnine, Simon 267
evo-devo 6, 180, 190, 200
exceptionalism, causal 235, 257n18
explanation: causal 2, 96, 202, 254,
   257–8; mechanistic 298, 313–15,
   317; metaphysical 212–13, 222–3,
   228–9; scientific 16, 28, 299

feeling 261–3, 277–8, 280, 282, 286
final cause 211, 213, 294, 307, 316;
   see also teleology
Fine, Kit 94, 267, 323n67
form: conjugate 295–7, 314–15; as
   principle of actuality 35; as principle
   of unity 15, 213; as structure 16, 120,
   126, 186–7, 199, 201–3, 228, 264–6,
   294–5, 297, 308, 314, 320n7, 321n11;
   substantial 4–9, 95–7, 101–2, 116,
   204n14, 213–16, 219–26, 228–9,
   230n18, 288n38, 297
Forrest, Peter 86
function 171–3, 176–8, 179n8, 219,
   275, 295, 300, 318–19, 324n85;
   see also quantum wavefunction
functional analysis 266–7, 272, 274,
   281
functionalism 5, 62, 70, 74, 76–80,
   82–5, 93, 275, 278
fundamentalism 15–19, 32–3
fundamentalist unification 3, 16–17,
   19–21, 24–6

Galileo 46, 215
general relativity *see* relativity
genetic regulatory network 191–2
Ghiselin, Michael 222, 231
Gibson, James Jerome 283–4
goal-directed activity 187, 200–1, 205,
   241
growing block theory 54
gunk 124, 134, 139

higher-order activity 187
higher-order dynamics 201
higher-order property 7, 266, 273, 275
higher-order structure 199–203,
   205n58
higher-order system 194
Hoffman, Joshua 100, 147–8, 153–4,
   161, 165, 220, 227–8
homeodynamic unity 169, 177–8
homeostasis 172, 175, 189, 213
Hume, David 7, 125, 215
Humeanism 2, 5, 123–5, 140
Humean supervenience *see*
   supervenience
hylomorphism: Aristotelian 15–16,
   33, 264, 267, 277–81; biological
   185–6, 188, 201–2; interpretation
   of quantum theory 96, 101, 111;
   New Mechanism 293–4, 300,
   308–11, 313–17, 319–20; real

essences 203n1; staunch 296–7, 312; Thomistic 55n1

incompatible outcomes 68
incompatible properties 154–6, 159
indeterminacy 63, 78, 80, 82, 86, 89
intentionality 1, 237, 268
interpretation function 69, 71–5, 77–8, 80–9
inverted goggles 285

Jaworski, William 294–7, 309, 311, 314, 320n3, 320n7, 324n74
Johnston, Mark 267, 323n67

Kim, Jaegwon 205, 207, 250, 257
Koons, Robert 15, 94–5, 258n28, 259n36, 267, 295–6, 312
Koslicki, Kathrin 228, 267, 288
Kripke, Saul 127, 142n16, 186

Ladyman, James 40, 48–9, 129, 137, 142, 161, 287
law of nature 110, 115, 118, 161, 163n57; contingent 213; explanatory 298; fundamental 1, 17, 123; grounded in powers 126; Humean 124
Leibniz, G. W. 119, 267
levels: of mechanism 275, 305–7, 311–13, 316–17; of organization 265–6, 294, 315
Lewis, David Kellogg 70–1, 73–6, 78–9, 85–6, 125, 214, 287n18
lichen 217–18, 230
Lockwood, Michael 35–6, 55–6
Lonergan, Bernard 295, 321
Lorentz, Hendrik 51
Lowe, Edward Jonathan 38, 212, 230, 252, 258

macro-object 95, 150–1, 157, 159–61, 162n28
Many Minds interpretation 4, 105–11, 121n4
Many Worlds interpretation 61–2, 66–70, 93, 97, 105–6
Marmodoro, Anna 116, 179n9, 204n14, 267, 295
Marr, David C. 282, 285
material cause 101, 235, 258n28, 316
materialism 15, 237, 262, 321
material property 309–10

matter 47, 54, 179n11, 192, 215, 251, 294; Cartesian view of 241, 256n10; continuous 42, 139; energy and 265, 267, 269, 280; form and 35, 120–1, 186, 188, 201–2, 222, 297, 315; functional 203; heaps of 111–13, 304; prime 228, 320n7; secondary 320n7; stability of 163n67
mental causation *see* causation, mental
mereology 228, 305, 311
mind: analytical functionalism 74–6; animals 115, 238–40, 255n9; immateriality of 47; origin of 115; philosophy of 15, 123, 140, 143n54, 237, 275, 296, 298; universals in the 214; *see also* dualism
mind/body dichotomy 7, 261–3, 282
mind-independent entities 2, 39, 94
Minkowski, Herrman 3, 35, 44, 51, 53
Minkowskian spacetime 36–7, 52–3; *see also* spacetime
modal interpretation of quantum theory 116–17, 121
modal logic 70–2, 125–8
monism 5, 142, 147–54, 157–63
morphological trait 191
morphology 6, 170–8
morphospace 174, 176–7, 180, 192–3, 195–6, 198, 201
moving spotlight theory 46, 50
multifaceted manifestation 177
multiverse 55, 106, 108, 110–15, 117–18

naturalism 7–8, 235, 266–7, 274, 287, 297
natural kind 93, 95–6, 126–8, 135, 171, 178n3, 206n70
natural necessity 1, 123–4, 126, 141
naturalness 77, 90
necessity 62, 123–4, 126, 141n13
neuroscience 262, 274–5, 282, 286, 293–320, 320n7
Newman, Maxwell Herman Alexander 57, 81
New Mechanists 294, 311
Newton, Isaac 28, 38
Newtonian mechanics 31, 43
Newton's Laws 28, 81
Ney, Alyssa 157–8, 160–1, 163
nonlocality 51, 130–1, 135, 162
non-mechanistic explanation 201
North, Jill 160, 163n69

observable 116–19, 130
Oderberg, David 56n5, 203n2, 230n19, 231n36, 267, 287n11, 295
ontic explanation 299–300, 307, 310, 318–19, 322n23
ontic organization 304–5, 314
orca, Orcinus 241
organ 7, 215–29, 230n13, 238, 265, 274, 303, 311; adaptation of 68–9; intellectual 288n45; sensory 47, 75
organism 70, 88, 97, 178n1, 201, 203, 217–29, 243, 263, 308–10; activity of 303; *versus* collectives 222–3, 231n44, 231n60; determinables of 117; development of 189–90, 271; mindless 115; as natural unity 1, 8, 169–70, 211–13, 215–16; powers of 15, 17, 252, 258n30, 269, 312; self-maintenance of 6, 171–8, 178n7; substantial forms of 9, 113, 204n24
organization 263–7, 273, 280, 285, 293–7, 300–2, 304–19

paradox, Putnam's 63, 83–5, 87, 93
Parmenides 35–6, 56
parthood 153, 250, 255n7, 288n33, 302, 308, 311–12
pattern 174, 198, 200, 226, 241, 264, **283**, 304, 309; real 76, 80, 93, 95
perception 39, 263, 278, 282, 284–7, 310; permanent possibilities of 72
persistence 311–12, 321n11; of activity 176–8, 243; of organism 169, 171–4, 265, 267, 269–70, 295; of peaks of quantum waves 160–1; of quantum branches 61; of substrate 37
personal/sub-personal explanation 309
phenomenalism 72–3, 76–7, 83, 86
phenomenon: cause of, mechanism for 299–301, 309, 317; classical 20–1; of generative robustness 173; higher-level 313; homeostatic 172; material 24; observable 130; occurrence of 245; predictable 198; of unity 211; *see also* substance, quantum entanglement
phenotypic plasticity 167–8, 172–5, 184, 186, 198, 200
physicalism 139, 266, 278–9, 294
pleonastic process 172, 176
pluralism: causal 247, 250, 308–10, 313, 315–16; *versus* priority monism

148–50, 153, 157–9, 161 (*see also* monism); property 273
possible world 14, 62, 68, 119, 127–8
potentiality 35–7, 54, 56n5, 185, 215, 268, 321n12; pure 203n3; second 220
presentism 4, 37, 50–4, 57–8
principled basis for quantum mechanics 61
principled model 20–7, 29–30, 32
priority monism *see* monism
probability: dynamic 176, 180n34, 201; quantum 68–9, 80, 83, 92–3, 97, 102, 107–9, 116, 121n4, 130, 150, 158; and simplicity 87; subjective 61–5; *see also* chance
process *see* causal process
process of development 185–6, 190, 192
process ontology 170, 194
property, emergent 273, 313; *see also* higher-order property
Pruss, Alexander R. 63, 78, 86, 90, 93, 121n4
psychology 90, 92, 278, 298; folk 75, 77, 79–80, 238, 261, 293–4; irreducibility of 275; teleological explanation in 319–21
Putnam, Hilary 78, 84, 93
puzzle of coincidence 244–6, 251, 254, 259

quantum mechanics: branching 97, 101, 108–9, 112; cosmic entanglement 86, 136–8, 149–50, 153, 159, 132–4; decoherence 62–3, 77, 84, 91–3, 97–8, 101, 150–3, 158, 162; entangled property 136; entangled state 125, 130–2, 162n21; entanglement 124–5, 129, 136–8, 142n36, 143n55, 147–50, 152–4, 162n21; holism 134, 136, 140–2, 144, 148–50, 152, 157, 161; many minds (*see* Many Minds interpretation); modal interpretation of 116–17, 121; quantum cosmology 77, 142; quantum Darwinism 153; quantum field 82, 121, 125–6, 130, 134–5, 140–1; quantum gravity 9, 43–4, 51–2; quantum wavefunction 61–3, 70, 76–7, 80, 86, 95–6, 101–3, 105–8, 110–22, 138–9; wavefunction collapse 105, 111, 120, 139, 149–50, 152, 158, 162

Ramsey, F. P. 64, 66, 70, 72–3, 75–6, 255n5
Rea, Michael 267
reaction norm 175, 180n29, 192
recombination 5, 125–6, 131, 135–8, 140
reduction: functionalist 4, 70; mereological 249, 253–4; relation between theories 33, 272, 288–9, 299, 317
reductionism 3, 186
relation: causal 48–9, 74, 85, 110, 127–9, 170, 191, 258n28, 305–7, 309–10, 315; constitutive 170, 305–7, 309–10, 315–16; spatiotemporal 125, 127, 149
relativity 3–4, 35–7, 41, 43, 46, 49–55, 57–8, 62
robustness 128, 135, 173–3, 176, 179–80
Rosenkrantz, Gary 147–8, 153–4, 161, 220, 227–8, 230
Russell, Bertrand 46, 57, 70, 74

Saunders, Simon 62, 64, 68–9, 80
Scaltsas, Theodore 116
Schaffer, Jonathan 142, 148–54, 161, 163
schematic unification 31
scholasticism 2, 185, 215, 221–2, 230, 293, 327
self-directed activity 171
self-directed power 179
Sider, Theodore 52–3, 56–7, 287
Sklar, Lawrence 17–18, 33, 45, 50, 57
slime mould 217–18, 219–20, 223–4
Smolin, Lee 43–5, 56
spacelike separation 52, 57, 130
spacetime: bounded region of 143n45; Minkowski 3; ordinary 5, 84–6, 138; realism about 63, 74, 77, 85–6; theory 42–4; as a whole 147–8
Spinoza, Baruch 215
Stern-Gerlach experiment 106, 129, 142
Steward, Helen 7, 240, 242–9, 252, 257–9
structuralism 70, 72, 76, 81, 85, 124

structural power 161, 163–72, 174
structural realism 49, 51–2, 59, 129, 134, 139, 283
subjunctive conditional 71–4, 76
substance 1–7; *versus* accident 320–1n7; act and potency of 35; Cartesian views of 238; causation 242–3, 246–54; *versus* collective 222–4; and hylomorphism 15, 214–16, 228–9, 265, 278, 294–7, 308, 313–16; *versus* organ 218–20; organism as 178; and quantum entanglement 147–8, 154, 157–9, 161; and quantum mechanics 95–8, 101–2, 111–14, 116–20; spacetime 51, 53–4; and substantial form 226–8, 314–15
substantivism 323n73
supervenience: Humean 5, 125–6, 134, 136–41; of mental on physical 97; and reduction 299; and substance causation 243–4, 246–7, 249, 254, 259n36; of value on being 102
synchronic unity 6, 8, 170, 172–3, 176

teleology 1, 8, 211–12, 216–17, 294, 317–20; *see also* final cause
theoretical unification 17, 20–1, 26–7, 31–3
Thompson, D'Arcy Wentworth 211, 229
Thompson, William 229
thought 238–40, 261, 263, 278, 280, 286
topological mapping 180, 198–201, 206
Traveling Minds interpretation 97, 109–12

unifying power 169, 220–1, 228–9
unity, biological 169, 172, 178, 216, 225, 227–8, 309, 323n73
utility 64–7

vagueness 87, 101–2, 216–17
van Inwagen, Peter 95, 97, 269–72, 274
verificationism 45–6, 50, 57
vision 282, 284

Wallace, David: emergence and decoherence 62, 91, 162n28; indeterminacy of interpretation

81, 85–7; probability 64, 66, 80; solutions to problems 102; spacetime realism 77

whirlpool 243–51, 253–4

Whitehead, Alfred North 45, 206

Wittgenstein, Ludwig 239–40

Worrall, John 57, 287

Zimmerman, Dean 66–7, 71–2

zombies 97, 109, 114–15